Science in the Modern
World Polity

# Science in the Modern World Polity

## INSTITUTIONALIZATION
## AND GLOBALIZATION

GILI S. DRORI, JOHN W. MEYER,
FRANCISCO O. RAMIREZ,
AND EVAN SCHOFER

*Stanford University Press*
*Stanford, California 2003*

Stanford University Press
Stanford, California

© 2003 by the Board of Trustees of the
Leland Stanford Junior University.
All rights reserved.

Printed in the United States of America
on acid-free, archival-quality paper.

Library of Congress Cataloging-in-Publication Data
Science in the modern world polity : institutionalization and
globalization / Gili S. Drori . . . [et al.].
    p.   cm.
Includes bibliographical references and index.
ISBN 0-8047-4491-2 (hardcover)—ISBN 0-8047-4492-0 (pbk.)
  1. Science—Social aspects.  2. Science and state.  I. Drori, Gili S.
Q175.5 .S3643   2003
303.48'3—dc21                                                      2002011862

Original Printing 2003
Last figure below indicates year of this printing:
12  11  10  09  08  07  06  05  04  03

Typeset by BookMatters in 10¼/13 Adobe Caslon

# Contents

# List of Tables and Figures

*Tables*

*Figures*

# Preface

The book reflects decades of sociological work at Stanford University on the development of the modern nation-state system since World War II. Nation-state structures and policies are often expected to reflect the great diversity in resources and cultures behind these societies. But in fact, our studies show that striking similarities increasingly characterize nation-state structures and policies. Across many different domains, countries appear to be enacting common models or scripts of what a nation-state ought to be. This is true with respect to national constitutions and state bureaucracies (for example, statements of purpose and goals, cabinet and agency formation), socio-economic progress models (for example, emphases on economic growth and population control), and egalitarian citizenship models (for example, the rights of women and children). Common trends are especially evident with respect to national educational systems. Protestations about uniqueness notwithstanding, these systems undergo similar expansionary trends at all levels. These growth patterns are rationalized by common principles of progress and justice, which are articulated by professionals and experts in increasingly similar organizations.

The resulting standardization is both underestimated and undertheorized in the social sciences. Much of the literature only stresses coercive mechanisms (for example, structural adjustment policies of the World Bank) or mimetic processes (for example, copying textbooks of the former colonial powers). The underlying assumption is that the outcomes are mainly due to the interests of the dominant economic and political actors. This is an untenable premise, given that standard models of progress and justice are at least as likely to provoke challenges to hegemonies as differentiated scripts and even such scripts could readily be used to justify all sorts of underdevelopments and inequalities. That is, alleged differences between peoples and cultures can be assumed to account for differences in economic and social outcomes. Instead, strong commonality assumptions about individuals (their needs and goals) and about humanity (the feasibility of progress and justice) give rise to standardization processes. We contend that at the center of this growing standardization lie the sciences: worldwide bodies of authoritative knowledge located in authoritative personnel defined by common professional standards. We further contend that the global spread of the standardized nation-state is clearly connected to the worldwide triumph of scientific authority and the higher education system within which scientific authority is prominently displayed.

The sciences, the scientific experts, and the educational system in general are central in two important ways. First, they are the central locus of the culture or knowledge system of modern globalizing world society. Second, they are the central means by which the world culture is transmitted to and installed within the particular national societies of the world. Schooled personnel are the most legitimated members and elites of the world. Their school-based knowledge is highly validated, in large part because it reflects rationalized understandings of a universalistic character, rather than unique local perspectives. The authority of science both presupposes and promotes a constitutive worldview which assigns much lawfulness to both nature and society and calls for much professionalization in order to successfully comprehend and apply the general laws and principles. From this perspective, science is not principally a set of sound techniques or smart technologies, but rather a powerful culture that pervades the world.

It would not have to be so. One could imagine a world society built around economic or political organizations dominating a culture made up of very specific instrumental techniques, a world with a culture involving limited science and no universities. And indeed, it seems obvious that a world really controlled by a single dominant political or economic organization would have little use for the sorts of free-floating knowledge systems that flower in our present world. By way of contrast, note that in our world, little groups of authoritative scholars wonder about discovering over here an ozone layer problem, over there the abuse of some children, and yonder corrupt organizational arrangements. Scientized perspectives addressing these problems arise and include environmental science regimes, human and children's rights agendas, and modern management and accounting systems and standards.

Gradually, the themes of this book took shape as we tried to make sense of disparate empirical findings. If schooled science is the core culture of the modern global system, it makes sense that its institutional arrangements spread everywhere, regardless of local functionality. They spread in a world society of organizations and discourse. They spread to and diffuse within every national society. And they work as a cultural system, not a technical one. That is, the effects of the globalization of science are diffuse, touching on every social institution; the effects are not narrowly instrumental or solely economic in character.

## ACKNOWLEDGMENTS

The work of this book directly followed, pinning down the different components of the overall story. In carrying it out, we have had a great deal of support and help. We are grateful for all the help and eager to acknowledge it.

First, we thank our collaborating authors, Ann Hironaka, Yong Suk Jang, Elizabeth H. McEneaney, and Christine Min Wotipka, who offered their own scholarly work for this volume and commented extensively on this book in manuscript.

Many additional people have contributed to our project. Again and again, the participants in Stanford University's ongoing Comparative Workshop—our core institutional locus—have commented on our work in progress and offered helpful suggestions. So have colleagues in seminars in many universities and professional meetings. Concretely, we have been aided by research support from Xiaowei Rose Luo, Miriam Abu Sharkh, and Kiyoteru Tsutsui. Valuable intellectual and editorial help on early versions of this manuscript has come from Lynn Eden, David Frank, Georg Kruecken, Gero Lenhardt, and Ian Malcolm.

Second, our project has depended on the resources provided by substantial grants from the following notable institutions: the National Science Foundation (NSF RED 92-54958), the Spencer Foundation (20000085), and the Bechtel Initiative on Global Growth and Change of Stanford University's Institute for International Studies (KEB608). For support and encouragement, we are grateful to Larry Suter of the National Science Foundation, Patricia Graham and Ellen Lagemann of the Spencer Foundation, and Coit Blacker and James Gibbons of Stanford University's Institute for International Studies. We are also pleased to acknowledge the Bechtel Initiative on Global Growth and Change and Stanford University's Institute for International Studies for providing a physical space for our project.

Third, we are thankful for the assistance and encouragement given by the editors with Stanford University Press, primarily Karen Hellekson, Patricia Katayama, Nathan McBrien, Norris Pope, and Mariana Raykov.

Fourth, we acknowledge specific individual contributions to the book. The book is the outcome of a full-fledged collaboration among the four authors and our associates. We employ a common theoretical perspective, addressing many issues on the institutionalization and effects of science. We also use common cross-national research design and data analysis strategies throughout the project. Particular authors took the lead in producing specific chapters. Chapters 5, 6, and 11 were, respectively, written by Yong Suk Jang, Elizabeth H. McEneaney, and Ann Hironaka. Chapter 2 is by Evan Schofer and Elizabeth H. McEneaney, and Chapter 9 is by Christine Min Wotipka and Francisco O. Ramirez. Beyond this, Gili S. Drori took the lead in producing Chapters 5, 12, and 13, which are adapted from her dissertation research (Drori 1997) and in developing Chapters 7 (with Francisco O. Ramirez) and 9 (with Evan Schofer).

Last, some parts of our book derive from earlier reports elsewhere. Chapter 3 is adapted from Evan Schofer, "The Rationalization of Science and the Scientization of Society: International Science Organizations, 1870–1995," in *Constructing World Culture: International Nongovernmental Organizations Since 1875*, edited by John Boli and George Thomas, 249–66 (Stanford, Calif.: Stanford University Press, 1999). Chapter 5 is a revised version of Yong Suk Jang, "The Worldwide Founding of Ministries of Science and Technology, 1950–1990," in *Sociological Perspectives* 43, no. 2 (2000): 247–70. Chapter 10 is a revised version of Evan Schofer, Francisco O. Ramirez, and John W. Meyer, "The Effects of Science on National Economic Development, 1970–1990," *American Sociological Review* 65 (2000): 877–98. Last, Figure 5.1 is adapted from Gili S. Drori's "Science Education for Economic Development: Trends, Relationships, and Research Agenda," *Studies in Science Education* 35, no. 2 (2000): 27–58. In each case, we acknowledge, with appreciation, the earlier publications.

GSD, JWM, FOR, ES

*San Francisco, October 2002*

# Science in the
# Modern World Polity

Science as a World Institution

Authority based on scientific knowledge plays a central role in modern society. It is obviously important in the developed world. It was also transparently crucial in the communist countries, whose ideological frame rested on the claims of a scientific Marxism. Authority justified in scientific terms is also evident in the aspirations of the developing countries, as the chapters of this book and many other studies demonstrate. No matter how distant national traditions are from scientific styles of thought, national policies emphasize science, not least in the education of the nation's citizens or planning for the future.

Instructions flow into the third world from global centers on the widest variety of issues: how to produce economic development, how to raise and educate children, how to preserve health, which aspects of nature require protection and how to protect them, or how to manage organizations. These instructions do not take the form of commands from core powers to conform to the interests of these core powers, although covert interests are undoubtedly involved. The instructions typically do not take the form of commands at all. They are carried along by communities of scientists and experts and by organizations that offer technical assistance, as if based on scientific knowledge divorced from the interests of these epistemic communities (Hall 1989; Haas 1992) and organizations (Finnemore 1996; Boli and Thomas 1997, 1999). They are influential because they are deeply grounded in the authority of science. The same processes, of course, routinely go on within more developed countries themselves, where scientific knowledge backs up social practice in an even wider range of domains.

Our work presents cross-national empirical research on the rise, nature, and impact of science as an authoritative world institution. We see science as spreading throughout world society as an expanded and intensive cultural package of ideas and assumptions about the lawful and comprehensible character of nature, including human and social nature. As a general cultural

1

model or framework, it spreads along with the modern emphasis on the competent and empowered individual, nation-state, and rationalized organization as the dramatic "actors" in social life. So, for example, with the spread of modern, democratic political arrangements worldwide after World War II, science (and the university system with which it is linked) went right along. Similarly, science and its academic credentials went along with the worldwide spread of modern corporate management practices.

As a general cultural model, scientific expansion and penetration have broad effects on society, far beyond specific outputs such as economic or technical growth. As our studies here show, science also encourages the incorporation of democratic practices, greater attention to environmental concerns, and expanded steps toward standardization of management and organizations. Science is a quite general rationalizing force in the modern system.

Our arguments contrast sharply with most views of the role of science in modern society. It is common to see science expanding as instrumental activity, particular to specific social goals or to the interests of certain social groups. According to some such theories, these goals, interests, and needs are perverse; in other theories, they derive from the propelling logic of scientific discovery itself. In neither case, however, is the expansion of science seen as a shift in general social authority or seen in terms of its quality as a general cultural package—ideas we emphasize here. The conventional theoretical emphasis on social needs, goals, and interests has an intuitive appeal because the main modern cultural assumptions about science stress its practical utility and its close instrumental links with social development. It is certainly easy enough to find examples of the practical impact of modern scientific activity, from medications to energy utilization to civil engineering. Yet, as we show in the following chapters, such views have difficulty accounting for the rapid worldwide diffusion of standard forms of scientific activity to societies with radically different perceived needs and power structures. They also have difficulty accounting for the diffuse effects of expanding scientific activity on national societies—on such matters as human rights and social rationalization, which are not typically thought of in scientific terms.

The chapters in this book thus emphasize science and its social authority as a general cultural model, spreading and affecting society in diffuse ways, rather than solely or primarily as means for achieving instrumental or technical goals. In this sense, we see science as playing a role in modern society analogous to the role of religion in more traditional societies.

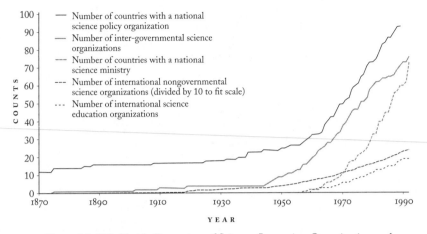

*Figure I.1.* Worldwide Expansion of Science: Increasing Organization and State Structure, 1870–1995

## CONCEPTUALIZING SCIENCE AS AN INSTITUTION

That science is rapidly expanding as an institution is a matter of fact—and a dramatic fact at that. To illustrate this, we provide in Table I.1 and Figure I.1 some examples of the explosive expansion of science at the world level and in the world's national societies. We show the expansion in numbers of world organizations, both governmental and nongovernmental, concerned with science and science education, as well as increasing number of countries joining the influential International Council of Scientific Unions. We show expansion trends in science enrollments in universities and increased numbers of scientists. We also show the dramatic expansion of publications in science journals. Science publications themselves have expanded dramatically during the modern period (Barnes 1985; Cozzens 1997; Schofer 1999b).

Perhaps because science is so central and so deeply institutionalized, there is great difficulty in defining its role and thus in explaining its spread, nature, and effects. We modern people take science so much for granted as constitutive of our worldview that we lose some capacity to analyze its centrality. This is what we mean by the term *institutionalization*—the creation of a knowledge system that is accepted as legitimate and valuable and that structures reality and activity for people and groups in society (Selznick 1949; Jepperson 1991). The institutionalization of science and the difficulty of distancing

TABLE I.1
Selected National Measures of Science Expansion:
Trends in Mean Values and Variation.[a]

| Variable | Year | Mean (SD) for All Countries | | CV |
|---|---|---|---|---|
| National enrollments | 1950 | 1.94 | (2.82) | 1.45 |
| in higher education | 1955 | 2.50 | (3.52) | 1.41 |
| (as percentage of 20–24 | 1960 | 3.50 | (4.72) | 1.35 |
| age group[b]) (n = 106) | 1965 | 5.66 | (7.31) | 1.29 |
| | 1970 | 6.96 | (8.47) | 1.22 |
| | 1975 | 10.06 | (10.55) | 1.05 |
| | 1980 | 11.85 | (11.50) | 0.97 |
| | 1985 | 13.79 | (13.13) | 0.95 |
| National memberships | 1954 | 1.71 | (3.44) | 2.01 |
| in International Council | 1959 | 2.31 | (4.33) | 1.87 |
| of Scientific Unions | 1964 | 2.84 | (4.91) | 1.73 |
| branch organizations[c] | 1969 | 3.59 | (5.80) | 1.62 |
| (n = 203) | 1974 | 3.89 | (6.23) | 1.60 |
| | 1979 | 4.47 | (6.23) | 1.48 |
| | 1984 | 5.38 | (7.53) | 1.40 |
| | 1989 | 5.60 | (7.57) | 1.35 |
| Paper publications in | 1973 | 0.07 | (.17) | 4.722 |
| scientific journals (hard | 1978 | 0.34 | (3.14) | 4.949 |
| sciences) per 100,000 | 1982 | 0.43 | (4.29) | 4.879 |
| citizens[d] (n = 134) | | | | |
| Scientists and engineers | 1960 | 682 | (828.24) | 1.21 |
| in R&D per million | 1970 | 1,098 | (1,278.86) | 1.16 |
| citizens[e] (n = 37) | 1980 | 1,549 | (2,022.95) | 1.31 |

SD = Standard deviation; CV = coefficient of variations.

[a]Table originally presented in Drori (1997:9)

[b]Data on the number of students in tertiary education are gathered from UNESCO statistical yearbooks, standardized by age group 20-24 years old, and provided for a constant set of cases.

[c]Data on the ratio of membership in ICSU organizations to the total number of ICSU organizations are compiled from ICSU directories and provided for a constant set of cases.

[d]Data on the number of papers published in scientific journals are derived from SCI data files, standardized by population size, and provided for a constant set of cases.

[e]Data on the number of scientists and engineers in R&D are gathered from UNESCO statistical yearbooks, standardized by population size, and provided for a constant set of cases.

ourselves from it show up strikingly in the work of the many intellectual critics of science, from the Frankfurt School to the Strong Programme to the postmodern studies of science. The attacks themselves tend to employ the language, claims, methods, and theories of science (see, for example, Mulkay 1983, 1991; Habermas 1993, 1996; Haraway 1976, 1996).

Two implications of the worldwide institutionalization of science may be noted here. First, as we show throughout the book, scientific activity flows throughout the world and through many domains of social life, considerably beyond obvious local needs or interests and independent of local resources. Science expands in rich and poor societies, complex and simple ones, in domains where it has demonstrated great utility, and in domains where positive uses are difficult to demonstrate.

Second, the worldwide social definition of scientific knowledge as the most authoritative kind of reality produces a situation where an independent definition of what is science and what is not is essentially impossible to sustain. A large body of literature addresses the question of objective or essential definition (for example, the chapters in McKnight 1992), but the enterprise tends to be fruitless. Science, highly institutionalized, is a sociocultural category with elaborate rituals of designation and definition of its own. In a study of an institution of this sort, one cannot usefully look at a set of activities and decide that they are or are not scientific; these decisions, sometimes highly contested, are a part of the modern system under investigation. So we define science in social practice as it is referred to, and thus defined, by the most legitimate organizations and discourse in the field. Science is what happens in major research facilities (for example, the European Organization for Nuclear Research or the Mayo Clinic) populated by Ph.D. researchers; these facilities and researchers are supported by organizations such as the National Science Foundation. Scientists earn their professional skills in universities and specialize by scientific discipline and degree. Thus, although laboratory activities, on examination, often do not map onto philosophical ideals and may occasionally be seen as incompetent or even fraudulent, consensual definitions clearly situate them within the domain called science. In studying highly legitimized social institutions, such as science, a commonsense definition seems most viable.

Thus, for the purposes of our empirical studies here, we define science as the activities that the relevant organizations and discourse patterns take to be science. During the twentieth century, the U.S. National Science Foundation (NSF), its analogs in different nations, and such international organizations as the United Nations Education, Scientific, and Cultural Organization (UNESCO), developed increasingly codified conceptions of science. So UNESCO now generates what might be termed "how-to" handbooks for developing nations to build "proper" scientific infrastructures. The mission statement of the NSF, for instance, focuses largely on the "applied" and "pure" sciences in the context of the research university that credentials scientists and social scientists and engineers, and their appropriate fields. Through this approach to science policy and the monitoring of current science trends, the NSF counts scientific personnel as a relevant output. It also counts publications and patents in established fields. Both these emphases provide our studies with concrete measures with which to identify and define science. In the logic of this system, Western science is kept sharply distinct from non-Western, premodern, or indigenous knowledge systems. In our analyses here, we also consider these systems as quite apart from Western modern science. From now-global conventions, one can clearly delineate a distinctive set of organizational and professional activities that become institutionalized as science in the past century.

THE AUTHORITY OF SCIENCE

The studies in this book do not look mainly at the internal roles and activities involved in science, such as the place of individual scientists in scientific discovery or the importance of social context to declaring scientific truths. The literature is filled with disputes about whether and to what extent scientific development is driven by evidence or social interests, and about whether scientific careers and organizations are shaped by considerations of merit or power struggles. We are not involved in assessments of this sort (although as researchers, we are centrally involved in sociology as a scientific enterprise). We focus, rather, on the social authority of science, as reflected in and created by decades of ever-increasing investment in science, in the obvious and growing importance of scientific discourse, and in the increase in numbers and power of scientific organizations. These trends have shaped the world and how we think about it. Thus, our studies examine science as an authoritative and deeply embedded world institution. Our studies do not take for granted the taken-for-granted authority of science, but treat it as a central phenomenon to be examined and explained. They see science as serving some of the same functions as religion and partially replacing it. They see it, in other words, as a cultural canopy covering the modern social world, more than as the instrumentally efficacious (or, in critical perspectives, arbitrary) knowledge that conventional analyses and debates conjure up (Berger 1968). Culture, in this sense, is a set of constructed social realities and models, not principally a set of values and preferences (Berger and Luckmann 1967). Our usage here contrasts with the view that science is an objectively definable set of activities, roles, and organizations shaped by value preferences, such as disinterestedness (the *locus classicus* of this line of thought is Merton 1973). Rather, we see the institution itself as an elaborate cognitive cultural model. Much modern value and policy are built up around the general concept of "science" rather than the specific bodies of knowledge, theory, and inquiry that nominally make up the general concept (such as the particularities of sociology, chaos theory, or normal distribution assumptions). The authority of science extends beyond its particular practitioners and beyond its particular claims and is awarded to the general concept, and to the institution, of science.

Our studies arise from the perspective of sociological institutionalism (Thomas et al. 1987; Meyer et al. 1997a). The core ideas are that wider cognitive cultural frames make up institutions such as science, and that these frames structure the identities and activities of such social subunits as national states. Global cultural frames, prominently including science, shape national cultures. To illustrate, modern notions of education and its value for both individuals and society have been established on a global scale, so that every sort of national society now has an expanded education system—and remarkably homogeneous education at that. Science is similarly institutionalized on a world level and similarly presumed to confer instrumental value. Offered a similar set of institutional arrangements by which to shape themselves, modern nation-states, organizations, and individuals are legitimated as social entities by the cultural frame of the modern world polity, or social order. This same order defines much of what these entities are to do and be while providing clear justifications for why such standardized and shared arrangements make sense. A central theme of this book is that science is central to the culture of contemporary world society and to the ways that this society structures national and local societies.

## Science as Global

Science is a worldwide institution in many senses. First, recognizably scientific roles and activities are found everywhere—for instance, in educational arrangements such as universities, which are now ubiquitous (Riddle 1989; UNESCO 1998). Second, as our studies show, explicitly global organizations, professions, and activities have expanded greatly, and it is routine for scientists and scientific discourse to make up global networks (Chapters 3 and 4; Schofer 1999a,b). Third, the claims of science are formulated as universally applicable and have standing everywhere. Whatever the personal views of political and economic authorities in the world, many of whom may be uninformed about, and hostile to, scientific matters, few claim public exemption from the various scientific laws or from the impact of scientific evidence (Meyer 1999). In all these ways, science is part of a world polity or sociocultural system.

## Science as Institutionalized

Science is deeply institutionalized in many ways, much more broadly than just as a profession within society. Science is built into the modern education system; science is an authoritative source of knowledge that informs policy and action. In order to become proper adults, essentially all children in the world, usually as a compulsory matter, pass through mass educational systems that emphasize science (Kamens and Benavot 1992). In order to become successful adults, it is usually necessary to obtain advanced or higher education, which again emphasizes science (Ramirez and Lee 1995; Kamens et al. 1996; see also Chapter 8).

Further, the science involved speaks with authority to the full range of public topics recognized in the modern system. Not only are some traditional technical issues involved, but so are the details of family and sexual relations; physical, psychological, and social health; economic, organizational, and political management; and so on. Almost all public issues now involve scientific testimony and evidence, from prenatal life (for example, the abortion question) to the definition of death (for example, the euthanasia debate).

## Science as a Cultural Frame

Over and above the often-clear bodies of knowledge and activity involved, science functions as a general cultural model of knowledge and power. Science is presented as method and orientation and even attitude (Chapter 6; McEneaney 1998), and it can be applied everywhere. Its essence is thought to lie in universal procedures, and its laws are understood to be similarly universal. Merton (1973) makes this a set of values underlying scientific procedures that actually take the appropriate form; but much criticism suggests that it is better to see universalism as a set of social forms or claims, or as a general cognitive model. Practical activity is decoupled from the abstract model, as we discuss below (and review in more detail in Chapter 7). In fact, we argue that the centrality of science as a cultural model generates so much expansion that actual linkages with practical activity are likely to be loose (see Chabbott 2002 for examples in domains related to national development).

## Science as an Ontology

Scientific knowledge and its value clearly transcend practical and mundane matters. Science, as a generalized cognitive cultural model, rests on the strong modern faith in the existence, operation, and accessibility of unseen general principles, even when these principles are not now known. The nineteenth-century founders of sociology, for example, received much social support that was based on assumptions that human social life was highly lawful and that humans could in the future command (and use) the laws involved. The same approach is taken now in relation to human disabilities, or collective economic problems, or any other public issue. The faith involved, which can carry on for a long time without experiential support, is impressive. As we suggest later in this book (Chapter 7 and elsewhere), the maintenance of this faith may involve some decoupling of scientific principles and policies from practical realities.

The conflicts with earlier religious models from which modern science arose are also impressive. There is the conflict about whether the world is all that standardized and objectively lawful (instead of fallen, arbitrary, or reflective of the will of God). Long battles over such issues, as in the case of evolution, result. There is also the conflict over the human hubris of trying to know and use the supposed laws involved, as in political, economic, or familial management. A reasonable way to interpret the long history of conflict between scientific and religious models is to see these institutions as competing on the same ground, rather than operating in different domains. For example, there is a striking contrast between the intensity of conflicts between science and more traditional religion on the one hand and the relative absence of similar conflicts between the domains of religion and mathematics on the other (Cohen 1982; Kamens and Benavot 1992).

## THE RESEARCH PROBLEM

The real issue in understanding the spread and penetration of science in modern world society is thus to comprehend the institutionalized cultural authority of science. Most social research on science does not confront this main issue; it focuses on the explanation of particular scientific activities, organizations, or professional stratification systems (see Zuckerman 1989 for a review). But the central problem, in our view, is not to understand science in terms of its particular uses or successes, its organizational and professional systems, or its actual or postured activities. We see such matters as following from, rather than causing, the extraordinary extent and depth of the authority of the institution. This is a key theme in this book: *the essential elements of science as it spreads across social domains and around the world lie in and result from its institutionalized cultural authority more than from any particular organizational and professional developments.*

Our answer to the question of the source of the spread and dominance of science is that the entire modern system, with its empowered and rationalistic individual, organizational, and nation-state actors, rests on and elaborates on basic cultural assumptions linked to science (see Chapter 1). Without these assumptions, the modern attribution of competence and responsibility to social actors—rather than, for example, tradition or nature or the gods—would make no sense. So all the processes that make the "rational actor" the central element of society also work to further expand science and its authority.

The most central of these assumptions are two crucial props to the cultural (and often legal) standing of the rational individual or state as "an actor" in history. First, sustaining the possibility of rational purposive and responsible action, there is the sweeping assumption that the visible natural world, including humans and their society, reflects basic underlying laws or principles that are ultimately universal, rational, and integrated or consistent. Thus, if one understands the underlying laws, one can understand and act rationally in the messy empirical world. The connection with Western religious traditions is obvious: as eloquently told by Geertz (1973), "People [perhaps particularly Western-influenced people] need to [and can] believe that God is not mad." Second, humans can, through rational procedures and institutional arrangements, discover, know, learn, and use these laws or principles, and they can act in terms of them. Specific rational techniques to be used vary from mental discipline and logic to empirical research, but the laws involved are knowable (and usable) by humans. Again, the connections with Western religious ideas, especially since the Reformation, are obvious. And again, the importance of the assumption to the modern perception of society as made up of effective individuals and organizations (rather, for example, than traditions or communities) is clear. Even in a lawful and orderly environment, if individuals cannot comprehend the laws involved, it is difficult to hold them responsible.

This view of science, as constitutive of cultural authority, contrasts with analyses of the role of science as a useful or perverse technical instrument in modern society. Our perspective helps explain the striking and conflicting observations produced by those analyses; it also helps explain a wide range of features of the scientific system not easily explained from other points of view. The studies in this book help explain surprising aspects of the origins and expansion of this system, its nature and structure, and its impact.

ISSUE 1: EXPLAINING THE ORIGINS, EXPANSION,
AND PENETRATION OF SCIENCE

Successful modern arrangements, once institutionalized, naturalize themselves as efficient and effective, and they support instrumental and functional accounts of their existence (Dobbin, forthcoming). Such cultural defenses are routinely used to support every modern arrangement, from rationalized forms of gender and racial inequality (and now equality) to the virtues of education or of modern organizational management. In this fashion, functional accounts of the origins and spread of now deeply institutionalized science are routine. Organizations such as the Organization for Economic Cooperation and Development, for example, can prescribe expanded investments in science as necessary for all sorts of progress.

So functional or instrumental theories of the usefulness of science are part of the folk culture of modernity. For the most part, their validity can now be taken for granted, and investment in science is a standard bromide for problems of economic growth and social development. This is all much easier because science has indeed plenty of successes to demonstrate in a wide variety of sectors. There are intellectual successes in discovering previously unknown matters from features of the universe to the world under the sea to subatomic structures; there are also great practical successes in discovering new technologies or cures for diseases.

As empirical explanations of scientific expansion, however, such functional ideas work badly. Scientific expansion occurred in the West for a long time with little evidence of functional effectiveness (Kuhn 1977). Science also penetrates the whole third world, across the spectrum of academic fields, with little evidence that local investments in science are actually instrumental for immediate developmental goals. Many aspects of social and technical life are penetrated by science with little evidence of direct instrumental effects, even long after the fact. It is unclear that developments in sociology, or economics, or astronomy, or many areas of medicine and biology have much to do with the instrumental efficiency of social actions and policies. Within academic fields, important questions are clearly far removed from social efficacy—it is hard to imagine much marginal utility of new theories and evidence of the big bang, the intelligence of a bird species, or the prehistory of humans. The modern defenses of such bodies of knowledge sometimes assert that they are basic or theoretical and thus serve as the foundation for more instrumental knowledge, but what practical applications follow from evidence of ice on a moon of Jupiter?

It makes more sense to see science not as efficiently instrumental knowledge, but rather as a cultural framework underlying and supporting the modern projects of a competent human individual in a rational society and of the tamed environment within which this empowered actor legitimately functions. It makes sense, that is, to see science as part of the cosmology underlying the modern instrumentalities, rather than to see science itself as an instrument.

One can then see the long period of expansion of science as the broad cultural or ontological component of modern rationalization—as supporting a meaning system rather than particularly effective in a technical sense. One can understand the flow of scientific analysis to all sorts of unlikely fields (for example, sociology) or questions (for example, ice on a moon of Jupiter) as part of this meaning system, rather than as primarily instrumental. One can understand the expansion of science in its broadest forms on a world scale as reflecting a penetrative meaning system at least as much as an efficient competitor with improved techniques.

This institutionalist point of view directly addresses the crucial question of the spread of the authority of science. This authority does not principally derive from proofs of instrumental effectiveness—for instance, for economic or political ends. Nor is it driven by true knowledge or the strategic plots of the sciences and scientists themselves, as some critical theories have it. The authority of science as a core cultural matter spreads with considerable independence from evidence of instrumental or technical effectiveness.

The modern system is culturally organized as made up of individual and organizational human "actors" (the term conventionally used to denote empowered, purposive, and fairly rational social units; Meyer and Jepperson 2000). Fundamental ingredients of these empowered identities are scientized pictures of the environment within which reasonable action can occur and scientized pictures of the actual or potential competence of the actor (with some presumably scientific advice) to comprehend the order of his or her environment. Wherever such cultural notions of dramatic rational actorhood spread, both the supply of and demand for science expand and penetrate (Chapter 1).

## ISSUE 2: THE NATURE OF INSTITUTIONALIZED SCIENCE

If we see science as a highly cultural institution—carrying along cultural assumptions, as much as theories, evidence, and bits of knowledge—we can better understand the two primary lines of thought about it in current social scientific discussions. Each line of thought has an army of supporters, and the field is substantially composed of the conflicts between these armies. In our view, both truths derive from the cultural character of science—and given this character, they are two sides of the same coin.

On the one side, associated with the name of Robert Merton, is a conception of science as highly structured around some general and basic and universalistic values. Professional and organizational systems articulate these values in terms of shared standards of rationality and merit (Merton 1970, 1973; Zuckerman 1989; Cole 1992; and others). Culture, in these conceptions, is more a matter of values and preferences than the cognitive models of reality that we emphasize in this book. The picture of science as rooted in some basic value preferences is often then linked to a functionalist conception of science as useful for, or needed in, modern society, but this linkage is not crucial. What is crucial for this rather realist picture of science is the emphasis on the magisterial professional and organizational clarity around general values and models, built into clearly bounded categories and stratification systems. The edifice is most impressive, and much evidence supports it as quite real, with, for example, worldwide patterns of scientific citation taking on orderly and stratified forms. The imagery carried along in this tradition suggests the existence of a controlling true natural reality and a set of sciences that increasingly comprehends it.

This elaborate professional and organizational form, extending dramatically across time and global space, is exactly what institutionalists might expect from a dominant world cultural system, especially when this cultural system confronts an empirical world that is often opaque. The actual physical and social environment, as we experience it, is messy, inconsistent, and variable even in local settings, let alone global ones. To impose a common cultural frame on it requires a great deal of ritual or ceremonial clarity. There must be clear hierarchies, from Nobel Prize winners down to mundane technicians. There must be clear boundaries around structures such as the doctorate, disciplines, and so on. The core activities must go on in special places, such as universities and laboratories far removed from the real world. The associated search for truth must take on cosmological issues far removed from immediate instrumental concerns (like the university, science should be universal—and thus for the most part impractical). Finally, there must be sharply defined standards of merit, value, and truth. All these things are, quite dramatically, central features of the very public pageant of science: once science becomes the central cultural frame, the pageant expands in length, dignity, color, and decibel (Wuthnow 1987).

Operating under such claims, science can obviously be seen as an emperor with no clothes: the high claims of the scientific parade necessary to hold up the modern system far transcend the muddy knowledge systems actually in place. So in dramatic opposition to the sociological analysts of high science come the contemporary students of science as it actually seems to happen backstage and offstage. Under labels such as "the sociology of scientific knowledge," or "science and technology studies," these researchers trace histories of scientific ideas and evidence and the details of day-to-day work and relationships in laboratories. They see the pervasive interpenetration of science with interests and powers in the larger society around science, and with the petty little social factions and schools within it (for example, MacKenzie 1981; Latour and Woolgar 1986; Bijker et al. 1989; Barnes et al. 1996). The imagery suggests a knowledge system little controlled by the nature supposedly under investigation, and heavily controlled by all sorts of social forces, from power in society and academia to the interests of internal and external participants to the small accidents arising from social events and networks. Again, the evidence produced in this research tradition here is impressive, and one can see much posturing in the pretenses of mundane science.

Between these extreme views come various practical compromises, emphasizing the simultaneous effects on the development of science, of evidence from the actual empirical world, and of the social networks of scientists studying that world (for example, Crane 1972; Latour 1987; Callon et al. 1986). These compromising approaches are valuable, and are recognized as such, in research on science as it happens. They are not especially relevant for our problem here, which is about the expanding social role of science, not its internal arrangements or effectiveness.

Our mission in this book is to analyze the authority of science in the modern system, not principally the activities of the sciences themselves. We are not in the business of assessing whether scientific knowledge is objective or constructed. The important thing for our purposes is to understand how two dramatically competing camps, both quite convincing in evidentiary terms, can arise and rather stably survive in the analysis of science.

This result is quite a normal one, according to sociological analyses informed by institutional theory. The term *loose coupling* describes it (March and Olsen 1976; Weick 1976; Meyer and Rowan 1977; Brunsson 1989; Krasner 1999). Highly institutionalized value systems, especially those carrying cultural authority, are located in clear and crystallized social forms that are sharply bounded from many practical realities. This permits great clarity in values and social authority. The activities within these schemes, and covered by them, are carried out backstage with much practical adaptation, covering up, and hypocrisy (Goffman 1959). The more inspiring and universal the values and claims at the institutional level are, the more bounding and decoupling at the practical level are likely to result.

Religious systems provide many examples of highly institutionalized yet decoupled social systems; they carry the most sweeping values known to humans in form, doctrine, and ritual, but in practice, they are accompanied by much disimplementation, not to mention squalor and corruption. In the modern system, an institution closely akin to science—namely education—has provided many research examples for similar traits (Meyer and Rowan 1978). National and worldwide standards of curriculum and knowledge, institutionally located in sweeping sets of standardizing diplomas and degrees, are accompanied by enormous (but often concealed) failure, inconsistency, and variability. We know in great detail exactly who is a college graduate and which occupations require a college degree, but we conceal the fact that some of these people are illiterate.

The cultural relevance of science to every sort of activity in the modern system is extreme; but the loose coupling involved is also extreme. The scientific training and research that are supposedly relevant to practical life go on in universities, quite segregated from this life. Science is organized around worldwide categories, models, standards, and communication systems, not local needs and activities. Many of the issues, topics, and theories discussed are far removed from any issues of instrumental importance and seem to be constitutive rather than instrumental.

If a scientized model of reality, and of the competence of individuals and rationalized states within reality, are among the central cultural principles of the modern system, it makes sense that they must be institutionalized in a generalized ceremonial system kept loosely coupled to the detailed practices involved. It would not be helpful to have the ordinary thoughts and activities of the crucial priests of the modern Delphic oracles routinely exposed. In no way could the mundane activities of scientists sustain the hubris of the modern human rational actor.

ISSUE 3: THE EFFECTS OF SCIENCE AS AN INSTITUTION

The success of institutionalized science has led to the assumption that it has direct instrumental values for the modern system. This happens in much the same way that the success of institutionalized education as the route to individual and social progress has led to doctrines that education has some sort of instrumental value as human capital. In other words, functionalist models are institutionalized in images such as a "research and development cycle." In standard sociologies of science, similar models occur. Even in the critical sociology of scientific knowledge traditions, there is the notion that however perverse, the scientific system functionally supports an array of activities, powers, and interests in society. As metaphors, such ideas are established and successful, and enough cases can be found to illustrate any theory with so much a priori status. But serious empirical analyses have found it unexpectedly difficult to support simple causal connections between scientific development and real-world progress. At the cross-national level, for instance, Shenhav and Kamens (1991) report negative effects on economic development, and other researchers find mixed evidence on the effects of science on economic growth (see Chapter 10).

If science is part of the constitutive culture of the modern system as much as an instrument of technical improvement, this situation makes more sense. The claimed domain of science is extreme and covers every aspect of the modern social environment. Even if science is seen as an instrument of power, the powers supported are broad ones. The idea that science helps make the world governable for centers of power (for example, the expanding national states) has as one component the notion that science provides techniques for social control across a broad set of domains (Foucault 1980, 1991; Rose and Miller 1992).

Science clearly affects every aspect of modern activity: no modern institution escapes its purview, and the scientization of any domain in a particular social setting clearly transforms it. Scientific expansion fuels modern environmentalism in polity and economy. It provides medical and social scientific evidence that changes the rules of family life, supporting penetrative analyses of both intergenerational and marital relationships as abusive. It shows ways in which long-distance economic relationships support inequalities that violate basic human rights. It can do all this governmentality on a world scale, affecting the policies of both dominant and peripheral powers and helping to mobilize a transnational system of organizations and professions (Chapters 11 to 13).

These effects support particular forms of development and may hinder other forms. In particular, they may limit the impact of science in the much-discussed area of economic development. If science functions as a broad cosmology, and if the forms of rationalization it supports cut across the board, its effects on economic development are not as simply positive as the literature supposes. Scientific authority supports economic rationalization—but it also supports more rationalized ideas about human rights, the environment, the requirements of world equality, cultural diversity, and the rights of indigenous people. These latter notions can support many constraints on purely economic development.

OVERVIEW OF THE FOLLOWING CHAPTERS

We see science as a constitutive cultural (or religious) part of a rapidly expanding modern world polity (Thomas et al. 1987; Boli and Thomas 1999). This conceptualization helps account for its expanded authority, its diffusion into both central world institutions and the world's periphery, and its expansion into new domains. The expansion and the authority of science grow as the emergent actors and organizations in the modern world polity depend on science for legitimation.

This general argument is developed in Part I of the book. Chapter 1 develops our account of the expanded authority of science in the modern system as a component of a world polity and culture, and Chapter 2 attends to the methodological problems of studying a process that is both worldwide and national at the same time.

Part II of this book analyzes the consolidation of a global field of science. Chapter 3 focuses on the rise of a global scientific organizational system, examining the rapid expansion in international associations in the domain. Chapter 4, in parallel, focuses on the discursive themes that produce and result from such organizational integration. We live in a world in which, despite much apparent cultural diversity and conflict, scientized communication on every dimension of public life flows easily and smoothly.

Part III describes and analyzes the impact of this global scientific system on national states worldwide. Chapter 5 examines the blossoming of national ministries of science in recent decades, reflecting the worldwide prominence of the institution, diffused through global linkages rather than local instrumental requirements. Chapter 6 shows the expansive meaning of, and striking global homogeneity in, science educational curricula as they developed over the twentieth century. It also shows the dramatic emphasis of science, as taught in mass education, on both the empowerment and the rationality of ordinary individual persons. Science, in the modern curriculum, is an appropriate form of thought for people in general, not mainly for disciplinary or functional specialists in particular. It is thought to be a natural way for people to think. Chapter 7 shows that the flow of all sorts of scientific arrangements from world society down to the national societies occurs on a loosely coupled basis, so that there is relatively little national-level coherence in the system. Order at the world level produces both conformity and decoupling at national settings, exactly as institutional theories predict (for example, Meyer and Rowan 1977). Chapter 8 shows worldwide isomorphism in the expansion of the participation of women in science training, again reflecting the status of science as culturally supporting universal individual rationality rather than specialized functional roles. Chapter 9 then shows the worldwide isomorphism in the spread of differing scientific fields—findings that reflect the general cultural rather than narrowly instrumental status of science; it also accounts for some of the variations in styles of emphasis on particular science fields. Such variations, much like the similarities across science domains and across world regions, reflect variations in political culture more than the technical requirements of a national economy.

Part IV shifts attention to the national-level consequences of scientific expansion. Chapter 10 shows that the effects of this expansion on contemporary economic growth rates are more problematic than most theories suggest; we propose in this chapter that this happens precisely because of the cultural character of science, which legitimates social rationalization across a broad scale. Positive effects of the expansion of scientific training are demonstrated, along with some negative effects of national research expansion; these latter effects are associated with scientific systems that emphasize a broad range of fields. Chapters 11 to 13 then demonstrate the effects of scientific expansion across this broad scale—on environmental protection, democratization, and organizational standardization and rationalization. The Conclusion provides an integrative review of our arguments and suggests new lines for future investigation.

COMMON THEMES

Throughout this book, some themes recur. First, science globalization generates increased homogeneity in form, or institutional isomorphism. This happens in science itself (Chapter 9) and also in educational curricula (Chapter 6). Large demographic changes in participation in science move in parallel (Chapter 8), and state organizational controls arise at the same time in radically different countries (Chapter 5; Finnemore 1996). Beyond science itself, this pattern characterizes the dominant entities of the modern system—nation-states, organizations, and even individual identities—as these are affected by standardizing science (Meyer et al. 1997a).

Second, although scientific authority diffuses worldwide and is incorporated in all national states, it has too much significance and too much a formulaic and standardized form to be incorporated in a coherent manner. Scientific practices expand, tied to a global system, but they are loosely coupled with each other and with national policies and practices, especially in developing countries (Chapter 7). They are, however, tightly linked with the general standards (Chapter 4) and organizational systems (Chapter 3) of world society. Naturally, science is closely linked to the university, itself better known for cultural authority than for functional efficiency (Riddle 1989).

Third, the transnational and cultural aspects of science and its authority that generate both worldwide isomorphism and local decoupling also produce effects on many dimensions of social life that are both massive and diffuse. The effects cut across many institutions. Scientific expansion under-lies the worldwide expansion in social organizational rationalization (Chapter 13), in human rights and democracy (Chapter 12), and in environmentalism (Chapter 11). Perhaps precisely for these reasons, its immediate effects on national economic growth are much less certain than is generally taken as given. The once-exotic scientist, as a cultural functionary, now seems to sup-port uncontrolled growth with one hand, but rationalized controls over this growth with the other hand.

CONCLUSION

The modern cultural frame supports much faith in human empowerment and agency, along with a highly rationalized and tamed picture of the natural and social environment within which social action is to occur. Science is a crucial part of the cultural frame that the modern system creates and depends on. It is crucial both in the creation of the human as scientist and the construction of nature as scientized. In the current period, the whole system has become worldwide in practice and global in ideologies and aspirations. At the world level, international professional and organizational expansion and structura-tion has been extreme (Haas 1992; Boli and Thomas 1997, 1999). The con-temporary overall result is a great expansion in the authority, domains, and activities of science, both internal to countries and internationally.

## Part I

# The Globalization of Science from a World Polity Perspective

## Chapter 1

# World Society and the Authority and Empowerment of Science

OVERVIEW

In this chapter, we put forward explanations of the expanded and institution-alized authority of science in modern societies. As we suggest in the Intro-duction, the contemporary worldwide expansion of scientific activity can be seen as reflecting this expanded collective authority as a cultural matter, rather than as producing authority through accretions of functional success and organizational power. And the crucial element of the expansion of science is indeed its broad authority, rather than its instrumental functions, powers, and interests. In this sense, science operates as the secular equivalent of a "sacred canopy" for the modern order, generating a modern, rational inter-pretation of world order and offering this logic as a secular interpretive grid for natural and social life.

Science arises and expands in close conjunction with the modern cultural invention and expansion of the model of the rational and purposive social "actor," whether individual, organization, or nation-state. For this highly agentic and competent actor, science serves functions far beyond those of an instrumental tool: it plays a broad role analogous to religious ones, providing constitutive and legitimating (that is, ontological and cosmological) supports for the hubris involved in claims to actorhood.

An explanation for the contemporary explosion of scientific authority and activity, then, may be the rapid recent expansion in the rights and responsibilities of a variety of social actors, from individuals to national states. And this expansion, in turn, is clearly a product of the rapid globalization of the current period, with the rise of a world public society, or polity. Worldwide, individuals, accepted as human persons, are accorded rights and capacities that were previously unrecognized or restricted to the citizens of a few core countries. National states, with greatly expanded rights and duties and attributed capacities, cover a globe in which previously only a few had much standing. And rational organizations, with elaborate responsibilities and capacities, are found everywhere, in every domain. The idea that all sorts of entities function as actors with exaggerated rights and capacities pops up everywhere in the modern world society. This idea implies that a wide range of actors enjoy an enormous capacity for a human agency that is conceived as a rational, purposeful, and ultimately predictable articulation of valid interests. Supporting this idea and its triumphant dissemination as a cultural frame, a world scientific culture expands and intensifies. The current structure/agency debates in the social sciences underestimate both the degree to which agency itself is a worldwide standardized social construction and the extent to which scientific authority supports and legitimates the agentic actor.

Taking this view to the authority of science in the modern system helps explain many features of science: its rapid spread around a world that in instrumental or functional terms is diverse, its spread across many different social domains despite limited effectiveness, the worldwide standardization of its forms and contents, and its tendency to focus on a variety of questions far removed from any instrumental significance.

Thus, to explain the modern global explosion of scientific activity, discourse, and organization, we need to understand the ways that the contemporary world polity has expanded conceptions of human social actorhood, as carried by persons, states, and organizations. Globalization has built up expanded sets of truths that are to apply everywhere, with science as the obvious cultural vehicle. And it has expanded and standardized the meaning, responsibilities, and rights of all sorts of social actors. Science comes into play to prop up the necessary model of the agentic actor, who with the proper approach and socialization can understand everything (Meyer 1986). So science props up the necessary model of the universalized and lawful environment, in response to which a scientifically understandable agentic actorhood is obviously correct and desirable.

## EXPANSION AND AUTHORITY

All the sectors of the rationalized society—the economy, political system, family and socialization, medicine, and others—rest on a cultural base that is substantially scientific in character. These sectors also change rapidly with scientific developments; for example, the introduction of modern ecological, economic, or human rights arrangements relies heavily on scientific advances and draws extensively from the scientific discourse. Or consider the raising of children: a variety of medical, biological, psychological, and social sciences define how it should best be done. These sciences generate clear cultural rules; they also generate categories of scientifically trained persons who can instruct (for example, in court) on correct or negligent forms of child rearing. The sciences have clear authority, but this authority is cultural in character. That is, the scientists do not raise the children. Rather, they instruct "real" social actors how it should be done. Individual actors are to do it. States are to provide legislation and resources. Organizations of various sorts (child care, medical, or educational) are to manage parts of it. All are to play their parts in a cultural drama laid out by the sciences.

The modern scientized society can be contrasted with more "traditional" societies, in which some religious or traditional authority over a limited range of social life is vested in a few elite interpreters, but in which much authority is embedded in mundane society. The priests may make some rules, for instance, about child rearing, but more are carried along by generations of grandmothers. The authority is now more concentrated. Authority has moved to science, which penetrates far down into society and spreads across all the different modern sectors and national societies. Direct reference to scientific authority routinely appears in legislative and courtroom hearings, in business decisions, or, for that matter, in conflicts between spouses. Religious leaders, too, now gain stature to the degree that they appear to be scientific; even grandmothers may try to appear so.

The scientization of society, or the penetration of science into everyday life, is extreme (Chapter 3). Science is thus no longer confined to exotic academic locales; it also resides in states and other organizations, and in the routine thoughts and practices of ordinary individuals (including children; see Chapter 6). Also, beyond physical nature, science addresses social and psychological worlds, and something close to the divine (as in theories of the big bang, or of extraterrestrial intelligence). The domain's expansion makes science a label of expertise and specialization available for a wide variety of cultural materials, so we can have "food science" or "sports science." The scientization of society extends to expressive issues as well: in order to resolve matters of observance of the Sabbath, rabbinical discussions focus on the technical details of the operations of elevators and the physics of gravitational forces; and in order to define death, authority goes to doctors, biologists, and psychologists. Finally, social concerns are defined in scientific and technical terms, from the naming of periods of time by their scientific advances (the Green Revolution or the Space Age) to the use of scientific methods in everyday discourse (the display of synoptic maps in weather reports or statistical error margins in polling reports).

CURRENT EXPLANATIONS OF SCIENTIFIC EXPANSION

Explanations for the modern expansion in scientific authority are less well developed than one might expect. One tradition sees expanded scientific authority as reflecting the expansion of science itself—the greatly increased numbers of scientific personnel, activities, research products, and investments in modern national societies (Ben-David 1990; Cozzens 1997). In this tradition, scientific expansion can simply be described, with explanations taken to be almost self-evident. Science expands and is frequently referred to because it "works" and provides functional advantages. The implied causal arguments here are that scientific expansion arises from the pressures, or needs, of powers and interests governing modern society. Science, in other words, is functionally necessary, either for the social system as a whole or for the dominating political and economic interests within it. Seeing science as an instrumental or functional necessity is clearly part of the modern culture itself. But science often spreads more as a faith than as a system of well-established and clearly known functional relationships (Wuthnow 1987). It has been unexpectedly difficult to document the functional necessity of, or contribution of, science in the modern system (see Chapters 4 and 10; Kuhn 1977; Shenhav and Kamens 1991). It is easy to show the important social consequences or benefits of particular scientific or technical developments, but it is much more difficult to show strong and consistent positive effects of overall scientization. Our own research reveals that some dimensions of scientization enhance economic growth, whereas others show negative effects (see Chapter 10).

Moreover, this faith in the utility of science in general obscures the obvious lack of functionality of many particular scientific ventures. Much scientific activity has always seemed to have few instrumental benefits for the modern system or for particular elites in it, as with the current and expensive exploration of various distant and receding galaxies, or the reproductive behavior of various organisms. Economists, mostly as a matter of theoretical commitment rather than strong evidence, subscribe to doctrines of the economic benefits of science but have not even convincingly demonstrated the marginal utility of an economist; for most social research, it would be most difficult to specify directly functional consequences.

A second, and more critical, explanatory theme in the field is that the expansion of science is produced by the interests of scientists themselves and the scientific organizational system (Merton 1973; Ben-David 1990). This line of reasoning supposes that functionally successful science can build up strong organizational and professional support, or it can imagine that the whole process is a kind of successful political strategy (Abbott 1988). Scientists clearly do perpetuate their own necessity to justify their professional standing, to encourage further sponsorship of their activities, and to support their social status (see, for example, Finnemore 1996). But unless scientists are seen as having extraordinary manipulative power and competence, and unless everyone else is seen as foolishly taken in, this explanation assumes what it sets out to explain. Why would members of the scientific elite have such extraordinary authority or legitimated empowerment to pursue their interests, given that they directly master neither the economic structures (capital) nor the political ones (the state) understood to control modern society? It seems clear that any form of power in the modern system depends on scientized conceptions of reality; understanding why this is so makes it clear that the scientists are much more than simply servants of power (see Chapters 12 and 13).

A third explanatory theme sees expansion as built into the nature of scientific activity itself. A less critical version sees science as reflecting a natural human curiosity, and as developing by creating endless new arenas for this quality. Science thus endogenously feeds itself, providing agendas and incentives for autonomous expansion (for example, Cole and Cole 1973). A more critical version—the "sociology of scientific work"—emphasizes the social embeddedness of this process, and the arational and arbitrary ways in which both social and scientific interests feed into a path-dependent expansion (see, for example, Callon et al. 1986). But this essentialist perspective on the motives for scientific expansion offers no explanation for the recent

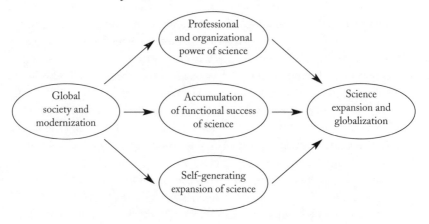

*Figure 1.1.* Factors Affecting the Global Expansion of Science

phenomenal rates of global proliferation or for the intense recent global structuration of the whole arena of science (Chapters 3 and 4).

Figure 1.1 shows the three dominant explanations of the global expansion of science in recent decades. A first theme emphasizes scientific functional success, aggregating across microcontexts to a macro–social effect, as a driving force. A second theme, directly macrosociological, emphasizes the professional power, organization, and interests of the scientists themselves. A third theme emphasizes the natural or cultural spirit of exploration and the self-reinforcing effects of scientific development over time. In a broad sense, all three themes perceive science as a natural (for good or ill) product of the long-term process of social modernization, becoming global as the modernization process itself becomes worldwide.

## THE PROBLEM

The primary explanatory issue has been poorly formulated in two ways. First, the real problem is to explain the legitimization, authority, resources, and attention given to science in the wider system, not the particular organizations and activities involved. Why does the extraordinary modern legitimated empowerment of science arise and expand? Why do all sorts of social forces—both populations and elites—pour attention, prestige, and resources into scientific activity? Scientific authority is to be found not in scientists and their particular activities themselves, but rather in wider societies that create, confer, expand, and concretely uphold this authority. It is built up by nations, states, and organizations that support scientific activity and that rest policies and activities on scientific knowledge. It is built by people—especially elite people—who, en masse in the modern world, send their young to study the sciences in universities, who help expand these universities and incorporate in them much scientific work in the name of various common goods, and who create the rules conferring social and economic advantage on the properly trained and credentialed persons so produced.

A central feature of every modern stratification system is that the greatest social prestige, and often status and income, are given to the schooled professionals. This is in sharp contrast to dominant theories, which stress the importance of state power and control over economic capital as the engines of stratification. In fact, the elite professionals, who dominate the prestige system, commonly do so through claims to scientific knowledge (Abbott 1988; Treiman 1977; Meyer 1994a,b). We need to explain why this is. It is unlikely to be the manipulative product of the professionals themselves because the behavior of others and of whole societies is involved in this conferral of status and authority.

As a second aspect of the explanatory problem, the broad social authority involved goes principally to science in general, as a broad cultural abstraction. It does not go mainly to specific (and possibly functional) particular activities and analyses. All sorts of scientists, from engineers to social scientists to astronomers, carry the prestige of science. All sorts of scientific roles, from educators to lab technicians to MBAs, are elaborated in the universities. All sorts of scientific training, from sociology to biology to abstract physics, are institutionalized and sought after. Instrumental theories try to explain variations in social investment among fields, and there are indeed such variations to explain; but in explaining the trees, the forest of a huge overall social and educational investment in science in general is underemphasized. For a parallel, one must imagine an economy in which profitable firms are greatly rewarded, less profitable ones are rewarded almost as much, and firms with no intention or hope of making a profit are rewarded only a bit less. Obviously, such a situation reflects a society attaching great meaning and value to the general category "firm." In the same way, a wide range of activities linked to the category "science" get great support, legitimacy, and often resources (as in the allocation of university positions) in the modern system.

Thus, science is to be seen as a broad and general authoritative cultural canopy covering an enormous domain of valued activity in the modern system. The problem is thus to explain the character of the institution as an authoritative one in modern society, not the specific activities, roles, and organizations involved. One cannot explain the modern general authority of science by looking mainly at the often-careerist and often-chaotic activities of the scientists and their organizations. The general and expanded authority of science is a property of the modern society, not mainly the scientists. An institutionalist perspective emphasizes that such a focus on scientists themselves and their interests is itself scripted and shaped by global myths, located in a polity that is substantially a cultural construction and that is worldwide in both claims and realities.

SCIENCE AND THE MODERN ACTOR

The global and cultural dimensions of modern science establish it as a central feature of modernity and as one axis of the modern, initially Western and now global, polity. Science's stand as a global cultural institution and its cultural qualities are central to understanding its social role and the impact of its globalization. And it is in its capacity as a global cultural institution that science alters all other modern institutions. We need to understand the forces that so strongly support this transcending cultural institution.

Science, seen as an authoritative and general cultural system, has close links with the construction of the modern agentic social actor. Science is thus not only socially employed for practical purposes but is also elaborated as a vital cultural frame (the institutionalist perspective), supporting the constitution or legitimacy of social entities as social actors. Actorhood is the principle that social life is built up of actors—human individuals, organizations, and national states with valid interests that others are to respect, and with the capacity (that is, agency) to validly represent those interests in activity (Meyer and Jepperson 2000). Rational and responsible human actorhood that everyone else is to respect only makes sense if, first, the world is a rather lawful and orderly (that is, scientizable) place, and second, if the human social actors can understand it (that is, be scientists). But if the world indeed has these properties, doing and having science is practically an obligation—and a highly valued one at that. In the contemporary world, the cultural authority is such that a great scientist—Einstein—can be defined as "Man of the Century" by *Time Magazine*.

It is generally understood that the rise of modern science and the rise of the rationally organized actorhood of individuals and states (and derived formally organized actors) go together (for example, Ezrahi 1990). Out of a medieval matrix in which order (and thus real actorhood and action) occurred only in a spiritual plane, and in which the natural world was seen as chaotic and "fallen," arose both concepts of strong human actors, empowered by their access to the wider spiritual and natural laws, and an orderly and scientized conception of the nature that these actors could comprehend and in which they could properly act. The Renaissance, Reformation, and Enlightenment progressively built modern individuals and states. They also, in subsequent revolutions, linked these together in the modern nation-state, which during the nineteenth and twentieth centuries led to organizational revolutions that created all sorts of organized actor structures in both state and society. Each step of the process also expanded the scientization of the environment within which the empowered actors take form: actorhood expands science, and science expands actorhood (Toulmin 1990).

Modern society is culturally constituted as made up of these social actors, with changing content and meaning over time. Foundationally, actors include human individuals. This conception has greatly intensified on a worldwide scale over time, and individual actorhood is increasingly seen as a matter of global human, rather than national citizen, rights (Soysal 1994). States (increasingly over time nation-states) are also foundationally seen as legitimate actors, both domestically and on a world scale; this is a central meaning of the principle of sovereignty (Krasner 1999). Many other recognized organized actors are built by individual actors in association with one another (as with the modern corporation) or by devolution from state actors (as with public bureaucracies). In contemporary societies, mixtures of both are common, as in public good, or nongovernmental, organizations in civil society (contrast Meyer 1994b; Coleman 1974; and Perrow 1991 for general perspectives).

The modern liberal society sees itself as produced and maintained by such actors, with a polity, economy, and culture almost entirely driven by their choices, rather than by nature, tradition, history, or spiritual powers. In this sense, actors are seen as bounded or as disembedded from wider spiritual, natural, and social forces: they carry an autonomous capability to act on their own purposes. Thus, the defining elements of actors include boundaries and purposes, coherent means-ends technologies using assessable resources to achieve the purposes, unified sovereign authority, and effective control systems over behavior. Such actors can understand and manage these aspects of their identity—and they do so with a comprehension of the lawful and orderly physical and social nature within which they are to act. This conception of the actor as a sort of all-competent "good guy" is of course unrealistic, particularly on the agency side, and much religious and cultural—and in the modern period, scientific—support is needed to sustain it. The myths of actorhood require that individuals, organizations, and states are understood to maintain elaborate information systems, extraordinary searches of the environment, extremely complex decisions, and highly effective control and implementation systems. Science, nominally taming the environment while nominally empowering the actor, is a great support (see Brunsson 1989; Brunsson and Sahlin-Andersson 2000).

Each of the specific properties of the modern actor is interdependent with a rationalized or scientized picture of the environment (prominently the natural environment, but increasingly also the social and psychological ones). Purposive human action makes little sense if the environment is utterly disordered, chaotic, corrupted, and unpredictable. In such disordered situations, means-ends technologies, and the rationalized use of resources toward goals, are similarly disabled, and the sovereignty of the agentic actor and effective control over the machinery of action lose meaning and legitimacy. Thus, in order to sustain actorhood, some science is vital. For example, if societies and individuals are to be responsible on matters of health, a coherent knowledge system must be assumed. Or as another example, a rationalized and universal educational system requires much scientized educational theory, and those who wish to put forward such a system generally support the sciences involved.

The scientific rationalization of the environment is thus crucial to the constitution of human agentic actorhood. If people can know the laws of disease, they acquire the capability and responsibility to control their behavior, as rational or rationalized actors, in light of these laws by such means as hygiene and immunization. If the technical and resource environment of economic production is scientized, states and organizations similarly gain legitimate authority and responsibility to take action, such as planning production cycles and manipulating currency exchange rates. Thus, the scientific rationalization of the environment of action creates the orderly uncertainties (making matters lawful and releasing them from the grip of arbitrary or chaotic forces) and calls for the consolidation of empowered actorhood.

In this sense, historically strong conceptions of the human (and state, and

organizational) actor arise in interdependence with the rise of the scientization of nature, including the nature of human actors themselves. And there is
a continuing cultural interdependence between strong models of actorhood
(for example, liberal individualist ones) and the scientization of nature (Frank
et al. 1999; see Schofer 1999a,b for discussions of the general relationship).
Scientized domains (for example, greater control over energy resources;
human sexuality) expand the authority, responsibility, and capacity of the
human actors involved. And such a call for the display of actorhood (authority, responsibility, and capacity) is translated into social mobilization around
claims of uniqueness. Thus, the scientization of human sexuality helps consolidate the claims for the rights of gay and lesbian people (Frank and
McEneaney 1999). Similarly, the scientization of the human life course for-
malizes the status and conditions of childhood and sets the foundation for
children's rights in labor laws, custody battles, and international educational
aid (Chapter 12).

Causal relations between scientization and the expansion social actorhood
run both ways and operate at both actor and cultural levels. At the cultural
level, expanded scientific rationalization creates expanded models of proper
human actorhood. For instance, the eighteenth-century development of
modern scientific analyses of the economy created an enormously expanded
model of the competencies and responsibilities of the proper human actor,
who became empowered and responsible to make the widest variety of
rational choices about consumption, productive investment, exchange, and
technical improvement. The same processes operate also at the actor level: the
new and scientifically legitimated market society created incentives for actorhood competencies that were previously much less important, or even stigmatized. In a similar process, the human capital and human rights revolutions
after World War II, with their scientific groundings, dramatize the triumph
of actorhood.

In the other causal direction, expanded religious and secular conceptions
of the role of human actors in society and state made culturally proper and
necessary an encyclopedic expansion of human scientific knowledge. They
also generated motives along these same lines for the emerging individual and
state actors. For example, the expansion of genetic knowledge can help gay
and lesbian activists refer to actual or potential genetic characteristics to
establish their claims for recognition. Similarly, consumer or labor interests
can refer to knowledge and analyses on their market positions because such
market relations are elaborated on by economists. Overall, then, scientization

supports the claims of actors, whereas the notion of actorhood encourages further investigation of knowledge on potential characteristics.

## SCIENCE AS INSTRUMENT AND ONTOLOGY FOR ACTORHOOD

Two kinds of ties link scientization and actorhood: instrumental and ontological interdependence. In conventional thinking, only the first is relevant. But to understand the authority and cultural significance of science in the modern system, the second is more important.

On the instrumental side, much modern theory imagines that actors are rather complete, given, and naturally evolving entities. These entities/actors may lack needed information, or may demand information that expands their capabilities. Science and scientization meet such needs and provide knowledge required for actor functioning. Knowledge may improve the specification of goals (for example, education, medicine, or economic survival), the analysis of resources and of means-ends relations (for example, for efficient production), the overall analysis of the social and physical environment (for example, mapping and taxonomies), and the ability to integrate and organize sovereignty and self-control (for example, economic, organizational, and psychological capacities). As rationalized and rational individual and collective actors behave and compete, demands for improved knowledge are crucial. Obviously, much modern scientific activity can be analyzed as responses to such social demands.

But it is a mistake to take competent agentic human actorhood as a given. Instead, it is necessary to recognize that cultural claims for this actorhood have been enormously inflated in the modern system. Political, economic, and social institutions rest on assumptions of actor capacity and competence beyond any plausibility. Similar expectations arise about organizations—and nation-states, which are now seen as responsible for all sorts of human progress and equity. Faced with this situation, there are great pressures for scientization, less as instrumentally valuable for given actors than for the validation of actorhood itself. This highly institutional or ontological (or expressive) function of science becomes overwhelmingly important in the modern actor-centric system.

Thus, throughout the whole modern period, and perhaps especially in recent decades, even the search for nominally instrumental scientific knowledge and competence clearly takes on more than instrumental meaning. Over and above the utility of great rockets going to the moon or delivering war-heads, dramatic and symbolic elements are present that transcend ordinary social interests (which might benefit more from a mundane new fertilizer). The universal knowledge and capability of human actorhood is being demonstrated, or highlighted, by such memorable quotes as "one giant step for mankind." This and other demonstrations are ordinarily seen as emotional or expressive, but in our view, they might better be seen as reflecting the ontological role of science in the modern system.

Beyond this, much modern scientific activity cannot be seen as substantially instrumental at all. The current literature imagines that science has something of a life—at least, an inertial force—of its own, in an organized "natural curiosity," or a suboptimizing drive of the scientific establishment itself (for example, Cole and Cole 1973). One can better analyze the global expansion of science by seeing science, in relation to human actorhood, as having an ontological role. During the modern era and more intensely during the twentieth century, science provides elaborated, rationalized, and more complete conceptions of the nature of things, including the nature of the human and organizational actors themselves. Science thus establishes a cosmos in which responsible and authoritative human actorhood is enhanced: first, an eternal frame within which human actors exist, and second, a demonstration of the all-knowing agentic competence of these actors.

This ontological dimension becomes more important if one sees the pretenses of human actorhood as deeply problematic. As much modern organizational theory emphasizes, human (and organizational) actorhood is constructed, fragile, uncertain, and incomplete. The first problem of any legitimated modern actor is to create and sustain actorhood itself—not to act rationally and effectively. Brunsson (1989), for instance, makes it a central theme that modern organizations, in order to maintain their actorhood, often make decisions and act in ways that seem highly arational (for example, in making decisions they avoid wide-scoped searches and concentrate on only one alternative; see also March 1988; Kahneman et al. 1982).

The link between science and human actorhood helps explain some of the distinctive features of science as it developed in, and integrated into, the modern educational system (Chapter 6; McEneaney 1998). Curricula and textbooks greatly emphasize—increasingly over time—the general all-purpose value of thinking scientifically, as well as the value of encouraging the broadest range of individual human participants. They do not emphasize narrowly instrumental, technical, or disciplinary structures; the idea is to build in the child a strong human actor who is a sort of "participatory scientist," rather than a passive student who knows a lot of science and accepts it as a higher authority. Science is more a part of the umbrella of understanding for proper actors than an instrument that they control (and that controls them). The form emphasizes participation.

An important role of scientization, in the modern system, is to sustain cosmologies—or abstract and general pictures of the wider universe—within which human social actors exist and maintain their universal status, or legitimacy, as agentic entities. Two aspects of this seem important.

First, there is a continuing concern with the location of the human actor in the wider lawful and rationalized universe. This involves, far beyond instrumental considerations, a concern with the fundamental nature of human social, psychological, and physical life, and of the contexts of existence. Such are the fascinations with the linguistic, evolutionary, and sociocultural prehistories, including prehistories of the Earth, the solar system, and indeed the universe. It becomes important to know about the big bang, the existence of life on Mars, or the potential location of other intelligent life forms in the universe. It also becomes important to trace the evolution of the human species and its cultures. Decisions about such questions seem to have meaning, less for rational and instrumental action and more for an understanding and stabilization of existence. It strengthens the ontological status of the rational human actor to establish the known and lawful character of all of nature stretching out from this actor to infinity on every horizon.

Second, derived from religious origins and spiritual concerns, legitimated claims for the modern human actor involve agency far beyond instrumental or realistic considerations. The human actor is thus assumed to hold the capacity for informed and rational knowledge, understanding, and action (Meyer and Jepperson 2000). The "full" actor, be it human or organization, is to have an extraordinary level of consciousness and the capacity for conscious rational action on a universal level. Science clearly reflects this concern. It shows up in commitments to universal theorization, to an extraordinary emphasis on logical and mathematical formalization, and to an investigation of all sorts of questions as relevant to the achievement of true and universal theory (Toulmin 1990). It is thus important in constructing agentic human actorhood to establish not only that the universe is a lawful and rational place, but that humans can and do figure it out. Gratuitous displays of knowledge, information, and analysis help to do this (see Chapter 6 on the importance of "science as fun" in the modern school curriculum).

Both these ontological foci of scientific activity help explain why so much of modern scientific expansion occurs in culturally focused, rather than instrumental, settings. Science is to be found in industry and in industrial research settings, to be sure, along with a variety of instrumentally oriented special institutions. But its main place is in the rapidly expanding university systems of the world (Riddle 1989) and in the curricula of universal mass education (Kamens and Benavot 1992; Benavot et al. 1991). These locales support a broad and ontological, rather than narrow and instrumental, conception of science.

GLOBALIZATION, ACTORHOOD, AND THE DEMAND FOR SCIENCE

The contemporary period, especially since World War II, has involved a great expansion in human social actorhood. Individuals have more recognized human rights (McNeely 1995, 1998). Nation-states (present in far greater numbers) take on greatly broadened rights and powers (Meyer et al. 1997a). And the formal organizations derived from these fundamental actors have greatly expanded their functions—to the rational management of personnel, legal matters, environmental concerns, safety matters, and research and innovation (Brunsson and Sahlin-Andersson 2000). Further, nation-state actorhood, organizational expansion, and the celebration of individual human status have become worldwide, extending to the whole territory and population of the world.

Behind this dramatic expansion in rationalized actorhood lie the forces that come under the common title "globalization." Many specific components of the general concept are involved. The disasters of World War II brought individual human rights to global prominence (Rauner 1998; Lauren 1998). The corresponding breakdown of colonial empires created a large number of independent and sovereign national states. Both the cold war and the nuclear age called attention to the importance of the responsibility of these states for the condition of the whole world. Economic integration, as reality and as myth, made the conditions of both states and human individuals matters of global relevance and significance. In all these ways, the situation called for responsible articulate actorhood, rather than raw Darwinian competition among mute powers. And the entire scenario was played out in the almost complete absence of anything like an integrating and controlling world state (Wallerstein 1974 makes this a defining condition for the whole system). If any order were to be found and maintained, the actors would have to do it.

In this situation, many immediate purposes of many different parties expanded demands and claims for the actorhood of states and individuals and all sorts of organizations in between. Groups internal to society had legitimate claims for equity and progress—the state was the obvious agent to respond to these. Similarly, in a world of high interdependence, states had every incentive to maintain their internal controls by assuming expanded powers. In the wider world, claimant social groups and states had every reason to insist on the expanded responsibilities of other states in the system to meet their obligations; few alternatives existed in the absence of the old (colonial) system or a new world state replacing it.

All these forces assembled in an exploding set of professions (Haas 1992 calls them "epistemic communities") and international organizations (Boli and Thomas 1997, 1999) during the period. These structures were not themselves agentic actors, with real line authority and with direct responsibilities for action. They can be discussed best as presumably disinterested "others" who tell actors what to do (Meyer 1994b; Meyer et al. 1997a). Their existence generates expanded actorhood rights and responsibilities in the putatively real actors of the system. And the language they speak is principally the language of science, as every expert or consultant knows.

The expansion in responsible actorhood has clearly been reflected in the expansion of rationalized and scientized knowledge systems: as new domains are tamed by scientization, new arenas of rightful and responsible actorhood are created (for example, in scientifically analyzed "economies"). All this has gone worldwide (see Chapters 3 and 4; Schofer 1999a,b).

There are therefore instrumental demands for scientific knowledge in all sorts of domains coming under the legitimate responsibilities of various actors, for everything from medical and psychological knowledge to population and reproductive control to information systems, environmental understanding, and production technologies. As more and more domains of human life are rationalized in terms of the rights, responsibilities, and powers of human actors, instrumental pressures for scientization process are increasing. Such pressures result in an explosion of scientization, with more intense penetration of particular domains and extension of it across domains. As rationalized, or scientized, society (and its related feature of collective actorhood) expands, science becomes socialized in a well-known sense and applied to the widest variety of issues, and scientific organizations proliferate at both national and world levels.

By itself, this instrumentalized use of science leads in some analyses, to a taming and debasement of true science. In other analyses, it leads to the penetration of society by a scientific leviathan and a subjection of society to instrumental reasoning. Either way, these lines of thought leave out the cosmological or ontological side of the science/society relationship and thus mistakenly describe what is happening. Pure theory does not disappear into application; it is reinforced by the demands of the empowered modern human actor for ontological status as an entity in the universe and as a rational "knower" of the universe. Thus, every step in the expansion of human actorhood produces both an expansion of instrumental science (the technical gadgetry of applications) and an expansion in cosmological, universal, and logical/mathematical theorization. The same human actors who now take an interest in scientifically managing the details of diet, administration, and poverty now also avidly support massive expenditures for the study of miles of Antarctic ice, of theories about the origins of both humankind and the universe, and so on. The instrumental scientific rationalization of the human actor and its environment is accompanied by the ontological quest for understanding of when the sun will burn out.

Key indicators of the ontological rather than instrumental status of modern expanded science, as noted above, can be found in the fact that so much scientific expansion occurs in educational systems, with its diffuse foci. For instance, it occurs in the educational and public attention given to scientific discoveries of extraordinarily little instrumental significance: the discovery that a species of crow can make and use tools may be relevant to ontological questions about the human condition, but it is not likely to lead to increases in the gross national product.

GLOBALIZATION OF SCIENCE

Above, we discuss ways in which globalization in the contemporary period expands the responsibilities and rights of agentic human actorhood (at several organizational levels), generating enormous demand for science of all sorts to provide accounts of both actor and the tamed environment permitting rational action. But of course, globalization also works to supply scientific expansion meeting—but also creating and enlarging—the demand. Thus, Haas's (1992) epistemic communities of experts both respond to and create international agendas for expanded actor responsibility (for example, for air and water pollution, or for global warming, or to control child labor). And the greatly expanded set of international organizations formed in the current period, often relying on scientific presumptions, works the same way (Boli and Thomas 1997, 1999), simultaneously generating supply of and actor demand for scientization. Through both broad routes, an explosion in globalized scientific discourse and activity is produced (Chapters 3 and 4).

Science, throughout its history, has a universalizing character. In the contemporary period, it has achieved effective global status. The abstract commitment to science in general transcends variations in time and space that might generate variations in rational or functional investment patterns. Science—and science in general—expands rapidly, not only in developed core economies and societies, but in the third world too. Scientific activity expands, in fact, in every sort of country worldwide (see Chapter 9). Scientific practices, policies, organizational forms, and ideas proliferate worldwide.

First, there is the globalization of the scientific system. Scientific data, laws, and organizations are developed to apply everywhere in the world (and beyond). Biological and social scientific knowledge, as well as physical scientific knowledge, applies everywhere: it is understood to be useful everywhere and to extend its ontology everywhere. The furthest tribe remaining has a society and economy that can be analyzed social-scientifically, and exists in a biological ecosystem that can be analyzed scientifically. And the knowledge involved should in principle be applied everywhere, with appropriate policies. Thus, even peripheral people and societies can and must develop properly within the "laws" of science. The dominance of this perspective helps explain the rapid world-level expansion, in recent decades, of all sorts of international (both governmental and nongovernmental) scientific organizations, with widespread participation from all sorts of countries (Chapter 3; Boli and Thomas 1999).

Second, there is an essentially universal extension of the scientific system itself into national-level societies. Almost every peripheral country now builds universities, and these handle the great and expanding bulk of postsecondary schooling, in contrast to particular institutes that might be relevant to local functional needs (Riddle 1989). National states, operating under world-level encouragement, create governmental scientific establishments, and they organize (at least formally or constitutionally) as if the principles of a universal knowledge system are locally empowered and authoritative (Finnemore 1996). Domestic nongovernmental scientific organizations also expand greatly in the current period (Schofer 1999a). And in every type of country—

from the richest and most economically complex to the poorest and most economically simple—scientific instruction is increasingly emphasized in mass education (Chapter 6; Benavot et al. 1991; Kamens and Benavot 1992). The researchers who study the question find essentially no economic correlates: the process is worldwide, and occurs in every type of country with only minor variants (see Chapter 9 and Lee 1990 on the development of "Islamic science" in Malaysia).

Given the extreme international differences in available resources, it is not surprising that world scientific activity and production are dominated by the core countries. What is surprising is how much such scientific activity goes on in the furthest peripheries of the world—and how rapidly this expands.

CONCLUSION

Our overall argument on the way in which the modern globalized system has supported the expansion of scientific activity is illustrated in Figure 1.2.

As we have argued, science expands in the modern system as a secularized version of a "sacred canopy" providing a cultural base for expanded and now globalized human actorhood. Far beyond instrumental considerations, the perverse influences of the scientists themselves or of the products of natural curiosities, science provides a cultural umbrella sustaining the ontological status of the rationalized human actor. It defines a lawful and universal nature (including human social nature) in which the rationalized human actor can make sense, and it provides a system of knowing that establishes the capacities of this actor for complete comprehension of, and access to, this nature.

This suggests that the science that spreads around the world is quite homogeneous in content or focus—because it relates to general abstract principles and cosmologies rather than to particular functional problems. The available studies of worldwide scientific expansion reinforce this conclusion. Despite enormous economic and cultural variation around the world, general abstract models of actorhood and scientized environment are remarkably similar. Science curricula and textbooks in mass education are remarkably homogeneous and change over time isomorphically (Chapter 6; McEneaney 1998). Standard models of science in the universities are worldwide. University enrollments expand in every type of country; this expansion involves growth in enrollments of both men and women, even in the sciences (Chapter 8). Similar general guidelines for science policy are institutionalized

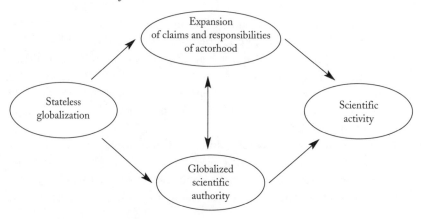

*Figure 1.2.* The Global Expansion of Science: An Institutional Perspective

across national variations (Finnemore 1996), and the same domains of scientific research are to be found everywhere (Chapter 9).

Some cross-national variations appear: these seem to be related to the extent to which national societies support scientific activity beyond the narrower "hard science" base that might link most tightly to immediate functional drives for national development (Chapter 10). Western liberal countries seem to support scientific activities in ways far beyond hard science, with large social scientific and medical research systems. Typical developing countries show a more limited range, particularly involving less social science research. Both communist countries and some newly industrializing Asian countries dramatically emphasize hard science research (and presumably instruction) in fields such as mathematics, physics, chemistry, and engineering (Chapter 9).

Overall, however, a broadly defined scientific system expands and proliferates worldwide. It expands as a global system, and also as a highly penetrative one into every type of country. With science carrying both instrumental and cosmological/ontological meanings, such global penetration is also highly consequential for nation-states worldwide. Science globalization alters the structure of nation-states by establishing scientific institutions locally. Most importantly, it changes their nature by allowing its rationale to become the axis of their social life—from defining the relationship between social and natural, to rationalizing social explanations and claims, and ultimately to empowering all forms of actorhood.

## Chapter 2

# Methodological Strategies and Tools for the Study of Globalization

EVAN SCHOFER AND
ELIZABETH H. MCENEANEY

Scholars in the institutional tradition have developed a broad research agenda addressing globalization and its effects on nation-states. This agenda involves two main focuses: first, documenting the rise of a "world society" consisting of organizations, institutions, and a shared world culture, within which nations are increasingly embedded; and second, studying the effects of world polity embeddedness on nation-state policy and behavior. Each research topic necessitates a different approach in terms of unit of analysis, types of empirical data, and statistical methodology. This chapter contrasts institutionalism with other perspectives on globalization, outlines the main research strategies involved, provides a detailed description of relevant statistical methods, and discusses some new research tools relevant for institutional research.[1]

## GLOBALIZATION PERSPECTIVES AND INSTITUTIONALISM

Institutionalism is a distinctive perspective on globalization, drawing attention to the formation of a world society and culture that transcends nation-states (for a full discussion, see Meyer et al. 1997a; Thomas et al. 1987). As such, institutional research on globalization is rather more specific than the congeries of topics denoted by the loose term *globalization*. Typically, globalization is used in a more descriptive sense, to characterize several related features of the contemporary world: increasing worldwide interconnectedness, which renders the world a "smaller" place due to new communication and transportation technology; the worldwide spread of capital, material goods, and various cultural forms resulting from increased interconnectedness (for example, the worldwide spread of McDonalds and rock music); and the increasing prominence of transnational structures, such as the United Nations (UN), the World Trade Organization, and multinational corporations. As a descriptive term, *globalization* encompasses a range of different worldwide trends, varying not only by intensity but also by their nature and the nature of their consequences.

Globalization, from an institutional viewpoint, is the rise and structuring (entrenchment in the form of institutions and organizations, or institutionalization) of a world society with a shared culture. This global "culture" is conceptualized as shared cognitive frames and understandings regarding the identity, goals, and activities of individuals, organizations, and states. Here, culture is not conceived as individual tastes or styles of dress, but rather as broad models or "maps" of how people and governments ought to behave. For example, the post–World War II growth of an international human rights "regime" supports a worldwide set of norms and views about human rights issues (to which nations and individuals adhere to varying degrees).

This world culture and society, often called the "world polity," evolved out of the West and reflects dominant Western values such as rationalism, universalism, progress, justice, and individualism (Boli and Thomas 1999). World culture is global not only in that it is built into transnational structures (for example, international organizations). It also embodies an explicitly global ideology or worldview, reflecting a global "imagined community," thus encouraging the identification of social problems and solutions on a global scale. The twentieth-century expansion of international associations and intergovernmental organizations reflects the increasing structuration of world society, rendering it more entrenched, stable, and consequential for other actors in the world.

The world polity provides norms and identity elements to actors (governments, organizations, and individuals) embedded within or linked to it, changing their behaviors, actions, and goals. In particular, nations are infused with cultural frames, or "models," and direct policy prescriptions that influence nation-states to be more rationalistic and progressive, and to be organized around and oriented toward universalistic perspectives. This results in great similarity (or isomorphism) of policies and behaviors worldwide despite the widely varying local circumstances that nations face. For example, researchers have observed that nation-states adopt strikingly similar education systems. Educational policy, organizational structure, curricula, and even the physical infrastructure of schools are similar throughout the world, even though national economies (and thus functional labor-force requirements) vary from subsistence agriculture to postindustrial societies (Meyer et al. 1992a; McEneaney 1998; Rauner 1998). Institutional researchers have argued that national strategies for organizing education systems are not created de novo in each country. Rather, they are imported as a package from the world polity—a package that contains cognitive models of, for example, how a "proper" school ought to be organized, what its specific goals are, and how it is to be implemented. Similar processes of policy diffusion and isomorphism have been observed in many instances, including national citizenship and suffrage laws (Ramirez et al. 1997), national environmental policy (Frank et al. 2000a), population planning (Barrett and Frank 1999), and science policy (Chapter 5). However, it is not just policies that spread. The actual cultural conceptions of "a student" or "a school" change in nations as global models diffuse and penetrate nations.

The distinctive contribution of institutionalism can best be seen in contrast to other perspectives, which focus mainly on economics: the impact of increased trade and capital flows across national boundaries. Economists and political scientists have taken substantial interest in global flows as a mechanism that changes national economies, alters the power of states compared with business interests, and consequently affects national policies. For example, it has been argued that increased world trade flows threaten workers in industrialized countries by forcing them to compete with low-wage workers from developing nations (for example, Alderson 1999). And the growing mobility of capital (and the consequent threat of capital flight) is thought to constrain the fiscal policies of governments, thus reducing the power of states relative to multinational capitalist interests (Keohane and Milner 1996). Economic globalization perspectives have relatively little to say about the social, organizational, or cultural dimensions of globalization that are the central focus of institutional analysis. At times, however, the two views of globalization make differing predictions about nation-state behavior and policy and thus can be construed as competing perspectives (see, for example, Frank et al. 2000a,b; and Buttel 2000 for the case of environmental policy).

A second tradition, one primarily associated with political science, conceptualizes globalization as the transnational agreements and organizations that nations create to solve collective problems, often referred to as international regimes (for example, Krasner 1983). International regimes range from relatively specific international agreements to share resources and avoid conflict (for example, over fisheries) to larger cooperative efforts such as the European Economic Community. International regime scholars frequently invoke the word *institution* to describe this process, but they retain a realist, actor-centric conception of institutions, resisting the rather Durkheimian imagery of a shared world culture that institutional researchers employ. International regime scholars typically envision institutions as the result of the aggregate behaviors of social actors operating under economistic or rational-choice logics. The assumption is that nations are self-interested actors that will cooperate only in circumstances where they stand to gain. The explanatory focus has been to understand the circumstances under which nations are willing to create, or submit to the demands of, transnational agreements and organizations.

There is little consideration in the international regimes literature of how world-level structures might have the capacity to fundamentally affect state goals and interests. Indeed, the notion of actors with predefined interests is the most basic assumption of economistic and rational-choice thinking. In contrast, institutionalists argue that political actors are not fully independent of this wider shared culture. Rather, they are constituted and empowered by shared understandings, models, and norms. This is particularly important when it comes to explaining why national governments adopt policies or behave in ways that do not seem to reflect local state interests, economic imperatives, or functional needs (as in the case of educational policy mentioned above). But even when national behavior seems to be "rational," it may be that the state is simply enacting world cultural models that are typically rationalistic in character. In sum, institutionalists view international institutions as more than the agreements among rational states. Rather, they are constitutive of national goals, interests, and behavior.

The research contributions of institutionalism have been quite distinct from other perspectives on globalization. Economistic views typically employ quantitative research methods to look for an association between increased global economic flows and changes in national policies or national economic circumstances. International regimes scholars have looked more explicitly at the history of intergovernmental organization and world institutions, generally by means of a historical or case-study approach (see, for example, Finnemore 1996). Institutional researchers have looked more broadly at the range of international structures that constitute a world polity. Typically, quantitative methods have been used to explain both the expansion of the world polity and its impact on nations. We now turn to a detailed discussion of this approach.

## THE RESEARCH AGENDA OF INSTITUTIONALISM

Institutional research focuses on the rise of "world society" and world institutions, and their effects on nations. This produces a dual research agenda. A first agenda is to study the origins, expansion, and characteristics of the world polity. To what extent can one observe a coherent world culture, society, and set of institutions that might plausibly influence nations? In which substantive areas are world society norms clearly worked out, codified, and institutionalized? And what processes caused the world polity to expand generally—or more particularly, around specific issue areas (for example, education, the environment)? Here, a longitudinal approach is required to study the historical building up of institutional structure in the world.

A second research agenda is to examine the effects of world polity structuration on nation-states. To what extent, and under what circumstances, does the world polity influence the behavior of social actors (particularly nations, but also firms and individuals) worldwide? Which nations are most embedded in the world polity and thus most subject to its influence? Such research focuses on nations, and the extent to which the world polity influences national policies. Here, a comparative, longitudinal approach is most often employed. The two elements of the institutional research agenda are substantially different in terms of research approach, data requirements, and methodologies. We discuss them separately.

## METHODOLOGICAL APPROACHES TO THE EXPANSION OF THE WORLD POLITY

The study of the world polity involves tracing the evolution of the international sphere over time. In essence, the "world" is the unit of analysis—that is, the transnational environment around nations, sometimes called the "world level" to contrast with the national, organizational, or individual levels of analysis.[2] Of interest are the properties of the world as it evolves over time: Is there evidence of a global shared culture and worldview that transcends nations? To what extent is there structural capacity at the world level to support norms, to prescribe policies, or to address particular issues or problems (for example, environmental or human rights issues)? Research methods used to study these processes necessarily focus on the historical variation over time. Case-based approaches have made substantial contributions by applying this approach to particular fields of the world polity, such as women's rights, development, or environmentalism (see Boli and Thomas 1999 for examples). However, we focus on quantitative and statistical approaches, which comprise the majority of existing research.

*Data on the World Polity*

An initial challenge has been to find data sources to document the existence and content of the world polity. Most commonly, institutional scholars have looked to concrete organizations or activities that indicate transnational interaction, cooperation, and structure: intergovernmental organizations, treaties among nations, international associations of individuals and professional groups, and international conferences and meetings. Many additional phenomena are suggestive of a world polity, some of which substantially predate modern international organizations. Historical evidence of widespread travel, personal correspondence, and intermarriage, not to mention European universities, monasteries, and Christendom itself, all indicate interconnections within the West and beyond. Also, the structures of communication, administration, and domination associated with European colonialism (typically linking metropole and colony) have, to varying extents, endured and linked non-Western nations to the West. The British Commonwealth, for example, supports strong ties among many nations, incorporating many quite poor and peripheral nations into the world polity. Additionally, nation-level and individual-level behaviors have been used to infer the existence of the world polity. Observations of individuals or countries conforming to world norms, in the absence of competing explanations, have been cited as indirect evidence of an external world polity. Finally, researchers have looked descriptively at the discourse of individuals, firms, and states to document the existence of world structures and norms (Chabbott 1999).

Institutional researchers typically focus on the most tangible, directly measurable, and face-valid evidence of the world polity: international conferences, international associations (INGOs), treaties, and intergovernmental organizations (IGOs). Conferences and associations reflect the participation of private individuals in a world civil society and common culture (although conferences held by official intergovernmental organizations are common as well). They address topics ranging from the interests of hobbyists to professional doctors and scientists to athletes to religious groups. Such associations are thought to reflect "world culture," the more diffuse shared cultural frames that underlie and help constitute the world polity.[3] The latter, namely treaties and intergovernmental organizations, represent the official institutionalization of agreements and cooperation among nations, sustaining both world norms and sometimes direct coercive power over states and individuals (for example, European Union constraints on national fiscal policy, or international tribunals for human rights violations). Treaties and intergovernmental organizations more directly reflect structure and capacity to address issues on a global scale and to influence the policies of nations. Both the diffuse associations that constitute world culture and the official structure of intergovernmental organizations and treaties have been shown to have independent (and sometimes disparate) effects on nations.[4]

The historical increase of conferences, associations, treaties, and intergovernmental organizations all suggest expansion of the world polity. Furthermore, the varying types of each (for example, environmental treaties versus human rights associations versus trade organizations) and the different discourses they produce reflect the different agenda issues dominant in the world polity. The number of organizations or treaties *of a given type* can be used as a measure of international attention to a given issue area compared with others or over time. For example, one can observe a tremendous increase in the number of international associations devoted to issues of economic development beginning in the 1960s as that issue became an important topic on the world agenda. The expansion of those organizations reflects increasing attention and structural capacity to support norms and action on issues of development. Also, the particular discourses expressed by international organizations at different points in time shed light on the various priorities of development professionals (Chabbott 2002). Discourse can be assessed through interviews or by examining the content of documents produced by international organizations (for example, annual summaries or conference reports; see Luo 2000).

To examine world polity structuration over time, institutional researchers commonly collect data on whole "populations" of similar organizations and their rates of founding over time—for example, all environmental treaties or all development organizations.[5] Indexes, compendiums, and yearbooks of treaties and associations are staple sources of data. In particular, the *Yearbook of International Organizations* deserves mention, providing the most extensive information on organizations in the world polity (UIA 1949–2000, published annually). The detailed keyword classification of associations and full descriptions of organizational goals and activities greatly simplifies the collection of data on world polity activity in different issue areas (see Boli and Thomas 1999).

## Describing and Explaining the Expansion of the World Polity

To describe the expansion of the world polity, institutional researchers have employed historical methods as well as descriptive charts and statistics to illustrate the increasing prevalence of international conferences, associations, treaties, and intergovernmental organizations. The most common way to operationalize world polity expansion is simply by counting—for example, enumerating the numbers of organizations, treaties, and discursive declarations in existence at different points in time. Graphs of organizational or treaty foundings (or cumulative foundings) versus time can powerfully illustrate increasing world organizational structure and attention to various issues (see Chapters 3 and 5). Periods when many new organizations or treaties are being founded can be interpreted as times of rapidly *increasing* world polity structuration in a given domain. Plots of cumulative foundings (total number founded to date) or organizational density (total number of organizations currently in existence) per year provide a sense of the overall level of world polity structuration at different points in historical time.

A second, more complex task is to account for the expansion of the world polity as a whole, or in specific domains (for example, the environment, human rights, economic development). The theoretical explanations involved are beyond the scope of this chapter, but commonly include the following: the expansion of the nation-state system, the rise of new professions and forms of expert knowledge, global economic changes, and the creation of the UN and other quasistate structures that serve as "nodes" around which further polity expansion can occur, and others (see Meyer et al. 1997b; Chapter 3). Explanations for the growth of the world polity can be tested by looking for statistical association over time between causal variables and measures of world polity structuration.

One may wish to test, for example, whether the growth of rationalized scientific expertise on environmental issues (measured, say, by the number of scientists addressing such issues worldwide) hastens world polity expansion around environmental concerns (measured by the number of international environmental associations), net of other causal factors (see Meyer et al. 1997b; Chapter 11). The "world" remains the unit of analysis, with both independent and dependent variables measured over time (for example, yearly or monthly).

Several related statistical methods are available to model such relationships among time-varying variables, where the unit of analysis is the "world" at different points in time. These include time-series analysis, various event count models, and event history models. The choice among these relies primarily on the nature of the dependent variable. If it can be measured at an interval level for a series of fixed time periods (for example, yearly), then time-series analysis may be appropriate. In contrast, if the dependent variable involves a count of relatively rare events over fixed time periods, Poisson or negative binomial regression approaches are often in order. Finally, event history methods are quite useful in modeling phenomena that are observed as instantaneous rates of occurrence of a particular event (for example, the date of creation of an environmental treaty). As such, event history analysis benefits from more precise information about the timing of the event: for example, to the day for analyses spanning shorter historical periods, or to the month or year for longer-term analyses. In practice, there is some overlap among these tools, leaving the choice of method to the discretion of the researcher (for example, yearly counts that occur infrequently can be analyzed by means of either count models or event history analysis). An overview of the methods discussed here is listed in Table 2.1.

*Time-Series Approaches*

Let us continue with the example above, in which we attempt to explain a particular aspect of the content of the world polity: the rise in environmentalism. On the basis of the general methodological approach outlined above, both the dependent variable and explanatory variables would be measured at the world level over time, either as an aggregate of measures at the nation-state level, or as some direct measure at the world level. Consider the following linear regression model:

$$Y_t = a + bX_t + e_t \qquad (1)$$

Here, $Y_t$ is number of the world's international environmental associations, measured yearly, from 1950 to the present.[6] For simplicity's sake, let us

TABLE 2.1
Overview of Selected Quantitative Models
for Globalization Research

| Unit of Analysis | Statistical Model | Type of Dependent Variable | Type of Explanatory Variables | Comments |
|---|---|---|---|---|
| World | Time series | Continuous, many fixed time points | Various, including lagged dependent variables | Can address problem of serial correlation; use Newey-West for heteroscedasticity |
| | Poisson, negative binomial, or GQL | Counts of relatively rare events, many fixed time periods | Various, Including lagged dependent variables | All handle censoring at 0; negative binomial and GQL address overdispersion; GQL adjusts for serial correlation |
| | Logistic regression | Whether an event occurred or condition existed during each of many fixed time periods (e.g., binary) | Various, including lagged dependent variables | Maximum likelihood estimation assumes independence of observations, yet serial correlation is likely |
| | Event history analysis | Rates of occurrence of particular events | Various, but must vary over time | Fairly precise data on timing of events needed; various specifications of the dependence of the rate on time are possible, although constant rate dependence on historical time is typical |
| Nation-State | Cross-sectional OLS regression[a] | Continuous, many nations measured at single time point | Measured at single time point | Inference regarding causal direction is difficult; convincing arguments can be made if reverse causality is implausible (e.g., on theoretical grounds) |
| | Panel analysis (OLS) | Continuous, many nations measured at single time point | Measured at two or a few points in time; lagged dependent variable included | Specification of lag time can influence results significantly |

| | | | | |
|---|---|---|---|---|
| | Panel analysis (latent variable approach) | Latent variable with multiple indicators, many nations at single time point | Latent variables with multiple indicators at two or a few points in time, including lagged dependent latent variable | Allows flexible handling of correlated error across time or across nations; identification problems possible; lag time can be influential |
| | Panel analysis (with Poisson, Negative Binary, GQL logistic) | Counts or binary occurrences of relatively rare events, single time period; many nations at single time point | Measured at two or a few points in time; lagged dependent variable included | Specification of lag time can influence results significantly |
| | Pooled time series | Continuous, latent variable, count, or binary at multiple time points or periods | Measured at multiple time points; lagged dependent variable included | Two-way random effects models can handle correlated error across time or across nations; data collection can focus on fewer nations |
| | Event history analysis | Rates of occurrence of particular events | Various, may vary over time, but not necessarily | Fairly precise data on timing of events needed; various specifications of the dependence of the rate on time are possible, although constant rate dependence is typically modeled |
| *Persons or organizations within nation-states* | Multilevel models | Continuous measures of individuals at one or more time points | Various types of measures at both the individual or nation-state level | World-level characteristics (including historical time) can be incorporated as a third level |

GQL = generalized quasilikelihood; OLS, ordinary least square.

[a]Latent variable, count, and logistic regression approaches are also possible at a single time point. See characterization of variable characeristcs listed separately.

consider only a single explanatory variable, $X_t$, the total number of environmental scientists in the world. The interval-level measure, $Y_t$, cannot be analyzed with a simple ordinary least-squares approach because undoubtedly, there is autocorrelation: the number of international environmental associations in one year is clearly related to that of the previous year or years, violating a basic assumption of ordinary least squares (OLS) regression.[7] The consequence of the likely positive autocorrelation in this example is that estimates of the standard error will be biased downward, inflating the $t$ statistic used to assess the significance of the estimated coefficient. This could lead us to reject the null hypothesis when we shouldn't. In practical terms, we might interpret the result to mean that the number of environmental scientists has a significant effect on the number of environmental organizations, when in fact the data do not support that conclusion.

To avoid false-positive findings, the statistical model may be modified in a number of ways. In the simplest case of year-to-year correlation, called first-order autoregressive processes, the error term $e_t$ is assumed to be linearly related only to $e_{t-1}$ and not to any earlier year or years:

$$e_t = \rho e_{t-1} + \upsilon_t \tag{2}$$

where $|\rho| \leq 1$ and the new error term $\upsilon_t$ is better behaved than the original error term $e_t$: it is unbiased (mean of $\upsilon_t = 0$) and has a constant variance $\upsilon_t = \sigma^2$. Such first-order autoregressive processes have been modeled by the method of "first differences," which assumes that $\rho = 1$. The first step is to calculate "change scores" by subtracting the values of $Y$ and $X$ across adjacent time points. Ordinary least-squares regressions that use these change scores can take advantage of the more tractable error term $\upsilon_t$. Somewhat less restrictive and thus more useful is the Cochrane-Orcutt method, a multistep process that first estimates residuals, $e_t$, in equation 1 by using an OLS model, then uses the residuals to estimate $\rho$ in equation 2.[8] In a modification of the first differencing approach, $X_t$ and $Y_t$ are transformed as follows:

$$Y_i^* = Y_t - \rho\, Y_{t-1} \tag{3a}$$
$$X_i^* = X_t - \rho\, X_{t-1} \tag{3b}$$

Finally, OLS regression is performed to estimate

$$Y_t^* = bX_t^* + \upsilon_t \tag{4}$$

Note that as in the first differencing approach, the aim here is to perform some type of transformation of the variables $X$ and $Y$ that produces a "cleaner" error term—one that meets the requirements of OLS regression. This is not always possible, and so it may be necessary to use models that allow for a more complicated error structure than OLS permits. We discuss a few of these models later.

These procedures address a particular type of autocorrelation—serial correlation—that is likely to be present in most historical analyses of globalization. However, they do not address problems of heteroscedasticity, the circumstance in which the variance of the error term is not constant. Typically, this problem is detected by plotting the estimated error terms—that is, the residuals from an OLS regression—against the dependent variable and each of the explanatory variables. Relating this to the earlier example, it could be that our model of the number of environmental organizations in the world is fairly accurate when there are relatively small numbers of scientists in the world, but is quite inaccurate when there are a lot of scientists. Inaccuracy is not the problem for OLS—although low $R^2$ will result—but the systematic swing from accurate to inaccurate is. Newey and West (1987) have suggested an estimator of the variance of the coefficient that is robust in the presence of heteroscedasticity (robust in the sense that the procedure does not require the OLS assumption of constant variance). It has the added virtue of correcting for autocorrelation up to a specified lag. Thus, the Newey-West procedure represents a significant improvement on other methods specific to first-order correlation. Relative to the example given, Newey-West estimates would allow us to explore whether the number of environmental organizations in a single year might be related to the number in the previous two, three, or more years. In addition, we would no longer need to assume that the model is as good at predicting the number of organizations in times with relatively few scientists as in times with a greater number of scientists.

*Analysis of Count Data*

Many world-level variables of interest are "count" variables (for example, numbers of treaties or organizations) that are always nonnegative integers. Linear time-series regression models are not ideal for count variables for a number of reasons. For one thing, linear functions with nonzero slope necessarily predict negative values of the dependent variable at some point, which makes no sense in the context of count variables. Also, the assumption of linearity is particularly problematic when the outcome variable is frequently zero. One such problematic measure is the yearly count of the number of multinational environmental conferences (such as the 1992 United Nations Conference on Environment and Development in Rio de Janeiro, Brazil). Modeling the number of these conferences year by year would help to explain the rise and institutionalization of environmentalism in world polity discourse. Because this is clearly not a continuous measure and subject to censoring at 0 (that is, it cannot be negative), the time-series approaches discussed above are not appropriate. Poisson regression is one solution (Long 1997). The phenomenon of interest is assumed to be generated through a Poisson process, with number of events at time $t$, $Y_t$, modeled as

$$E(Y_t) = \lambda_t = e^{X\beta} \tag{5a}$$
$$E(Y_t) = Var(Y_t) = \lambda_t \tag{5b}$$

where $\lambda_t$ represents the mean rate of occurrence at time $t$, and $X$ is a vector of observed explanatory and control variables, whereas $\beta$ is the set of "effect" coefficients to be estimated. One advantage of using the exponential function in equation 5a is that $e$ (in fact, any positive real number) raised to any power is a nonnegative value—which corresponds to the fact that count measures are never less than zero. However, Poisson regression requires a strong assumption in equation 5b that is rarely empirically justifiable: that the variance of the dependent variable is equal to the mean. Negative binomial regression is a more appropriate technique in the presence of overdispersion—that is, when var($Y_t$) > $E(Y_t)$. It adds a random error (that is, stochastic) component to the Poisson specification in equation 5a. Overdispersion may result when there is time dependence in the rate if the rate of occurrence of events within a time interval is positively related to the number of previous events in the interval (for example, intrainterval contagion), or "unobserved heterogeneity" (for example, variation in the dependent variable that is not accounted for in the model, an omnipresent problem of varying severity) (Barron 1992; Carroll and Hannan 2000:144–49). Recent versions of the software package Stata can estimate both of these types of models.

Neither approach addresses the problem of serial correlation (for example, year-to-year correlation), which in this context can be thought of as *inter*interval contagion. As in the case of time-series data, historical modeling of global counts, such as the number of international environmental conferences in any given year, may well involve this type of autocorrelation; in nontechnical terms, the number of international environmental conferences in which a country participates in a particular year is highly related to the number of conferences in which its delegates participated in previous years. Yet both Poisson and negative binomial regression techniques use maximum-likelihood estimation, which relies on independence of individual observations in order to derive a likelihood function for the full set of observations. Zeger's (1988) generalization of quasilikelihood estimation has been recommended as a method for modeling serially correlated count data (Barron 1992; Cameron and Trivedi 1998:240–42). The basic approach is to leave the underlying probability distribution undefined and to derive an estimator based solely on the first two moments (for example, $E(Y_t)$, var($Y_t$), no equality assumed).

Zeger (1988) begins with the Poisson model in which autocorrelation is assumed to operate through a latent process $\varepsilon_t$ that is continuous in nature. The latent process might be thought of as a propensity to generate observable events, for example. A second assumption is that the variance–covariance matrix of the latent process $e_t$ can be written strictly as a function of time elapsed. This allows us to relax the strong constraint that there is no correlation in the latent processes at different points in time. We replace this strong constraint of zero correlation with a weaker constraint: we specify that this correlation takes a particular form—namely, that it depends only on the duration between the two time points. It can then be shown that the unconditional expectation and variance of $Y_t$ are

$$E(Y_t) = \lambda_t \tag{6a}$$
$$\mathrm{Var}(Y_t) = \lambda_t + \gamma \lambda_t^2 \tag{6b}$$

Consequently, Zeger's approach attends to the nature of the data as counts (because it is grounded in the Poisson process); to the possibility of overdispersion since mean and variance need not be equal; and to autocorrelation in the form of a latent process. This class of quasilikelihood models can be computed by the COUNT module in the software package Gauss.

*Event History Analysis*

Event history techniques may also be used to model a phenomenon occurring repeatedly at the world level, such as the creation of new international associations or the incidence of armed conflict in the world. In this case, the outcome is conceptualized as a series of discrete repeatable events occurring sequentially at particular points in historical time, rather than as an aggregate number of counts within a given time period (for example, per year). If, for example, we know the time at which international environmental associations were founded (preferably to the exact day, but usually cruder units such as month or year are used), we can model the development of the environmental sector of the world polity by means of event history analysis.[9] We may believe that international environmental organizations are spun off from more general international scientific organizations, so that the greater the density of scientific organizations, the more frequently environmental organizations are founded.

In this case, the founding of an environmental organization marks the end of one episode, or "spell," and the beginning of another. The dependent variable is the founding rate of international environmental organizations, or the hazard rate of founding. Explanatory variables can be operationalized in a number of ways, such as lagged yearly worldwide counts of scientific organizations; moving averages of these counts in lagged multiyear intervals; or the instantaneous founding rate of scientific organizations (perhaps lagged). In each case, however, these explanatory variables must be time varying because the world is the unit of analysis, and so change over time is the only source of variation.

A typical formulation is the proportional hazard model:

$$h(t,X) = \theta(X)q(t) \qquad (7)$$

where $X$ is a vector of explanatory (and control) variables. In globalization research, $t$ typically represents historical time, with the occurrence of an event marking a new episode. We often assume that the process that drives the transition rate from one state (for example, having experienced $N$ organizational foundings) to another ($N + 1$ foundings) is the same regardless of $N$. As a result, the multiepisode, multistate nature of the foundings data can be modeled in a way analogous to the single-episode, one-transition case (Tuma 1994:159). Choice of the above model is based on an assumption that the hazard rates for two different values of $X$ are proportional to each other simply as functions of $X$ and not of $t$. In other words, the assumption is that the effects of time dependence do not vary according to values of $X$. Often, the effect of the explanatory variables on the rate is specified as multiplicative, such that

$$\theta(X) = e^{\lambda X} \qquad (8)$$

The specification of the rate as a function of time in equation 7 can be fairly complicated. The most common approach in globalization research has been to assume that the time dependence is constant—that it does not change for different values of $t$. The focus of these constant rate models then shifts to adequate specification of $\theta(X)$. In essence, one is allowing variance to be absorbed by time-varying independent variables, rather than being absorbed by some (usually arbitrary) function of $t$. The constant rate dependence assumption is a strong one, based on the notion that the history of the process prior to a given episode can be accounted for in $X$. This assumption is not always plausible, in which case one may employ a range of functional forms of $q(t)$. The choice of a specification for $q(t)$ depends, as a first step, on whether the hazard rate is monotonically related to time (that is, the rate is ever increasing or ever decreasing as time passes) or a nonmonotonic function. Common specifications for monotonic time dependence are the Gompertz-Makeham, the Weibull, and the gamma function. Log-logistic and log-Gaussian specifications are often used to model nonmonotonic time dependence (see Tuma 1994:152 for details).[10]

Once the hazard rate $h(t, X)$ is specified, a so-called survivor function can be derived on the basis of $h(t, X)$. One can then write a likelihood function based on both the survivor function, which is a formulation for a unit's chance of survival remaining at risk of having an event, and the hazard rate, which is the probability of an event occurring given that it has not happened previously.

*Summary*

Taken together, these three types of approaches can model phenomena occurring at the world level. To quantitatively analyze the world polity as a whole, we must rely on slices of time to differentiate "cases," albeit usually autocorrelated ones. For time-series and count data analyses, time is typically divided up into equal-sized chunks. In the case of event history analysis, the variation in the lengths of the episodes is essentially what is modeled. In theory, the choice of the particular model typically depends on the exact nature of the dependent variable to be analyzed: time series for continuous measures, count models for count variables, and event history for discrete-time events. In practice, however, many social phenomena can be measured in a manner as either a count or an event, and may also closely approximate a continuous measure as well. Thus, the frequency of the event often determines which statistical model is used. Highly frequent events, generating high numbers of counts in a given time interval, are often modeled with time series. If there are many zeros in data that are collected over fixed time periods, event count models are used. If the event is fairly infrequent, event history analysis is often used. In any case, the results are often consistent across these various modeling approaches.

In all these approaches, the emphasis is on analyzing a phenomenon that depends in some way on previous history. This is also true as we consider the historical effects of the world polity on individual nation-states, but we must also consider variance across the units of analysis at single points in time as well.

## METHODOLOGICAL APPROACHES TO STUDYING THE EFFECTS OF THE WORLD POLITY ON NATIONS

A second part of the institutional research agenda attends to the impact of the world polity on nation-states. In fact, the cornerstone of institutional theory and research has been to show that the structures of the world polity affect the behavior and constitutive identity of nation-states through cultural frames, cognitive models, policy advice, norms, and (on occasion) coercive power. Initially, research in this area relied on rather indirect evidence of world polity effects. Worldwide similarities among nations that could not be accounted for by other theories (for example, the adoption of similar education policies despite very different labor-force requirements) were seen as indicative of world polity influence—meaning that a certain policy had become taken for granted. S-shaped diffusion curves were interpreted similarly (for example, Ramirez et al. 1997). These techniques have given way to more direct measures of world polity structure and the use of statistical models to identify its effect on nations.

Here, the research methods employed by institutionalists have been resolutely quantitative and statistical, with few case-based contributions. This is the result of the inherently comparative nature of the research question. Single-country case studies provide little basis for generalizing about the cause of national policies and structure. Comparative case study approaches have the potential to prove more fruitful, although choosing representative cases remains a sticking point.[11] Case-based approaches are proving extremely useful, however, in clarifying the mechanisms involved in institutional policy diffusion and gauging the depth of national policy penetration. Nevertheless, the institutional research literature remains primarily quantitative in orientation.

*Measuring World Polity Influence*

How does the world polity affect nations, and how can we measure the process? The core understanding of institutionalism is that culture and ideas, typically thought of as cognitive frames and schema, constitute actor identity and construct behavior. Frames and schema that are entrenched (that is, institutionalized) in the world polity will become more prevalent among actors that are exposed to those views, leading to changes in actor behavior (see Meyer et al. 1997b). Institutional researchers have increasingly begun to specify the particular mechanisms through which world culture influences nations, in response to repeated criticism on this issue (see, for example, Chabbott 2002; Meyer et al. 1997b; Frank et al. 2000a).

The essential point is that the influence of world culture occurs via *linkages* of various sorts that introduce or expose certain cultural frames and schema to nations, organizations, and individuals around the world. Often, linkages are quite specific and identifiable. For example, Finnemore (1996) and Barrett and Frank (1999) observe that international organizations frequently send out personnel or consultants who literally go to different countries and teach government officials to see the world in a particular way and to implement policies and behaviors that conform to world polity norms. At times, policy changes are immediate and specifically traceable to particular international conferences or organizations. However, linkages may also be diffuse and sometimes difficult to measure. For example, norms are carried through the media; individuals are enculturated by participation in Western educational institutions as exchange students; and political leaders learn world norms about "proper" economic policy or human rights legislation from UN and World Bank discourse. All these processes would be difficult to operationalize for many nations and over time.

The strength of world polity influence over a nation's behavior in a given issue area (for example, education or the environment) is a function of two factors that operate interactively.[12] The first factor is the level of world institutionalization—that is, the amount of world polity structure supporting cultural frames and norms in a given issue area. It is only reasonable to expect effects of the world polity in domains where world polity structuration has occurred. As an example, national adherence to world norms of environmental protection was rather inconsistent before the creation of the UN Environment Programme and other important international organizations that address environmental issues (Frank et al. 2000a). Once world structure and clear world norms and standards came into existence, national adherence increased. The second factor is the degree to which a given nation is linked to the world polity via either diffuse or specific connections. Effects on national policy and behavior should logically occur most rapidly and most dramatically in nations more tightly linked to the cultural frames and norms of the world polity. European nations, which participate actively in the world polity and are members of many international organizations, tend to conform to world policy models and norms quickly. Likewise, non-European nations that participate heavily in international organizations also tend to rapidly conform to world norms. However, remote or politically isolated nations do so less quickly and consistently, even controlling for overall level of national development. Nations that are explicitly autarkic or opposed to the Western capitalist system (for example, North Korea and nations behind the iron curtain) adopt world cultural models least of all.

The *level of world polity structuration* and *the degree of national linkage* reflect the "supply" of world culture and the size of the conduits carrying world culture to nations, respectively. World polity structuration can be measured by the number of international organizations addressing a given issue area, as described above. Linkage is often gauged by national participation or membership in the associations, conferences, treaties, and intergovernmental organizations of the world polity. For example, data can be collected over time (often yearly) on how many memberships in intergovernmental organizations a nation holds or how many treaties a nation has signed. Likewise, citizen memberships in intergovernmental associations or conference attendance can be counted as a measure of world linkage.[13] Again, measures can be constructed generally, or to identify linkage to a specific domain or issue area of the world polity.

These "participation" or "linkage" measures also reflect world polity structuration. The fact that a country can have many organizational memberships implies the existence of many organizations and thus of substantial world structure. So such measures are useful in that they capture both of the processes mediating world polity influence—level of world polity structure *and* strength of national linkage. Linkage may also be measured in many other ways: flows of students or professionals between a country and the Western nations that form the center of the world polity; participation in projects run by international organizations (for example, development projects); and diplomatic, political, or even colonial ties to the West. Even economic flows (for example, trade) can be seen as a possible measure of linkage to the West and the world polity, although such measures are more commonly seen as indicative of purely economic, rather than cultural, globalization.

## Explaining National Structure and Behavior

Employing the kind of measures described above, institutional researchers have used statistical analyses to determine if the world polity influences the structure or behavior of nations, controlling for other relevant factors. Competing explanations of national policies typically include functional and modernization theories (still commonly invoked to explain national policies), economic theories, world-system dependency theory and domestic Marxist views, and international conflict theories.

Here, the nation-state is the unit of analysis. A variety of different statistical models may be used to test competing explanations, ranging from simple OLS regression to event history analysis. Independent variables often include world-level indicators reflecting the structuration of world polity or national-level linkage measures, as well as variables reflecting competing explanations. The most important distinction among the available research approaches is between cross-sectional analyses, which compare nations at a single point in time, and dynamic analyses, which look at change in nations over time.

## Cross-Sectional and Panel Designs

Suppose we wish to analyze the effect of world polity linkage on national government spending on environmental protection in a particular year. One hypothesis from a cross-sectional perspective is: nations that are members of many international associations may devote more resources to environmental protection, after controlling for per capita gross domestic product and other relevant variables. One approach for modeling a continuous variable such as this is ordinary least-squares regression:

$$Y_i = a + bX_i + e_i \qquad (9)$$

Here, $Y_i$ is the amount of environmental spending in a particular year $t$ by individual country $i$, $X_i$ is the number of memberships in international associations during the year, and $\varepsilon_i$ is an error term.[14] The term $b$ represents the effects of a change in $X_i$ on $Y_i$. In this example, $b$ is the expected change in the amount of environmental spending in a country for each additional international association membership.[15] However, we are often interested in change over time, rather than just cross-sectional covariation. There are two reasons for this: first, the effects of the world polity arguably take effect over a substantial period of time, and statistical models should reflect this point. It may take many years for a given set of world polity linkages to transform national environmental spending. Second, a focus on change over time helps resolve issues of causality. Cross-sectional analyses may leave ambiguity—for example, that national spending on environmental protection may generate more linkages to the world polity, rather than the other way around. In such a case, it is preferable to show that a given level of world linkage leads to future change in environmental spending.

Panel analyses commonly incorporate lag time between a change in $X_i$ and a change in $Y_i$. One form of a panel analysis is to simply measure the explanatory variable at an appropriate lag time before the measure of the dependent variable. In equation 9, if the existing literature suggests that a five-year lag is appropriate, then $X_i$ could be measured in year $t - 5$. More commonly, panel analyses also attempt to control for the initial value of $Y$ at the time that $X$ is measured. In this case, models assess change in the dependent variable (environmental spending) by controlling for initial environmental spending and other variables such as world polity linkage.

Chapter 10 in this volume employs a form of OLS panel analysis. The dependent variable can be computed as a change score:

$$Y_{it} - Y_{it-1} = a + bX_{it-1} + e_i \qquad (10)$$

Or the lagged dependent variable is entered as a control on the right-hand side of the equation:

$$Y_{it} = a + bX_{it-1} + cY_{it-1} + e_i \qquad (11)$$

Note that equation 11 is a special case of equation 10 where $c = 1$. Models of cross-panel rates of change can also account for initial conditions of the dependent variable:

$$\frac{(Y_{it} - Y_{it-1})}{Y_{it-1}} = a + bX_{it-1} + e_i \qquad (12)$$

In this case, the dependent variable represents a percentage of change over the two time points. Hence, panel designs require that data be collected over all individual units (for example, nations) at two or more points in time. Such analyses can be more convincing than single-time-point analyses, particularly with regard to issues of causal direction. However, their proper implementation can be complex for a number of reasons: high levels of autocorrelation over time make it difficult to get meaningful results; "floor" and "ceiling" effects can produce odd trends in the data (see, for example, Meyer et al. 1992b). Careful analysis of residuals and outliers is important.[16] In addition, the choice of lag time is important in all of these panel models. Although it is not uncommon for the lag time between waves of data to be determined by the availability of data, a strong theoretical basis for such decisions is far more preferable.

## Pooled Time Series

Pooling time series of several or even many nations has the advantage of allowing more flexibility to specify the lag structure, handling historical period effects, and so forth. As noted above, however, data may not be available in time-series form for all of the nations of interest. Although data availability may be good in central measures of economic activity and political systems, especially in recent decades, data may be somewhat scarce in other areas, such as scientific infrastructure or cultural production. The lack of data may lead, in both pooled time-series and event history analyses, to substantial sample bias. In the conclusion below, we suggest a possible remedy to this difficulty.

Hicks (1994) differentiates between "temporally dominant" pools (for example, data on ten countries over fifty years) versus "cross-sectionally dominant" pools (for example, a hundred countries over twenty years). Given the network diffusion imagery of many globalization arguments, cross sections that encompass all nations are typically sought. Returning to our example, if we had data for sixty-five nations on environmental spending and number of international organization memberships at five-year intervals between 1950 and 1995, a pooled approach might be helpful. Individual country-level time-series analysis is not possible solely on the basis of the ten time points, but pooling the relatively short time-series produces a data set with enough data points to be analyzable. However, this does present particular statistical challenges.

In addition to the typical problems of heteroscedasticity and autocorrelation discussed in previous sections, a major complication for model specification of pooled time series is that the error term has both temporal and cross-sectional components. That is, all cases deriving from a single country (albeit in different years) are likely to have correlated error. Likewise, all cases from within a given year may have correlated error (for example, due to worldwide trends or events such as the "oil shocks" of the 1970s). To the extent that there is autocorrelation, the interdependence may follow a different pattern among cross-sectional components than among temporal components. A single error term, as in equations 9 to 12, does not allow this more complicated process to be modeled.

For cross-sectionally dominant data where $i$ is the number of countries and $t$ is the number of time periods, Hicks (1994:176–78) recommends the error components model

$$Y_{it} = a + bX_{it} + \varepsilon_{it} \tag{13}$$

where

$$\varepsilon_{it} = U_i + V_t + W_{it} \tag{14}$$

The summary error term $\varepsilon_{it}$ is decomposed here as $U_i$, which varies across countries but is constant over time, $V_t$, which varies across time but is constant across countries, and $W_{it}$, which is a residual that varies across countries and time. Each error component is normally distributed with mean zero, and although the summary error may be correlated over time, across countries, or both, the individual components are not. An alternative specification is to replace the intercept $a$ in equation 13 with $T$ temporally distinct intercepts, one for each time period. This fixed effects approach allows the cross-national means of $Y$ to bump up or down across time, but it calculates an estimate of $b$, the effect of $X$, which is fixed over time. Random effects models build a variance component into $b$ rather than the intercept $a$, allowing the effect of $X$ on $Y$ to vary depending on the time period or the nation-state.

## Structural Equation Models with Latent Variables

It is fair to say that in virtually all quantitative social science research, and perhaps especially research on globalization, it is appropriate to think of the phenomena of interest as latent variables. That is, the characteristics we wish to analyze are broad and unmeasured (and therefore latent), and the measures actually used in the analysis are understood to reflect or serve as indicators of the characteristics under study. In all of the methods we described previously, there is an implicit assumption that the observable indicators are perfectly correlated with the latent variables they measure. For example, the number of international environmental associations to which a nation belongs is not necessarily of interest in itself, but rather is of interest as an indicator of a latent variable—for example, a nation's participation in a particular segment of the world polity. The indicator does not perfectly measure the latent characteristic, but methods such as ordinary least-squares regression and the others we have described provide no method for taking measurement error into account. In addition, other indicators are possible: the number of international environmental treaties signed; official participation in international environmental conferences; the number of publications about the environment in international professional journals. Even if we had complete data on all of these indicators for all countries, the four indicators taken together (for example, in a principle components factor analysis) would still not measure the latent quality perfectly.

Structural equation modeling is a statistical method that allows researchers to use multiple indicators of latent variables as well as to account for measurement error of both the explanatory (or exogenous) latent variables, as well as those that are the "effects" (that is, the endogenous latent variables). One step of the approach involves the specification of a model relating the endogenous latent variables with the exogenous latent variable or variables. Often, this relationship is specified as linear:

$$\eta = \mathbf{B}\eta + \Gamma\xi + \xi \tag{15}$$

Here, $\eta$ represents an $(m \times 1)$ vector of the $m$ endogenous, "effect" latent variables, and $\xi$ is an $(n \times 1)$ vector of the $n$ exogenous "causal" latent variables. $\mathbf{B}$ is an $(m \times m)$ coefficient matrix showing the influence of the latent endogenous variables on each other, whereas $\Gamma$ is the effects of the exogenous variables on the endogenous variables, in the form of an $(m \times n)$ matrix. Finally, $\zeta$ is an $m \times 1$ vector of unbiased error terms such that the mean of each element of the vector equals 0 (Bollen 1989). Equation 15 represents a general case in which there could be more than one "effect" of interest, and in which there is some relationship, perhaps even reciprocal relationships or feedback loops, between effect latent variables. Often, there is ultimately only one latent variable that we wish to explain (that is, $m = 1$). Chapter 12 of this volume, for example, uses a latent variable approach to model a single endogenous latent variable, popular mobilization (as indicated by the mean number of strikes, demonstrations, and riots in a nation over a five-year period) as a function of prior levels of scientization, an exogenous latent variable measured by a set of indicators such as the number of science books produced by a nation and the number of science citations. In such a case, $\eta$ is a single element and the $\mathbf{B}\eta$ term on the right-hand side essentially drops out of the equation.

In addition, structural equation modeling involves specification of the relationship between each of the latent variables and their respective measurable indicators. Once again, this relationship is often assumed to be linear:

$$y = \Lambda_y \eta + \varepsilon \qquad (16a)$$
$$x = \Lambda_y \xi + \delta \qquad (16b)$$

In this set of equations, $y$ and $x$ are $(p \times 1)$ and $(q \times 1)$ vectors of the observed indicators, respectively. $\Lambda_y$, a $(p \times m)$ coefficient matrix, shows the relative strength of the relationship between each of the indicators in $y$ and $\eta$ itself. Similarly, $\Lambda_x$ is a $(q \times m)$ matrix of coefficients showing the relationship between the indicators in $x$ and $\xi$. An important feature is that both equations have error terms: $\varepsilon$ is a $(p \times 1)$ vector of the errors in measuring $\eta$, and $\delta$ is a $(q \times 1)$ vector of the measurement errors with respect to $\xi$. Means of the error terms are assumed to be zero. The terms in $\varepsilon$ are allowed to be correlated among each other, as are the terms in $\delta$. This is an important feature in globalization research because indicators are quite likely to be intercorrelated. High correlation among explanatory or control variables (that is, multicollinearity) in OLS models can lead to parameter estimates that are highly unstable. A latent variable approach can avoid this problem. One restriction, however, is that $\varepsilon$ is assumed to be uncorrelated to $\delta$. In other words, measurement errors related to the exogenous latent variables are not permitted to be cross-correlated with measurement error on the endogenous side.

Which set of estimated parameters for equations 15, 16a, and 16b represent the "best" solution? Unlike ordinary least-squares regression and its variants, which minimize the residuals of each individual case, latent variable modeling minimizes the differences between the observed covariances among variables and the estimated covariances based on the estimated model (Bollen 1989).

As a concrete example, we might investigate whether a nation's participation in the world polity pertaining to the environment is related to concern about the environment within each nation. These two latent variables could be measured in various ways. Participation in this aspect of the world polity could be measured by using the four indicators noted above. Concern about the environment within each country could be measured by the following: (1) the percentage of opinion poll respondents in each country who agree that the government should take steps to protect the environment; (2) the percentage of the primary school science curriculum devoted to the environment; and (3) government spending on environmental protection as a percentage of gross domestic product.

Estimation of this model by software packages such as AMOS, LISREL, or EQS would produce, first of all, estimates of the sets of coefficients $\Lambda_x$ and $\Lambda_y$ showing the relationships between the indicators and the latent variables. Usually, this relationship is normed to one of the indicators of each of the latent variables. That is, the $\Lambda$ coefficient of a "reference" indicator is set equal to 1, so that the coefficients for other indicators show the strength of the relationship between the indicator and the latent variable *relative to* the reference indicator. The $\Lambda$ coefficients can therefore be thought of as the relative contributions of the indicators to the latent variable. The estimation of $\Gamma$ is substantively particularly important. In this case, because there is only one exogenous variable (linkage to the environmental world polity) and one endogenous variable (concern about the environment), $\Gamma$ is a single number. This coefficient represents the effect of national linkage on environmental concern. Whether it is significantly positive or negative can be determined in the usual way with a *t* test that uses the standard error of the coefficient.

One common problem in structural equation modeling stems from a real strength of the approach: its flexibility in model specification. Many of the possible specifications of the relationships between and among latent variables and indicators are underidentified. That is, the set of observed covariances in the sample do not provide enough information to narrow all of the possible solutions down to a single best solution. Proper identification of a model usually requires that the analyst add information by constraining previously free parameters to equal 0, another constant, or another of the parameters to be estimated. This might involve assuming independence between a pair of indicators that had been previously specified to allow nonzero correlation, or setting the impact of a set of indicators of a latent variable equal to each other. Bollen (1989) summarizes a set of rules to help determine whether a model is identified. Some software packages, such as AMOS, list possible modifications of an unidentified model that might solve the problem.

## Event History Analysis

Analysis of the effect of the world polity on event histories at the nation-state level follows much of the previous discussion of event history at the world level. Once again, precise information is required about the timing of the events, but we are no longer constrained to the world as the single unit. Instead, nation-states are the units, each able to experience a particular "event," such as the adoption of a particular policy. Chapters 5 and 11 use this type of approach. As with pooled time series, research designs in institutional globalization studies have often sought wider cross-sectional coverage, without sacrificing too much in terms of temporal coverage.[17]

Suppose we had exact dates in which individual nations ratified an important international environmental treaty. We may model this "event," national ratification of an environmental treaty, as a function of both nation-level variables (time varying or constant), time varying world-level measures, or some combination of these. We might hypothesize that the rate at which countries ratified the treaty depends on a nation's geographical region, number of national memberships in environmental associations (indicating national linkage to the cognitive models and frames of the world polity), as well as the cumulative number of international environmental associations in existence that might put pressure on nations to ratify treaties. A proportional hazard rate model would be much the same as in equation 7, except that now the vector $X$ could include explanatory and control factors that vary across country, time, or both. Also in contrast to the previous discussion of event history at the world level, we can model one-way transitions in addition to repeatable events. Thus, we could analyze either national adoption of a specific treaty, a single event that occurs within a nation only once, or we could analyze national ratification of any environmental treaty, an event that may happen repeatedly. In contrast, one cannot model a singular event at the world level.

The choice of modeling strategy depends a great deal on the type of outcome to be analyzed. Discrete events occurring at a specific time, such as national adoption of a policy, are best analyzed by event history analysis. Interval outcomes are better analyzed by panel or pooled panel regression. Cross-sectional analysis is often a last resort, as it is more open to concerns regarding the direction of causal effects. However, it is often useful where cross-time data are limited or when cross-national differences dwarf temporal shifts, leaving little variance to analyze in a panel or change-score analysis.

## FUTURE DIRECTIONS FOR INSTITUTIONAL RESEARCH

We now turn to a brief discussion of some analytical techniques that we believe hold significant promise for future research on globalization as grounded in institutional theory. Until recently, most of these techniques were not readily available to social scientists, but they are rapidly being incorporated by widely used statistical packages such as SPSS, SAS, and Stata.

*Correction for Sample Bias*

As discussed previously, institutional investigations of the effect of the world polity on individual nation-states typically require historical measures for all nation-states. It is virtually impossible to secure complete data sets in most cases—the data simply do not exist for all countries at all relevant time points. The problem is that the missing data are not randomly distributed: less developed, smaller, or more isolated countries often fail to collect data, particularly more intricate social welfare measures. This kind of sample selection bias can often lead to bias in the parameter estimates. One possible solution is to incorporate a correction procedure for sample selection, as described by Heckman (1979). The first step is to specify a binary model predicting whether or not a nation-state is included in the sample—that is, whether a nation-state had a full set of measures or whether missing data caused it to be dropped from the analysis. Here, globalization researchers have a genuine advantage: we have a well-defined population. In other words, we know at any given time all of the nation-states in existence (with few disputable cases). The task of modeling the likelihood that a nation makes it into our sample is made easier by the fact that at least some rudimentary data (for example, population, geographical location, economic output) are available for virtually all nation-states. In the second step of the correction procedure, the predicted likelihood that a nation-state is included in the sample itself becomes a control variable in the model of the dependent variable of interest. Although the coverage of international data collection projects seems to be improving in recent decades, relatively sparse data will continue to plague studies of historical globalization processes. Sample selection approaches such as this one may ameliorate this difficulty.

*Multilevel Models*

Much research has focused on the nation-state as the social actor, deemphasizing (often with solid theoretical support) subnational variation and process. Concretely, this has meant that nation-states are taken as the units of analysis, with some national-level measures aggregated up from subunits (for example, mean science score for third-graders in international achievement studies, or mean spending by private firms on research and development). With the development of various kinds of multilevel modeling techniques, globalization research can begin to take subnational variation across individuals, firms, or agencies into account as theoretically warranted. In this type of approach, sets of subnational units (for example, individuals, firms) are understood to be nested in a group, such as a nation-state. The "outcome" is measured at the subnational unit level, but explanatory (and control) variables are measured at both the subnational and national level. In essence, two or more separate sets of equations are specified: one for each nation-state, estimating the effect coefficients strictly at the subnational level, and another in which the constant or variable coefficients for each nation-state are used as dependent variables for analysis at the nation-state level (see Bryk and Raudenbush 1992 for more details). Specialized software is available to estimate these models, such as Hierarchical Linear Models (HLM) and MLn, but they can also be estimated by PROC MIXED in SAS.

Suppose, for example, that we wanted to test the proposition that linkage to the world polity is positively related to per-pupil expenditures in science and technology education. Because many educational systems around the world are quite decentralized, we can imagine that much of the variation in expenditures occurs within school districts. If detailed school district data were available for a number of countries, we could use a multilevel model to test whether there is any effect at the nation-state level of world polity linkage after controlling adequately for school district characteristics. Multilevel models closely match the imagery of institutional approaches to the study of globalization—actors embedded in social contexts such as the world polity. As more direct measures of the world polity are developed, it will be possible to incorporate "the world" at different points in time as a third level in these types of analyses. We can then imagine more complete investigations of how changes in the world polity differentially shape policies and behavior of nation-states, which in turn differentially impact individual organizations or people.

*Network Analysis*

Institutional arguments frequently employ network imagery. Notions of "linkage" and "embeddedness" within the world polity or "exposure" to ideas and cultural frames are central to institutional arguments. However, until recently, little work has combined network analysis and institutional arguments. This is perhaps because the relevant networks are often extremely simple—with a central node (the world polity) connected to each nation with varying strength. Rather than using the formal tools of network analysis, institutional researchers have typically opted to measure network linkages as a property of actors. For example, the strength of a nation's linkage to the world polity may be measured by the number of memberships a country holds in international associations. In this case, the imagery employed is that of network analysis, but standard regression-type analyses are used to model data.

However, in order to address more complex networks, such as dyadic influences among nations or complex influences of many international and regional associations, explicit network models are be needed. Most notably, Greve et al. (1995) have incorporated network effects into event history models by specifying the effects of network matrices (for example, reflecting the strength of dyadic links among nations), in addition to ordinary unit-level variables, on the hazard rates of a given event. Additionally, such models allow the specification of unit-level variables that reflect "level of influence" or "receptiveness." For example, one might argue that the adoption of an environmental treaty by a wealthy industrialized nation is likely to increase the rate of adoption worldwide—more so than would adoption by a poor, peripheral nation. In addition, one might expect certain sorts of nations (for example, newly independent nations that lack internal legitimacy) to be particularly easily influenced by others or by world norms. Both types of hypotheses may be explicitly modeled by using event history diffusion techniques (see Greve et al. 1995 for a full discussion).

CONCLUSION: TRENDS IN INSTITUTIONAL RESEARCH

Over the past decade, institutional researchers have developed increasingly rich and nuanced analyses that incorporate more complex mechanisms and processes. The sparse world diffusion stories of the 1970s and 1980s have been replaced by more elaborate depictions of the development of a world culture and polity linked to nations by multiple mechanisms that influences nation-state behavior. First, there has been more direct study of the world polity itself and the world culture transmitted by it. Institutional diffusion is not simply a story that nations copy each other (simple mimesis) but can now more clearly be viewed as the result of national embeddedness in a world polity and culture. Second, the linkage mechanisms carrying world culture and discourse are becoming more clearly specified. Nations do not adopt policies out of thin air; rather, they are influenced by specific structures—international organizations or networks of organizations, epistemic communities, prevalent transnational discourses, and so on, to which nations are directly (or diffusely) linked.

Although the future is difficult to predict, three trends appear on the horizon. The first is simply a continuation of the last decade, toward increasingly rich analyses of the world polity and its effects on nations, with greater emphasis on context and mechanisms. A second growing trend is the study of the broader societal effects of the world polity on nations—economic, social, and environmental. The last section of this book serves as an example, tracing the second-order effects of science globalization on such things as national economies, national environmental activity, national politics, and political participation. Such studies are not completely new. Previously, for example, institutional researchers argued that there can be economic costs of isomorphism (for example, Shenhav and Kamens 1991) resulting from national conformity to world polity models. Recent work goes a step further, acknowledging that the worldwide spread of policies and models of governance are transforming societies, often in ways far removed from the original intent. However, these processes are only now just being studied, so the full scope of world polity influence is far from clear.

Third, a few scholars have begun to examine the depth of world culture penetration within nations. It is clear that the world polity influences governmental policy and law. To what extent are the practices and beliefs of individuals or firms affected? And to what extent are government policies and laws actually effective, as opposed to being decoupled from practice? Although the state is clearly an extremely important locus of social activity, it is useful to extend institutional analysis to the behavior of individuals, firms, and other social actors within nations. Such topics are difficult to study due to limitations on the availability of empirical data, especially for a large number of countries. Also, before the availability of simple software tools for multilevel models (for example, Hierarchical Linear Models), analysis of such data was nontrivial. Despite the complexities involved, some encouraging steps (for example, Boyle et al. 2002) suggest a nascent trend toward examining the impact of the world polity at the subnational level.

All of these are encouraging trends that represent substantial improvement and innovation in world polity research. They suggest continued growth in the level of sophistication of arguments and empirical studies within the institutional research tradition.

*Part II*     The Global Field
of Science

# Introduction to Part II

Complemented by the diffusion of science practices to nation-states world-wide—the focus of Part III of this book—a central dimension of science globalization is the consolidation of a global field of science. This world-level field of science extends as an umbrella, or a canopy, of science over global affairs and global processes. The two chapters included here in Part II provide us with an overview of the cultural and organizational dimensions of science in world society. Together, they demonstrate how central science is to the modern world polity.

Current literature addresses such global intensification of science-based work as an outcome of needs, either professional or social. Science, it is argued, expands as a result of a professional necessity to coordinate actions, share information, and follow a normative code. Yet science expands into domains that are less and less professional in nature and into spheres that have no direct consequence. Our study of international science organizations has led us to conclude that the current trend of science expansion is the growth of socially oriented science organizations (Chapter 3). Also, thematically, the expansion of science is justified by its appeal to a greater social good, such as national economic development (Chapter 4). Such global expansion thus defines science in reference not to professional interests, but rather in the context of broader societal concerns. It is this broader context—the themes of world society rather than the interests of a particular professional group—that sets the background for the consolidation of a global field of science.

The overarching presence of the global field of science sets both an organizational basis and a thematic rationale for further science globalization. It serves as a source of legitimacy for further science expansion and a pool of models for new science organizations to emulate. In this sense, international science organizations, such as the United Nations Education, Scientific, and Cultural Organization (UNESCO), encourage the establishment of additional science international organizations and of national science organizations while also infusing them with the justification for such expansion and imprinting them with a uniform message of their structure and social role. Moreover, the density of the global field of science offers further support for science globalization. In this sense, the more science disciplines organize international disciplinary organizations, the more additional science disciplines feel the need to organize internationally.

The global field of science has two main dimensions: organizational and discursive. First, the globalization of science means the emergence, consolidation, and expansion of an organizational field of science. Such an organizational field is the network of organizations, a web of organizational ties among science-related organizations. Second, the globalization of science involves the consolidation of discursive themes that define science and its goals. These themes offer a justification for science globalization, mark the boundaries of the institution of science, and identify its relationship with society. Although the organizational and discursive dimensions of science globalization are analytically distinct, they are inextricably intertwined; moreover, the discursive and the organizational are coconstitutive: the discourse defines the parameters for organizational expansion, whereas the organizations serve as the carriers of the discourse of science. Finally, the broadening of the discourse of science results in the broadening of its organizational basis beyond international science organizations. Once science became synonymous with national development, not only science organizations, such as UNESCO, but also development-oriented international organizations, such as the World Bank and the International Monetary Fund, contributed to its globalization. The full world canopy of science therefore extends beyond the narrow, professional meaning of science to include its broader, socially relevant meaning. In this sense, global science overlaps with other global fields or institutions (for example, economic development), yet it also usurps their domains as its own.

The intensification of this discursive and organizational density of global science, or the amassing of this global canopy of science, is the result of several factors. First, as the first wave of professionalized science international organizations came into being, such formation only became possible once science was transformed from a gentlemen's hobby into a professional activity during the nineteenth century. Indeed, the term *scientist* was coined only circa 1840 (McClellan 1985). As mentioned in Chapter 3, such professionalization of science involved the definition of a certain set of practices as scientific, the strengthening of professional identities through the formalization of professional boundaries by relying on credentials, and the establishment of formal locations, such as the research university, as the organizational infrastructure of scientific professional activity.

Second, the consolidation of a scientific ethos, to mark the normative expectations for the emerging profession, greatly contributed to the solidification of global science. With universalism, disinterestedness, and communality as its main norms (Merton 1942/1973), the ethos of science describes scientific knowledge and laws as universal (that is, without boundaries), not affected by particularistic interests (for example, national security), and therefore inherently shared (that is, transferable from one context to another). These cultural guidelines mean that international scientific cooperation is natural and preferable. In other words, because "natural laws" are valid worldwide (the laws of gravity are similar in Europe and Africa; human psychological needs are similar regardless of racial affiliation; economic principles are equal in different economies), science is applicable worldwide, and its application should be achieved through cooperation.

Finally, world events affect the rates of expansion of global science. For example, the two world wars and the Depression temporarily inhibited the rates of establishment of new science international organizations. Recently, the general tendency of "globalization of globalization" (the expansion of legitimacy of "globalization" and of "the global") provides an added boost to this process of global science expansion. These days, under the new canopy of legitimating globalization in general, science is further energized to globalize.

As a result of the combination of these factors, science is increasingly consolidated as a global field. Moreover, science is increasingly present in international policy and governance affairs. Such presence is evident and obvious when the matters are scientific in essence, yet increasingly, science speaks in the language of social concerns, and social matters take the form of scientific laws. Many are the examples for these parallel processes of "scientization of society" and "socialization of science." Most notably, science is positioned in reference to human progress. As described in Chapter 4, the tight link between science and development sets science as an integral part of human—particularly national—progress. Similarly, the tight link between science and environmental protection positions science in reference to human, or social, welfare (Chapter 11; Frank et al. 2000a,b). In its capacity as the authority on social means, science is hence called on to comment on social matters and social priorities. This position further enhances science's stand as a central feature of the modern world polity and encourages its globalization. As such, the consolidation of a global field of science is expanded.

## Chapter 3

International Science
Associations, 1870–1990

Subsequent chapters trace the global spread of Western science and its consequences. Here, however, we begin with a discussion of the rise and organizational expansion of science in the international sphere. How and why did science expand internationally to become an important part of a growing "world society"? The history of international scientific associations provides a powerful lens to gain insight into this process.

As a first exploration of science globalization, we trace the growth and changing character of international science associations (also called science INGOs because they are international nongovernmental organizations). Many international science associations sustain lofty and universalistic goals, visions, and ideology—often hoping to unite scientific knowledge or bring science to bear upon pressing global problems. They serve as potent carriers of universalistic, rationalistic Western culture, and their growth in the international sphere reflects (and directly constitutes) the institutionalization of that culture on a global scale. By examining the evolution of international science associations, we gain insight into the changing character of science and scientific authority. To foreshadow our main findings, we observe a major shift from science associations organized as an inward-looking profession to science associations organized around social problems and societal concerns, often sustaining advisory positions (or other authoritative roles) in relation to the United Nations (UN) or particular national governments. The history of these organizations thus chronicles the expansion of science as a social authority over society and governance.

## BACKGROUND

European science had transnational pretensions that substantially predate the rise of international science associations. Early scientific societies, although tied to specific nations or city-states, supported an international network of communication since the seventeenth century (Ornstein 1928). Journals, personal correspondence, emigration, and travel linked scientists throughout Europe and beyond. The very emergence of Western science was clearly a Europewide phenomenon, with deep cultural roots in Western visions of universalism and rationalism (Wuthnow 1980; Toulmin 1989). Yet it was not until the latter part of the nineteenth century that scientists (and nonscientists concerned with science or its applications) came together to form international associations—that is, associations not based solely in one city or nation, and with members from several other countries (Crawford 1992).

Since the turn of the century, international science associations have grown rapidly, and they are now fairly common. Several hundred exist, and new ones are being created at an increasing rate. Much international scientific activity is professional in character, supporting international communication and collaboration among scientists. Associations such as the International Union of Biological Sciences produce journals and hold meetings that bring together scientists from around the world. More recently, collaborative laboratories have been created to aggregate the resources of many nations, supporting scientific projects of immense scale.

However, the majority of international science associations are not narrowly oriented around the interests of the scientific profession. Rather, they reflect a broad progressive vision that science will benefit all humankind. For instance, the International Organization for Chemical Sciences in Development is a collection of scientists, research consortiums, development professionals, and individuals who support chemistry not for its own sake, but for the purpose of improving the economies of developing countries. Similarly, the International Network of Engineers and Scientists for Global Responsibility focuses on issues of peace, environmental protection, and development. Their goals are fundamentally social.

The growth of scientific activity in industrialized Western nations provided a foundation for the creation of professional science associations in the international sphere. But the broader cultural centrality of science in Western societies (see Chapter 1) provided a strong basis for linking science to progressive visions of social amelioration. With the growth of the scientific profession came new cognitive models linking scientific activity to societal development. That vision proliferated within the postwar intensification of the liberal world polity, leading to unprecedented levels of international organization around science.

## FOR "SCIENCE" AND FOR "SOCIETY": PROFESSIONAL AND SOCIALLY ORIENTED SCIENCE ASSOCIATIONS

By the late 1990s, there were well over 300 active international science associations. The earliest was founded in the mid-nineteenth century, although such early organizations were relatively rare; more than 95 percent of the associations were founded in this century, more than 70 percent of them after 1945. Like most sectors of international activity, scientific associations have been growing in number at an accelerating rate.

Figure 3.1 shows the cumulative number of science INGOs founded in the world over the period 1870–1990. This graph shows that organizational activity began in the mid-1800s, accelerated in the 1920s only to be interrupted by World War II, and grew rapidly in the postwar era. However, the appearance of smooth, continuous expansion shown in the figure is misleading. When one examines the characteristics of individual science INGOs— such as their stated goals, activities, and structure—it becomes clear that the population of international science organizations is composed of two distinct types of organizations.

Professional science associations were the first type to appear. Here we use the term *professional* broadly to include organizations focused on the following: (1) the professional interests of a specific scientific field; (2) scientific standards and nomenclature (for example, the International Commission on Zoological Nomenclature); and (3) the production of scientific knowledge (for example, the International Statistical Institute). Essentially, these organizations are made up of scientists working to benefit either their research or their profession. Indeed, many such associations emerged directly out of professional activities, such as annual meetings or congresses. A typical example, the International Union of Biological Sciences, publicly states the following

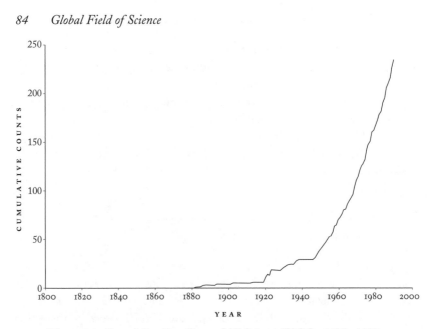

*Figure 3.1.* Cumulative Foundings of All Science INGOs, 1870–1990

aims: "To promote the study of biological sciences; initiate, facilitate, and coordinate research and other scientific activities that require international cooperation; ensure the discussion and dissemination of the results of cooperative research; promote the organization of international conferences and assist in the publication of their reports" (UIA 1994–1995:1076).

Figure 3.2 shows the cumulative number of professional science INGOs founded over time, showing much activity well before World War II. In addition to their early origins, professional organizations as a group share many of the same characteristics. They tend to be small organizations, their memberships consisting of active scientists, national branch organizations, and some interested amateurs. Larger bodies have memberships in the thousands, whereas associations based on obscure scientific subfields may have only a few dozen members. Another shared characteristic of professional science INGOs is their relative isolation from other organizations in the international sphere. They rarely establish formal relations with other associations, with the exception of other science associations in related subfields. Similarly, they rarely have formal relations with intergovernmental organizations (IGOs) such as the UN.[1] Finally, they only rarely advise nation-states on

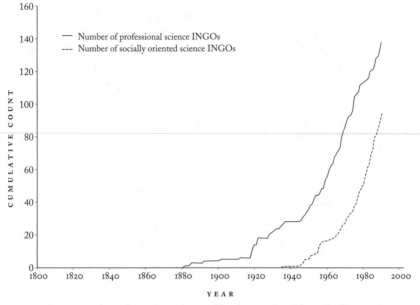

*Figure 3.2.* Cumulative Foundings of Professional and Socially Oriented Science INGOs, 1870–1990

matters related to their expertise. In short, they are associations of and for scientists.

In contrast, the recent period is typified by increasing numbers of international science associations focused primarily on issues outside the narrow professional interests of scientists (Figure 3.2). These associations support science in order to address social problems such as economic development, environmental degradation, war, nuclear weapons, and ethics. We refer to such organizations as "socially oriented" science INGOs (see Moore 1996 for a related discussion of public interest science organizations). Activities frequently include the following: (1) bringing scientific information to the citizenry or policy makers (for example, the International Network for the Availability of Scientific Publications); (2) promotion of science or science policy that directly ameliorates social problems (for example, the Research and Development Forum for Science-Led Development in Africa); and (3) promotion of ethics in the application of science (for example, the International Network of Engineers for Global Responsibility).

The character of these organizations is evident upon examination of their professed aims, activities, and membership. For example, the stated goals of the International Organization for Chemical Sciences in Development are to "encourage cooperation among chemists . . . ; endeavor to harness chemical sciences to work towards solutions of socio-economic problems; . . . to assist in the determination of priorities for development with respect to the chemical sciences in less developed countries; focus attention, effort, and resources from developed countries onto the needs and problems of developing countries by enlisting the help of selected specialists and institutions" (UIA 1994–1995:983).

These goals clearly extend beyond the narrow professional interests of scientists. Rather, this organization hopes to utilize the expertise of chemists to help solve a particular social problem: economic underdevelopment.

These socially oriented science associations differ from professional associations not only in their aims but also in their memberships and linkages to other international organizations. Scientists are not always the sole or even dominant members of socially oriented science associations. These organizations typically include policy professionals and interested citizens, sometimes even to the exclusion of scientists. Also, in contrast to professional organizations, which are relatively isolated, socially oriented science INGOs tend to have many links to other international organizations, particularly IGOs such as the United Nations Development Programme and the United Nations Education, Science, and Cultural Organization (UNESCO). Finally, socially oriented science INGOs are almost exclusively a postwar phenomenon. They were quite rare before 1945 and were founded in significant numbers starting only in the 1960s. However, their rate of growth has accelerated and continues to accelerate up to the present (Figure 3.2).

The dramatic rise of two such different populations of science organizations suggests long-term historical changes in the nature of science in the international sphere. To understand these changes, we must look both to historical changes in science itself and to the changing characteristics of the international sphere.

THE RATIONALIZATION AND PROFESSIONALIZATION OF SCIENCE
AND THE RISE OF PROFESSIONAL SCIENCE INGOS

When one considers the term *science*, one cannot help but think of highly rationalized activities—the Baconian experimental ideal, Popper's logic of hypothesis testing, the image of a team of researchers methodically searching for a cure for cancer. It is easy to forget that the organization of science into a clearly defined set of activities to be practiced by professional trained personnel is, in fact, a historically recent development. The modern scientific profession was a product of the nineteenth century (Bates 1965). Even as late as the early twentieth century, science remained an activity often performed as entertainment by or for the wealthy gentlemen of society. The subject of a scientific study was often an interesting curiosity rather than a systematically planned experiment designed to augment scientific knowledge. In scientific journals of the late nineteenth century, it is not uncommon to see articles about chemical experiments juxtaposed with ones concerning newly invented farming tools, a recently sighted meteor, a debate about whether or not humans are innately aggressive, and even reports of oddities such as the birth of a two-headed goat on a farm near London (*Nature* 1876).

In other words, exactly what constituted science was not well established—indeed, the word *scientist* was not commonly used until around 1840 (McClellan 1985). There are those whom we look upon with hindsight as scientists, such as Kelvin or Huygens, because their work conformed, more or less, to modern conceptions of scientific activity. However, there were also hundreds of tinkerers, collectors, and observers of nature whose work was routinely included in the scientific discourse of the time but has since been relegated to the realm of "invention," "amateur observation," or, most harshly of all, "pseudoscience" and "quackery" (Russell 1983; Slotten 1994). In other words, although some apparently "modern" scientific activity went on before the nineteenth century, science as a whole was not a modern, rational enterprise (Ellul 1964).

The historical process that yielded modern science was a complex one, involving worldwide historical movements toward modernity, secularization, and rationalization (Ellul 1964; Habermas 1987; Toulmin 1989). At the same time, these broad trends were paralleled by local struggles of scientists hoping to increase their status, share their findings, or improve the quality of scientific research (Bates 1965; Slotten 1994). The rationalization and professionalization of science involved several related phenomena: (1) the establishment of refined methodologies and practices that were "scientific"—that is to say, the rationalization and, to some extent, routinization of scientific practice; (2) the creation and strengthening of professional identities and credentials (professional boundaries), thus establishing a formal scientific community; and (3) the creation of organizational infrastructure to support routine scientific practice—for example, the rise of the research university, industrial research labs, and so on (Russell 1983; Slotten 1994; Abbott 1988; Ben-David 1990).

A rationalized professional infrastructure in developed nations provided a strong base for the transition of scientific professional activity and practice to the international sphere, especially compared with the weaker, less professional scientific societies of earlier centuries. However, this factor by itself is not sufficient to account for the expansion of science into the international arena. Local- or national-level activity can become highly rationalized and organized without necessarily leading to organization at the international level.[2]

One reason that science organizations first made the transition to the international sphere lies in the culture of science itself, which can be described as rooted both in universalism and a strong community of practitioners (Merton 1942/1973). The essence of the modernist scientific project is the search for knowledge that is universal rather than embedded in local context (Toulmin 1989). This universality is part of modern science and is evident in the attitudes of individual scientists and in professional discourse (Zuckerman 1989). Consequently, the work of scientists in any part of the world is seen as relevant to scientists everywhere. This provides impetus and legitimation for organizing beyond national boundaries.

But as Chapter 1 points out, the progressive, rationalistic, and universalistic culture is not limited to scientists alone. It is a central feature of the modern system—a system in which the "actorhood" of states and individuals depends on a scientized, rationalized cosmology. This environment encouraged not only the creation of professionally focused associations, but also associations with grandiose progressive aims—that is, socially oriented science associations.

This discussion leads to the following propositions.

PROPOSITION I.    *The expansion of professional science activity (which is rationalized, universalistic, and communitarian in character) within core nations increases the founding rate of professional science INGOs in the international sphere.*

However, we do not expect the population of professional science INGOs to grow boundlessly. The same universalistic impulse that initially led scientists to organize internationally also makes their associations inherently competitive or preemptive in nature. By this, we do not mean that they literally compete to undermine each other. Rather, they compete in the sense that professional science associations claim to offer universal, abstract knowledge over a particular domain of science. Once the International Union of Analytical Chemistry was founded, scientists had little incentive to found additional associations of analytical chemistry. Instead, analytical chemists would typically join the existing organization. In fact, historical evidence suggests that in areas where multiple organizations exist in the same scientific field, mergers commonly occur (UIA 1986). Thus, the very expansion of professional science INGOs limits subsequent growth, suggesting the following proposition:

PROPOSITION 2.    *An increase in the number of professional science INGOs has a negative effect on the rate at which new ones are founded.*

In addition, the rise of science INGOs was influenced by a series of historical processes that have affected all international activity. The most straightforward of these factors are the world wars and economic depression, which disrupted international activity and reduced the rate of formation of science INGOs. Another factor was the rising number of nation-states in the world. The creation of new nations provides additional bases for generating international organizations, either around newly created regional identities (for example, physicists in Asia, ecologists in the Caribbean) or as the result of regionally specialized scientific interests (for example, studying African fauna). For example, virtually no African science associations existed before the decolonization of Africa, where several now exist.

Furthermore, science organization would be rare in a time when any form of international association was difficult to arrange or hard to imagine. On the one hand, factors such as the difficulty of international travel and slowness of long-distance communication inhibit transnational organizing. On the other hand, without the models and experiences of a wide array of international organizations that could be copied or used as guides, and in the absence of cognitive cultural models of "the world" as an important unit of activity and social progress, even imagining an international science association becomes unlikely. Since the late nineteenth century, however, both of these inhibiting factors gave way to conditions fostering the creation of new international organizations.

An obvious indicator of the increase in facilitating technical means and motivating cognitive structures for international science organization is the cumulative number of existing international associations. This variable captures the aforementioned processes, including raw growth in international communication, legitimation of INGOs as an organizational form, and expansion in the number of governmental organizations that fund INGO activity, all of which support the creation and survival of science INGOs. Thus, a series of control variables are suggested:

PROPOSITION 3.    *Periods of war and economic depression decrease the founding rate of professional science INGOs, but increase in the number of independent nations and in the number of INGOs in general increase the founding rate of professional science INGOs.*

MODELS OF SCIENCE AND SOCIETY AND INSTITUTIONALIZATION OF THE LIBERAL WORLD POLITY

The rationalization and professionalization of science was accompanied by new visions of science as deeply valuable to all members of society, a view promoted earliest and most vigorously by scientists themselves. Many early scientists were also social reformers who saw science and the scientization of society as the solution to many of society's ills (Mackenzie 1981; Russell 1983). In the late 1870s, science professionals first gained significant financial support from the state in Britain and the United States. That support was dramatically augmented in the 1910s and after World War II (Alter 1987). Active science professionals proclaimed that they, rather than inventors, entrepreneurs, or engineers, held the keys to progress and economic growth. Furthermore, they argued that the rational application of scientific logic to many domains of social life (for example, the workplace, management, production) would solve social ills ranging from labor disputes to social anomie to war (Alter 1987; Moore 1996).

This model of science as a primary source of progress and social development was incorporated into many nation-states, as well as many IGOs, in the years around World War II (Finnemore 1991; Drori 1997; Haraway 1989; Chapter 4). A variety of factors prompting this trend have been cited, among them functional (Merton 1938/1970; Ben-David 1990), organizational (Russell 1983), and cultural (Haraway 1989; Ramirez and Lee 1995; Chapter 1). Functional and organizational perspectives typically presume that the "obvious" social utility of science, along with the strength of professional science organizations, led to the general acceptance of science as the source of social progress. Cultural perspectives depict the "science for society" model as contingently emergent from interwar and postwar liberal, democratic cultural discourse—either preoccupied with the war experience, the bomb, decolonization, and progress (Haraway 1989), or as a product of the triumph of liberal models and the expansion of the world polity (Finnemore 1991; Ramirez and Lee 1995; Chapter 1).

Although the cause may be disputed, it is clear that liberal nations such as the United States and Britain—and organizations they created, such as UNESCO—became crucial carriers of the "science for society" vision (Chapter 4). UNESCO is the consummate example of the process. Its primary activities include encouraging scientific activity and the diffusion of scientific knowledge not for its own sake, but rather for the benefit of society as a whole (Finnemore 1991).

By the postwar period, nation-states, scientific professionals, and important IGOs all viewed science as socially valuable. Scientists were urged to address topics of social import, whereas other social actors were encouraged to adopt more scientific approaches. This environment supported the founding of socially oriented science INGOs.

At a basic level, the existence of socially oriented science associations crucially depends on the existence of organized, professionalized scientific activity. If science were not coherently organized, the task of pursuing "science for development," or of forging links between science and policy makers, would be problematic at best. However, as science was organized around consistent, rationalized field categories (for example, physics, hydrology) with professional identities, it became possible to routinely consult appropriate scientists for advice on problems of social import, such as nuclear radiation or environmental degradation. Furthermore, professional organizations themselves were the strongest supporters of this conception. Their increasing presence in the international sphere provided support for the creation of socially oriented science INGOs. This suggests the following proposition:

PROPOSITION 4.    *The rise of professional science INGOs increases the founding rate of socially oriented science INGOs.*

Unlike professional science INGOs, these socially oriented science INGOs are not inherently preemptive or competitive. Scientists are bound by a universalistic discourse that encourages them to organize collectively but monopolistically: one organization in a given domain is sufficient. Socially oriented science INGOs, in contrast, organize around social problems and social action. There is no expectation that any single organization will or can hegemonically devise solutions to a given social problem. For example, a social problem such as environmental degradation generates a tremendous amount of organization with varying substantive foci (pollution, deforestation, ozone depletion), varying regional focus (Brazilian rain forests, Mediterranean marine dumping), and varying methodologies (grassroots mobilization, local projects, lobbying organizations). This logic applies directly to socially oriented science INGOs, which organize around many substantive areas, emphasize different regions, and employ varied methodologies. Indeed, rather than competing, new associations serve to legitimate the application of science to social problem and thus promote the creation of additional science INGOs. This suggests the following proposition:

PROPOSITION 5.    *The number of socially oriented science INGOs in existence will not negatively affect the rate at which new ones are founded.*

However, socially oriented science INGOs do experience certain forms of competitive pressure. It has been observed that the creation of IGOs with jurisdiction in a given substantive area can outcompete or preempt INGO activity in that area. In the case of environmental INGOs, for example, the expansion of IGO activity and the founding of a single dominant environmental IGO (the UN Environment Programme) reduced the founding rate of environmental INGOs (see Meyer et al. 1997b). By that same logic, "official" socially oriented science IGOs might have sufficient status and resources to undercut their unofficial INGO counterparts.[3] Hence the last proposition:

PROPOSITION 6.    *The formation of socially oriented science IGOs has a negative effect on the founding rate of socially oriented science INGOs.*

Finally, we expect that socially oriented science INGOs will be affected by the same factors (described above) that affect professional science INGO foundings: periods of war and economic depression, the increasing number of nation-states in the world, and the general expansion of INGO activity.

A QUANTITATIVE ANALYSIS OF THE RISE OF SCIENCE INGOS

By use of data on the foundings of international science associations, we modeled the historical expansion of professional and socially oriented science

INGOs as separate organizational populations. The source for organizational foundings is the 1995 *Yearbook of International Organizations*, published by the Union of International Associations, which attempts to catalog all international organizations that have ever existed. International INGOs are defined in the *Yearbook* as organizations with members from at least three different nations. We use the categories provided in the *Yearbook* to select the superset of all science-related organizations and examine the description of each organization to determine which organizations fit the concepts employed in this study.[4] We also use the organizational description to determine whether an organization is professional or socially oriented in character. The data encompass the full population of professional and socially oriented science INGOs currently active.[5]

The event of interest in this analysis is the founding of new science INGOs. The rate of such events over time (the hazard rate) can be modeled by means of event history analysis methods (Tuma and Hannan 1984; models were estimated by RATE [Tuma 1992]; see Chapter 2). We employ a constant rate (exponential) model because it assumes a constant rather than time-dependent hazard rate. Any time dependence is of fundamental substantive interest and will be accounted for by independent variables, rather than being specified in the functional form of the model itself.

Independent variables for these analyses are collected yearly. We use two variables to test hypotheses regarding foundings of professional science INGOs: first, the rationalization and professionalization of scientific activity in core nations, as measured by the cumulative number of professional science associations in the United States[6]; and second, competitive and preemptive pressures among professional science INGOs, as measured by the cumulative number of professional science INGOs that have previously been founded.

Three variables are used to test hypotheses regarding the founding of socially oriented science INGOs. Professional scientific activity in the international sphere is measured by the cumulative number of professional science INGOs in the world. Competition among socially oriented science INGOs is indicated by the cumulative number of socially oriented science INGOs previously founded. Finally, competition and preemption caused by science IGOs is measured by the cumulative number of socially oriented science IGOs in the world.

Additionally, the following control variables are included in both models: wartime and economic depression are indicated by dummy variables reflecting the years of World War I, World War II, and the Great Depression.

TABLE 3.1
Constant Rate Model Showing the Effects of Covariates
on the Founding Rate of International Professional Science INGOs,
1875–1980 (n=114 events)[a]

| Variable | Parameter | Standard Error |
|---|---|---|
| Constant | −1.61 | 2.23 |
| Professionalization of science in core nations | 0.0035* | 0.002 |
| Competition: Density of professional science NGOs | −0.003* | 0.0016 |
| Expansion of the international system | 0.289 | 1.43 |
| Wartime and world economic depression | −0.953** | 0.407 |

[a]$\chi^2$ compared with constant-rate model, no covariates, is 93.48***.
* $p<0.1$, ** $p<0.05$, *** $p<0.01$

To conserve degrees of freedom, these three time-period dummy variables were combined into a single variable.[7] Measures for increasing numbers of nation-states and increasing INGO activity are simply the cumulative numbers of those in existence. Because the latter two variables are highly collinear and reflect related underlying processes, they were combined into a single index via factor analysis.

Table 3.1 shows results from a constant-rate model predicting the founding rate of professional science INGOs. The model strongly supports the propositions outlined above. Professionalization of science at the national level has a significant positive effect on the founding of professional INGOs. The density of existing professional INGOs has a strong negative effect on the rate of new organizational foundings, suggesting preemption or competition among professional science INGOs. Expansion of the international system has a nonsignificant effect on the founding rate of professional science INGOs. Finally, the combined wartime and world economic depression variable has a negative but nonsignificant impact.

Table 3.2 shows results of a model predicting the founding rate of socially oriented science INGOs. International professional activity has a significant positive effect on foundings, but the insignificant effect of cumulative density of socially oriented science INGOs suggests that competition and preemption are not occurring. However, the existence of socially oriented science IGOs does have a significant negative effect on new foundings, suggesting

TABLE 3.2

Constant Rate Model Showing the Effects of Covariates on the
Founding Rate of Socially Oriented International Science INGOs,
1875–1980 (n = 48 events)[a]

| Variable | Parameter | Standard Error |
|---|---|---|
| Constant | −2.57 | 2.26 |
| International professional activity: | | |
|   Cumulative density of professional science INGOs | 0.124* | 0.0736 |
| Competition: Cumulative density of socially oriented | | |
|   science INGOs | 0.005 | .006 |
| Competition from IGOs: Cumulative density of | | |
|   science IGOs | −0.370*** | 0.125 |
| Expansion of the international system | 1.33 | 1.84 |
| Wartime and world economic depression | −2.08* | 1.12 |

[a] $x^2$ compared with constant-rate model, no covariates, is 101.29***
* $p<0.1$, ** $p<0.05$, *** $p<0.01$

clear preemption by these more official science organizations. Finally, the war/economic depression dummy has a significant negative effect, but the factor variable combining expansion of the nation-state form and total INGO creation has a nonsignificant positive effect.

These results support the description of the rise of science INGOs developed above. In the nineteenth century, science professionalized in the dominant core nations, producing local organizations with strong international linkages. Combined with the expansion of the international system, this professional activity emerged at the international level in the form of professional science INGOs. This process in turn legitimated the application of science to society, yielding socially oriented science INGOs.

IMPLICATIONS AT THE NATION-STATE LEVEL

Within core countries, the impact of professional science INGOs is eclipsed by the strength of national-level science organizations. In the United States, for example, professional science associations are typically larger and wealthier, and they support more prestigious conferences and journals than their international counterparts. In peripheral countries, the situation could not be more different because national scientific infrastructure is often weak or absent. Professional science INGOs, in conjunction with elite journals, become the primary link for scientists to current activity in their field.

For scientists in peripheral nations, participation in professional science INGOs is likely to be more important to their research than participation in local science organizations. Local organization is typically poorly funded in the periphery, and in some fields, it may not even exist. A (nonsystematic) comparison of the number of national-level professional science societies to the number of memberships in science INGOs suggests the following pattern: individual scientists in a given peripheral nation are typically connected to dozens of international science associations, yet national-level organizations are often rare—sometimes nations have fewer than five (UIA 1989; Sachs 1990). Whether or not such participation in science INGOs actually precipitates national-level organizations cannot be determined without a rigorous longitudinal analysis. Nevertheless, it is clear that peripheral scientists often participate in INGOs in a given scientific field before any significant organization of that field in their home countries.

Thus, for the periphery, the transnational network of professional science associations is likely to be of consequence. Such a network may help scientists keep up on developments in their fields, even in the absence of local scientific communities. At the same time, science associations facilitate communication and participation by peripheral scientists, potentially reducing the massive inequalities in world scientific research output that exist today.

The impact of socially oriented science INGOs is harder to gauge, in part because they are a rather heterogeneous group of organizations. Some focus on scientific ethics and primarily affect the work of scientists themselves. Other socially oriented science INGOs focus on specific social issues such as environmental protection. As international lobbyists, they have the potential to affect policy in nation-states around the world. However, many socially oriented science INGOs focus specifically on economic development or the support of science in underdeveloped nations. They typically bring resources, technologies, training, and policy advice to developing nations, intervening via programs, conferences, and grassroots efforts. One result is that peripheral nations almost certainly have more scientific activity and infrastructure than they would in the absence of socially oriented science INGOs.

BROADER IMPLICATIONS: THE SOCIALIZATION OF SCIENCE
AND SCIENTIZATION OF SOCIETY

The rise of socially oriented science INGOs, as well as their counterparts at the national level, is indicative of a broad social trend: the advent of science and rationality as a dominant model for organizing social activity. By this, we mean the following: (1) scientists increasingly shape and define social issues and the identification of problems worth solving; (2) scientific expertise and information is increasingly integrated into governmental organization and decision making; (3) scientific discourse increasingly infuses policy discussion and debate.

As we have argued above, this trend toward the "scientization" of social activity is particularly pronounced at the international level, where world-polity values of rationality and universalism are strong (Boli and Thomas 1999). John Meyer (1994b) introduces the notion of "rationalized others"—professional or scientific bodies that have no formal power but offer information and policy advice to nation-states and bodies of international governance. Socially oriented science INGOs are one organizational embodiment of this phenomenon. Peter Haas (1992) summarizes the notion of "epistemic communities," collections of like-minded scientists and policy professionals operating at the international level to provide information to nation-states and thereby influence or even shape state goals and interests. The many linkages that socially oriented science INGOs form among themselves and with IGOs suggests that they form a network and are part of the organizational basis for epistemic communities. If the rapid, continuing growth of socially oriented science INGOs is any indication, "rationalized others" and "epistemic communities" are proliferating.

The increasing authority and dominance of science in social decision making has another important effect: nation-states and IGOs begin demanding scientific input and information in decision-making processes. They increasingly invite scientists and science INGOs to serve as advisers and consultants. More recently, IGOs and nation-states have encouraged the creation of new science associations, in part to gain policy information. As of 1994, more than twenty science INGOs had been founded by IGOs such as UNESCO, the European Union, and the UN. One example is the European Pollution Science Society, an INGO concerned with the effects of pollution on the world ecosystem, founded by the European Union in 1988. The founding of this organization preceded the professionalization of "pollution science" as a field. Science has become such a routine solution to social problems among IGOs and states that it is encouraged and even demanded, irrespective of the amount of scientific knowledge available. This circumstance, the hyperdemand for rational action plans, is described in detail for one arena, development, by Chabbott (1999). It likely exists in many areas where pressure for governmental action is acute but no existing structures provide routinized strategies or solutions.

There is a tendency in both academic and nonacademic discussion to treat the scientization of social planning and governance as a purely instrumental response to the efficacy of science. Although this is certainly the case in many domains, it hardly tells the whole story. Much scientization takes place in domains where there is little scientific consensus or the efficacy of science is questionable—for example, complex environmental problems such as global warming, or issues of economic underdevelopment. The fact that science "works" in many contexts is not sufficient to account for the intense and often unrealistic level of scientization observed. It makes more sense to view scientization as driven in part by the rationalized culture of the Western-derived world polity. Moreover, scientific rationality was a foundational ideology that coevolved with and justified modern forms of organization and governance (Shapin and Shaffer 1985). As a consequence, it is unsurprising that states and branches of international governance tend to organize activities and solutions in terms of rationalized, scientized policy solutions.

Finally, science is not left unchanged by its increasing role in social policy formation. There is evidence that both IGOs and science INGOs are working to support new scientific fields that can cope with the social problems of society. As resources are increasingly allocated to these new types of scientific activity, the very disciplinary structure of science may be transformed. Many new or rapidly expanding scientific fields have clear socially oriented components: environmental sciences, genetic engineering research, and so on. At the same time, so-called big science is losing support now that governments are less willing to spend billions to support arcane research with little social utility. Thus, there is evidence to argue that science itself is "socializing"—that is, becoming more organized around social issues and problems.

The increasing presence of science in policy and governance is becoming apparent. However, it is less clear exactly how science is being incorporated into governance and what implications it will have. One could imagine science being directly incorporated by a strong central world state. However, this seems unlikely given the highly decentralized nature of the world polity.

Research on "rationalized others" and "epistemic communities" has begun to theorize the process, but much work remains. At this point, even basic knowledge is lacking. How often do decision makers use or cite scientific knowledge? Does this vary across substantive domains? What sort of scientific organizations provide policy information? With which governments do they interact? Further study of international science associations would provide insight into some of these questions.

## Chapter 4

The Discourses of Science Policy

In the previous chapter, we examined the organizational aspects of the expansion of science in world society over many decades. We now turn to the discursive or cultural dimensions of the same expansion. In this chapter, we review a wide range of policy documents and other texts generated in the international system over the last half-century—particularly by intergovernmental organizations in the broad United Nations (UN) system—to discover the ways in which they conceptualize science and its values for society. Cultural perceptions during this period defined a broad social role for science and produced both the consolidation of science as a global field and a wave of nation-state activities incorporating and expanding scientific activity.

Our review of the international discourse on matters of science during the period shows that instrumental or technical conceptions of the value of science for human activity have been utterly dominant. Science is seen not as a general cultural framework (the view emphasized in this book), but rather as a means related to the achievement of important social ends (see also Finnemore 1996; Chabbott 2002).

Principally, expanded science is seen as instrumental for the achievement of national development, seen mostly as an economic enterprise. We call the vision here the "science for development" model. It depicts science in a most positive way as promoting economic growth through improved labor-force skills and technical discoveries. Because national development has been a most highly legitimated goal for national states, and because science is increasingly depicted as a central means to achieve development, waves of policies for the national and international expansion of science have resulted.

A second, and equally instrumental, conception of the role of science has been much more negative. We call it the "science and human rights" model. Here the vision is that scientific expansion produces various threats to the achievement of broader human goals by subjecting more and more domains to central rationalized controls or by misuse of scientific achievements. Policy focuses here are on needed controls over scientific development, rather than enhancements of it. This set of concerns has been marginal through the past half-century, overwhelmed by a world emphasis on economic development. But in recent years (after the period covered by our review), it seems to have become more prominent, evidenced by world movements concerned with biotechnology and related fields.

In this chapter, we describe both variants of international discourse on science, noting the texts that exemplify these models and the agencies that promote them. We emphasize that both models see science and its expansion in a highly instrumentalized way, as useful for the most narrowly economic human goals (development) and as dangerous to broader and more expressive ones (human rights). In this book, we correct this instrumentalized picture of science, and our arguments and evidence emphasize science as a cultural framework and form. This reframing helps us analyze ways in which scientific expansion may be much less relevant for narrowly economic development than is commonly supposed (see Chapter 10)—and much more supportive of broader human rights and development than critics suggest (see Chapters 10 to 12). We suggest that our broader view of the nature of modern scientific authority produces a more realistic view of the effects of expanded science.

THE "SCIENCE FOR DEVELOPMENT" POLICY MODEL

Science is praised as a requirement for any nation-state aspiring to become modern and prosperous. Much like the institutions of modern education and technology, science is assumed to pave the path for national development. Whether through introducing people to modern beliefs[1] or by causing a change in the structure of the local labor force and thus of the economy, science is closely associated with national plans for development. Indeed, national development policies often cite science, and the related alteration of the skills of national populations, as major factors in attaining progress.

Adopting this vision, policy makers and lay people regard economic growth as dependent on the scientific and technological capabilities of the labor force.

## The Model

The "science for development" policy model describes national economic growth as dependent on the scientific and technical capabilities of the labor force. Such capabilities rely on advanced scientific and technical training, whereas such training depends on the foundations of science education in primary and secondary schools. Simultaneously, the skills of the labor force directly set a basis for technology development (or the encouragement of local technical innovation) or for effective technology transfer. In this scheme, the role of science is twofold. First, science is used as an education and socialization mechanism to shape positive attitudes toward modernization and to train candidates in science and technology, in this way preparing them for higher education and more sophisticated production roles. Second, science is used to create a knowledge base for technology, either transferred from core economies or locally produced. Technology is assumed to be a basis for an internationally integrated, advanced, and prosperous economy. In general, then, the effects of science on national development are mediated by the skills of the labor force and hence resting on a foundation created for the use of sophisticated technology and advanced manufacturing practices. Figure 4.1 graphically illustrates of the principles of the "science for development" conceptual model.

In an international atmosphere of economy-based developmentalism, the "science for development" model is often at the core of national self-definitions. The publication of the results of cross-national testing in science and mathematics, such as the Third International Math and Science Study, in the news media elicits waves of national panic, reflecting the widely held belief that schoolchildren's scientific achievements bear on national economic standing. Similarly, figures predicting an inadequate supply of scientists and engineers in the future labor force are regarded as a serious economic problems. These public reactions show how central science is perceived to be in national planning. They also reflect how science is judged by its production of, or at least relations with, economic performance. As Kenneth King writes, when describing the interpretation of the role of science education in the phenomenal economic success of the newly industrialized countries: "It is

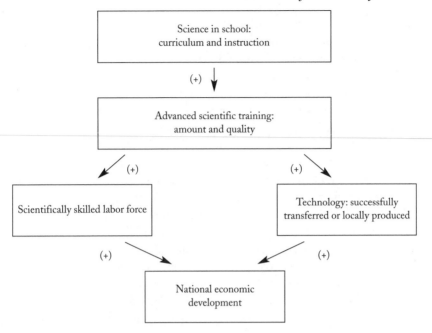

*Figure 4.1.* The Image of Science: The "Science for Development"
Policy Model

widely *assumed* that the human resource development strategy involving science and technology education must have played some part in the break-
through" (1989:108; emphasis added). When studying American science policy, Daniel Sarewitz (1996) refers to this belief in science's utility as "the myth
of infinite benefit."

## ASSUMPTIONS

The "science for development" model specifies that the benefit from science
(science education, scientific research, or technical training) is national development, most often described in economic terms. In the late 1990s, science
and development policy texts introduced broader notions of national development by including references to such issues as environmentalism and
income inequality,[2] but for most of the past half-century, such policy texts

explicitly cited economic goals as the main consideration, and measure, of national output. In this sense, the economy-based discursive regime of "development"[3] dominates any discussion of science and its social role.[4]

The "science for development" policy model has several related assumptions. First, the model regards science as a *national* project: science policy is aimed at providing benefits for the nation as a whole, and it relies on national financial support and societal legitimacy. Second, it sets a systemic plan for the achievement of national development, thus offering an explicit program linking science to national economic development. This may seem like an obvious trait of rational social planning, but it is not a given for all social policies (for example, toward the expressive arts). Third, it considers science to be a "real" social institution, with much shared content, culture, and organization, rather than a socially constructed phenomenon. This permits the expansion of science in general structures, prominently universities. Finally, this economy-based vision is also utilitarian in essence; it judges science by its end products or consequences. In science policy circles, this utilitarian stand is a discursive move away from the Vanaver Bush ideal of science as the endless frontier, with its future and general benefit to humanity and an implied concentration on basic research. The pinnacle of this move in science policy terms is the hyperutilitarian position of the "new production of knowledge" approach (Gibbons et al. 1994) and of "triple helix" modeling (Etzkowitz and Leydesdroff 1997), with their focus on the applicability of each type of research and on a changed role for academia in relationships with government and industry. In summary, the "science for development" policy model promotes a vision of science as (1) national, (2) systemically planned, (3) realist, (4) centered on economic development, and (5) utilitarian.

Obviously, the "science for development" policy model is rooted in the functionalist assumptions of modernization theory, which is related to liberal socioevolutionist perspectives. Such theories envision an "ideal form" of society[5] and regard development as a process of progression from lower, more backward stages of development to advanced modernity. It further supposes that situations and events may be replicated from one nation to another. As a policy guide, it thus suggests that less developed countries follow in the footsteps of developed ones, in order to achieve economic prosperity and social liberties. On the basis of such assumptions, missions of scientists and development experts, sponsored by international organizations, visit less developed countries and promulgate science aims for them to conform to those of developed countries (Finnemore 1993, 1996; see also McNeely 1995). The "science for development" model, although laden with these assumptions, like other expressions of the Western models of rational utlitarianism (especially in policy circles), is taken for granted as a neutral and context-free paradigm.

The "science for development" policy model is ultimately rooted in the related political notions of the nation-state as a goal-oriented, rational, authoritative, interlinked, yet autonomous actor. This is in effect an expression of the dislocation of the authority of governance away from the world polity, with its discursive regimes and organizational carriers. This discursive dislocation is misleading because the "science for development" model is formalized and is intensely propagated by international organizations.

### The Role of International Organizations in Propagating the "Science for Development" Model

Although science and technology have been agenda items for the UN and United Nations Education, Scientific, and Cultural Organization (UNESCO) since their formation, more concentrated and organized efforts to promote "science for development" started in the early 1960s. Since then, all development and science-oriented UN agencies have conveyed similar messages that present science as a means for national economic development. There was a steady increase in the number of policy declarations by UN agencies promoting "science for development," and there was a parallel increase in the attention given to this issue.[6] Science and technology were also highlighted as focal concerns for progress-seeking nation-states during one of the UN's Development Decades.[7] The peak of the efforts to promote "science for development" were the 1979 Vienna Conference on Science and Technology for Development and the declaration of 1980 as the International Year of Science and Technology for Development. These 1979 events are viewed as watershed points in the UN promotion of "science for development" (see Rittberger 1982).[8]

The efforts of UN agencies to promote "science for development" reached far beyond the assembly of periodic international conferences to include organizing regional conferences and conferences directed at target countries[9] and sponsoring and widely distributing a multitude of publications on "science for development."[10] Great efforts were also invested to create an organizational basis for these ventures. Because the issue of "science for development" is of concern for several UN agencies,[11] international organizations were founded to coordinate "science for development" activities[12] and to fund related activities.[13]

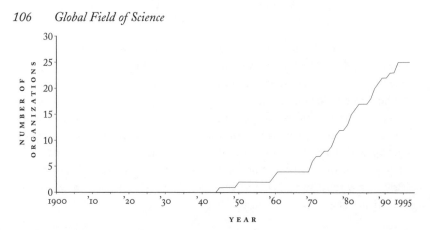

*Figure 4.2.* "Science for Development" International Organizations, 1990–1995

In addition to the institutionalization of the organizational basis for "science for development" model within the UN system, parallel trends occurred in the general field of science international organizations. Development-minded science international organizations have been established at far greater rates since the early 1970s. As presented in Figure 4.2, the first "science for development" international organization—UNESCO—was established in 1946, but the dramatic change in establishment rates of such organizations occurred in the early 1970s. By the early 1990s, there were twenty-five such international organizations, namely science international organizations that set development as a core goal. Such organizations include the International Association of Science and Technology for Development,[14] the International Organization for Chemical Sciences in Development,[15] the Islamic Foundation for Science, Technology, and Development,[16] and the Research and Development Forum for Science-Led Development in Africa.[17] Although there are more science international organizations over time in general (overall, some seven hundred science organizations in this data set; see Chapter 3), the trend displayed in Figure 4.2 shows the expansion of "science for development" international organizations in particular. Moreover, although the expansion of the field of science international organizations starts early on, with a dramatic rise after 1945, "science for development" international organizations are mainly a post-1970 phenomenon. Overall, the "science for development" policy model is currently a fully institutionalized global organizational field, connecting both international governmental and nongovernmental in a web of organizational ties.

*Transmitting the Image of Science in Texts*

In addition to the organization of the "science for development" field in international agencies, these organizations also imprint science policy texts with a particular picture of science. Figures of speech, which are used to convey the social role of science as the basis for national progress, are found in policy statements of UN agencies, national policy bodies, nongovernmental organizations, and other agencies and institutions. Such statements tend to carry similar themes: they emphasize the role of science and technology in the achievement of modern, economic, national development. Statements tend to vary only in the degree of their explicitness: although some planning organizations stylize the "science for development" as a clear guideline for action,[18] others show only general commitment to this scheme.[19] Similar statements are echoed in declarations of national policy organizations.[20] In additional, other institutions (not necessarily governmental or international organizations) reiterate "science for development" themes and place themselves within its scheme.[21]

Finally, additional contributions to the globalization of science imagery are provided by international nongovernmental organizations. International scientific associations (both professionally and socially oriented; see Chapter 3) display both the scientific ethos of universalism and the notions that science is useful for growth-seeking nation-states. Professionally oriented science organizations use a specialized jargon to convey the usefulness of their professional disciplines. Socially oriented scientific associations, which explicitly promote the social responsibility of scientists and scientific disciplines, establish a strong link between the vocation of science and the setting of national and global goals, with development as a central concern. Of course, the texts of more socially oriented science international nongovernmental organizations are more explicit than the texts of professionally oriented science organizations in their definition of their goals as having a social import. Also, general science organizations convey a stronger emphasis on development issues than do disciplinary scientific associations.[22]

Overall, "science for development" themes are imprinted on the operations and plans in countries worldwide—developed as well as developing—and on various environments—governments, public institutions of science, and private corporations. This imprinting is channeled through both advising and funding (see King 1989). UN agencies and other international organizations are instrumental in the diffusion of these science policy texts worldwide, and hence encourage the isomorphism of policy texts (Finnemore 1993, 1996; McNeely

1995). Most importantly, these international organizations serve as sites for the expression and articulation of the dominant "science for development" discourse. It is in their texts that the discourse is most explicitly demonstrated.

In summary, both governmental and nongovernmental international organizations promote the "science for development" model and the image of science that it carries. They set the agenda, formalize the discourse, organize the field, and carry the responsibility for the diffusion of scientific image to all nation-states. The interconnections among these science organizations—and between them and development-oriented and commercial organizations—establish a dense web of institutional ties, which are immersed in, and simultaneously promote, the image of science as a vehicle for national economic development.

### SCIENCE AND HUMAN RIGHTS

The international discourse of science, although dominated by the "science for development" model, offers another vision of science's social role. This vision is formulated in a model of "science and human rights." This model is—like "science for development"—evident in numerous international declarations and policy texts and is presented as an alternative to "science for development." On the basis of this premise, the models seemingly diverge on a number of dimensions. Although the "science for development" model carries a hopeful (or meliorist) perspective on science, the science and human rights model regards science as a source of potential infringement on human rights.[23] In other words, whereas in science and human rights texts science is mentioned in relation to world peace, security, freedom, and independence, these texts reflect a concern with the *infringement* of such ideals through scientific means and a call for securing scientific advances so they will not be used for ill aims. For example, the 1975 UN "Declaration on the Use of Scientific and Technological Progress in the Interests of Peace and for the Benefit of Mankind" boldly states that nation-states should "*Refrain* from any acts involving the use of scientific and technological achievements for the purposes of *violating* the sovereignty and territorial integrity of other states, *interfering* in their internal affairs, *waging* aggressive wars, *suppressing* national liberation movements or pursuing a policy of racial discrimination."[24]

A similar message is conveyed in the goal definition of the Pugwash Conference on Science and World Affairs, which convened in Pugwash,

Canada, in 1957: "Bring together scholars and public figures concerned with reducing the *danger* of armed conflict and seeking cooperative solutions for global problems. Organize international conferences of scientists to discuss: *problems* that have arisen as a result of the progress of science, and particular *dangers to mankind* from the development of weapons of mass destruction; *problems* of disarmament, international scientific collaboration, aid to developing countries."[25]

In this sense, the science and human rights model describes science as affecting human rights negatively: scientific advancements provide the technical tools to inflict harm on human populations. Technical advances in recording devices (such as computerized data banks), for example, infringe on one's right of privacy; psychiatric advances are employed for political torture; and advances in the biomedical sciences, which enable artificial insemination and genetic engineering, go against some notions of human evolution. Whether painting science as a demonic tool or as "bad science," the science and human rights model describes science and its effects on human development in negative terms.

The science and human rights model envisions the relationship between science and human rights as direct. Whereas science-development relations are seen as mediated by labor-force skills and technology, science has direct effects on human rights. The implementation of scientific advances by "the wrong people" or "for the wrong purposes" results in the infringement of rights. Thus, there are few mediating elements between science and human rights; rather, there is a general factor that shapes the relationship. In this sense, although the "science for development" model assumes that the science–development relationship is stable across times and locations, the science and human rights model identifies conditions for the relationship between science and human rights.

Overall, science is seen as a resource capable of violating a variety of legal, political, medical, economic, and cultural rights. Mentions of the positive effects of scientific and technological advances on human rights are rare. Texts do mention, for example, the contribution of science to sustaining and extending life through eradication of illnesses and control of plagues; to improving human relations through the development of communications technology or via enabling of popular elections through computerized national data management; to securing life via the technology of weather and natural disaster prediction. Few of these positive effects are, however, codified into international formal programs of action. Again, the positive benefits of science in securing human rights are overshadowed by the attention given to the hazards to human rights by the ill uses of scientific advances.

## *International Action on Science and Human Rights*

Although the science and human rights model addresses one of the main concerns in modern times—fear of science and technology's evil uses—this issue was marginalized in international discourse and policy. Overall, the science and human rights model is overshadowed by the "science for development" model. Moreover, science and human rights discourse never consolidated as a *policy* model. In other words, unlike "science for development," which was translated into policy action, science and human rights does not serve as the guiding principle for a program of action by international organizations or nation-states. Rather, science and human rights are used as an additional parameter in the general evaluation of human rights matters.

A few factors contribute to this overshadowing of "science for rights" and its dismissal as a policy guideline. First, the discourse of development is more dominant in international affairs than the discourse of human rights. Moreover, developmentalism tended to regard rights as subsumed under "development." For example, the latest wave of UN-sponsored studies of science that focus on the effects of science on human rights share this reductionist perspective by referring to "third-generation rights" (Weeramantry 1990, 1993). Their definition of human rights is peace, development, and healthy environment, and their focus is on the rights to work, to proper health care, and to certain minimal standards of living. They therefore consider both human rights and human progress within a confined economic framework. This corresponds with the implicit link between economic rights and human rights in other UN texts,[26] and the peak of such reductionist efforts was the 1981 initiative to institutionalize a "right to development"[27] (see Ogata 1990:5). UN texts reflect the dominant attitude that human rights are embodied in, or promoted by, the deliverance of economic prosperity.

Second, the prevailing broad definition of human rights served as an obstacle for the creation of a supportive international lobby. The expanded definition—which ranges from the right to privacy, right to work, right to vote, and right for life—resulted in the lack of international consensus and inhibited international governmental cooperation. The differences between countries in their definition of human rights resulted in, among other things, the abstention of different blocs of nation-states from voting on international action to secure such rights.[28] Cold war political alliances blocked initiatives and action on science and human rights,[29] and a smaller organizational basis was created.[30]

Third, the science and human rights themes were limited by their inherently transnational claim. "Science for development," by focusing on national development, defines science in national terms and therefore appeals to governmental formats (as in international governmental organizations). Science and human rights, on the other hand, appeals to international nongovernmental organizations because it refers to science in terms of transnational values. Overall, then, science and human rights was left to the domain of international nongovernmental organizations, whereas "science for development" is mainly the concern of international governmental organizations.[31]

As a result of these three intertwined factors—the dominance of developmentalism, political alliances, and transnationalist claims—science and human rights did not gain prominence in science discussions, nor did they consolidate as a global organizational field. This is clearly evident when we compare the number of human rights–minded science international organizations in comparison to the rise of development-minded science international organizations (as in Figure 4.2). Applying a similar definition to categorize science international organizations as human-rights minded, our analysis reaps only two such science and human rights organizations. These organizations are UNESCO and the Arab League's Educational, Cultural, and Scientific Organization. Although numerous other science international organizations mention human rights in their goal statements, they do so in a rather diffuse manner, most often as an associated social concern.[32] Overall, the lack of global structuration of science and human rights organizational field is evident: we find that only two science and human rights international organizations exist.[33] Moreover, these few international organizations lack the teeth to stop violations of international formal norms. With this weak policy base and a shakier organizational basis, science and human rights is placed at the periphery of international policy circles and overshadowed by "science for development" themes.

## SUMMARY: ON THE INSTRUMENTALIZATION OF SCIENCE

The globalization of science practices, clearly an accelerated process, is supported by the discourse of science. Organizations proliferate because of the understanding that science is a useful, or beneficial, enterprise. These benefits are defined primarily in general economic terms, most recently emphasizing economic competitiveness in a global context. Governments and national institutions rely on the prospect of benefits when making decisions to incorporate science into their practices. But the whole process is highly scripted: national science policies are prescribed to national bodies by science, education, development, and other related international organizations. A global polity, with its core themes carried by a web of international organizations, shapes national institutions. As a whole, science is inextricably linked with economic development and the nation-state system. Therefore, as much as they exalt science and technology, science policies also tend to celebrate nation-statehood.[34]

The international discourse regarding science, whether reflecting "science for development" or science and human rights, presents science as a strategic resource. Science is viewed as an instrument in furthering social goals, such as economic development, or as an instrument for infringing on human rights. Hence, both "science for development" and science and human rights models are products of the same utilitarian mold.

Because developmentalism reflects a technical, problem-solving perspective (Escobar 1995:44), the approach of "science for development" instrumentalizes science: it tends to reduce science to its uses, and it conceptualizes science as a tool or a technical solution for social problems.[35] In doing so, it also confines the investigation of the social role of science to its "material" dimension and understates the cultural accounts of this role (see Part IV of this book).

"Science for development" is the overwhelmingly dominant model and has been translated into worldwide policy and action. Hence, it is the utilitarian undertone of "science for development" that most clearly reflects the discourse of, and about, science, eclipsing conceptual alternatives.

The instrumentalization of science implies a narrow definition of social goals. It overemphasizes the goal of economic progress and marginalizes other social goals, such as social equality. As discussed earlier, this perspective on social goals resulted in the cooptation of human rights discussions into an economic-centered regime of developmentalism. This limited the investigation of the cultural consequences of the institutionalization and globalization of science.

The current discourse on science juxtaposes the discourses of "science for development" and science and human rights with a "good science/bad science" debate: "science for development" propagates the "good" consequences of scientific advances by focusing on economic growth, whereas science and human rights emphasize the "bad" consequences of science by focusing on the violations of human rights. The good/bad axis offers a critique of, and seemingly poses the alternative to, each discourse: the critique of "science for development" is the dependency theory's arguments on the enslavement to the Western production mode and logic (Sagasti 1973; Nandy 1988; Alvares 1992), whereas the critique of science and human rights is the description of the benefits of scientific advances, such as those mentioned earlier. Hence, the criticism of the current discourse of science is itself instrumental in nature and does not offer an *alternative* perspective.

As an empirical matter, results of research on the economic impact of science are inconsistent and mostly suggestive rather than conclusive (see review in Chapter 10; see also Drori 1998, 2000). Such is also the case regarding human rights: national violations of human rights through scientific means are today more sensitive to international scrutiny and thus more subtle, yet they do not cease altogether. Regardless of their validity and proof, both discourses fail to question the powerful instrumentalist approach to science and its ultimate current dependent variable: economic development. Moreover, the effectiveness of science policies for both economic development and human rights has little relevance to the discussion of the power of science to command global attention and to guide international and national action. The worldwide authority and legitimacy of science, relying on its image as a progressive tool, go beyond both functionality and false consciousness.

This instrumental approach to science—reflected in both "science for development" and science and human rights—disregards the expanded authority of science in modern society. Rather than viewing science and its worldwide expansion as results of its instrumental benefits and interests of related social groups, science is usefully regarded as a general rationalizing force, which, much like religion, offers an interpretive scale for world order and enjoys great legitimacy to comment on social priorities. Thus, it is science in general—as an abstraction rather than only a set of disciplines or discoveries—that is regarded with such esteem. In this book, science's social role is seen as both cosmological and ontological, rather than instrumental.

## Part III

# Cross-National Incorporation of Science

# Introduction to Part III

Shifting attention to the level of the nation-state, we now focus on globalization as the process of diffusion of an institution to the relevant units of the global system, namely each and every nation-state. Globalization thus takes the added meaning of a process by which nation-states worldwide are imprinted with the global institution. Specifically to the institution of science, nation-states worldwide incorporate the characteristics of the global model of science. They do so by locally institutionalizing such widely accepted organizational forms as governmental ministries or by drafting national policies of science and technology that carry globally accepted themes.

Our redirection of attention to the nature of the globalization process of science is more than a shift in the level of analysis from the global to the national. We also draw attention to the multilayered functions of globalization. Globalization is thus not merely the consolidation of a global model, which defines both an organizational format and discursive themes, as described in Part II. Globalization is also a powerful process in shaping nation-states[1] by setting and offering the global model of science.

The global model of science advocates the role of science in society (constructing the "need" for science) and offers the organizational formula (somewhat of a prescribed receipt) for nation-states to adopt; it prescribes science its themes and the legitimate range of disciplines, policy texts, products, and organizations. Thus, by the middle of 1990s, all nation-states publish at least ten papers in scientific journals, 95 percent of all nation-states established a local university, 99 percent are linked to the Internet, about 80 percent have at least one national scientific organization,[2] and 35 percent have a governmental ministry for science affairs. Overall, local science is a variant of the global model of science, more than a truly indigenous science form.

The common source and "pick-and-paste" strategies result in cross-national similarities in science. This isomorphic quality of science globalization and the general trend toward national-level expansion occur on various dimensions of scientific activity: in science education,[3] the scientifically trained labor force,[4] the production of science,[5] the participation of nation-states in scientific exchange,[6] and, finally, in the field of national science organizations.[7] In summary, these different measures of growth in science, representing various dimensions of scientific activity, indicate the institutionalization of a national-level field of science worldwide. This process of national-level institutionalization is marked, however, by loose coupling (most dramatically in less developed countries; Chapter 7) and by variations due to isomorphic pressures (Chapter 9).

Heavy support for this cross-national isomorphism, and therefore for the phenomenon of loose coupling, comes from the global arena. International organizations—most forcefully such development- and science-oriented organizations as the United Nations Education, Scientific, and Cultural Organization and the World Bank—convey similar messages and conduct similar promotion programs to advance the theme of science for national development and to boost the institutionalization of science worldwide. The contribution of international organizations to the diffusion of science to various nation-states is tremendous: they set the agenda, formalize the discourse, organize the field, and, most importantly, are responsible for the incorporation of similar programs to all nation-states. This global machine greatly expedites the process of globalization, as is evident in the exponential rates of global science institutionalization during the post–World War II era.

Alternative explanations exist to theorize the cause of such rapid global growth and of the astonishing similarities across nation-states. All such explanations attempt to answer the question, why do nation-states, which differ greatly in their needs and cultures, incorporate a similar form of scientific activities into their practices? First, theories of social movement offer a bottom-up explanation and argue that the local institutionalization of the field of science occurs in response to local needs to organize a growing local community of scientists (for example, Ben-David 1990; Schott 1993). Yet as described in Chapter 5, although such explanations describe the institutionalization of governmental initiatives in science among Organization for Economic Cooperation and Development (OECD) countries and during the first stage of the global diffusion process, they do not account for the incorporation of science ministries in the later part of the global process in the less developed countries. Second, world hegemony explanations suggest that pressures from the global hegemon and its coercive power cause such global similarities of form (for example, Cerny 1997; and specifically about science, Sagasti 1973; Nandy 1988). Yet although during the post-1945 era of great expansion, the global scientific core is clearly the United States, it is not necessarily the American scientific format that is diffused worldwide (see Chapter 9). The United States, because of its civic society structure and culture, is among the very few countries not to have a governmental basis for its scientific activity (such as a ministry or government-sponsored initiatives for science and development). It is true that the neoliberal global wave since the 1980s brought such American-like themes to science policy initiatives worldwide; yet this more recent policy fashion does not account for the process as a whole. Last, postmaterial theories argue that once basic human needs are satisfied, societies devote attention and resources to such peripheral matters as science and learning. Yet science is institutionalized in nation-states that are extremely poor, in a state of war, or face plagues. These countries are clearly far from satisfying their citizens' basic needs; nevertheless, national resources are devoted to the advancement of scientific projects.

The cause for the national incorporation of scientific practices is therefore to be found outside of, and beyond, national boundaries. The force behind the globalization of science is beyond local initiatives, international superpowers, or the gratification of basic needs. Rather, the globalization of science is supported by the image of science as a force of salvation and by the authority of science to comment on social priorities. It is the expanded authority and legitimacy of science that propel science to its dramatic rates of globalization.

Chapter 5    The Global Diffusion of Ministries
of Science and Technology

YONG SUK JANG

As the rationales of science and its impact on national development have emerged and spread globally, nation-states have become increasingly more involved in developing the institutions of science. Evidently, although science institutions flourish in both private and public sectors, state structures incorporate science into their set of organizational formats and institutional activities. For example, many states expand financial support for, and investment in, scientific research and development; most states employ a growing number of scientists within state bureaucracies; all states incorporated science as an integral part of school curricula; and finally, most states established organizational bodies to oversee local science sectors. This chapter explores the latter aspect of increasing state involvement in science, in particular the establishment of a new cabinet-level ministry in charge of science and technology. The central question is, when and under what conditions do central governments of nation-states directly address themselves to the business and activity of science?

Several theoretical approaches can offer explanations for this expansion in central governments. Functional approaches locate the causes of science ministry foundings in the national needs and conditions—economic, political, and military, internal or external to local science. Alternatively, institutional approaches emphasize the impact of global circumstances, relating to the world-level environment involving cultural norms and to the operations of science-related international organizations.

*120*

This chapter confronts these two approaches through an empirical investigation of the historical process of the worldwide establishment of science and technology ministries. Although I refer to the complexity of national governmental structures, the aim of this chapter is to comment on the founding pattern of newly emerging science ministries in particular.

## EXPANSION AND ISOMORPHISM OF CENTRAL GOVERNMENTS

In 1949, New Zealand established a ministry for science affairs, thus starting the 1950s trend of countries organizing ministries specifically concerned with science and technology. This trend has especially accelerated since the 1970s (Figure 5.1), so that by 1990, more than seventy nation-states have such a ministry.

This trend, although one that focuses our attention on governmental consideration of science affairs, is part of a broader trend of increased size and greater complexity of central governments. Two features connect the establishment science ministries with the general trend of expanded governmental structures. First, the brunt of the general trend is the establishment of ministries for social sectors and demands, formerly private or professional in nature. Second, central governments worldwide take form similar to that of other governments, taking charge over a similar range of social affairs (Kim 1996). These days, central governments formally attend to the "traditional" issues of foreign affairs, defense, and finance, but increasingly also newly fashioned social issues, such as welfare, trade, environment, and, most to my point, science.

The principle that science and its management need to be incorporated into the nation-state was developed in the seventeenth century (Wuthnow 1987), and central scientific academies spread among the European countries (Schofer 1999b). The linkage between science and national goals became systematically tighter in the nineteenth and the early twentieth centuries with national activities of science expanding exponentially and globally since World War II (see Chapters 3 and 4). Models of national development, which were formed by economic theories especially, have contributed toward an even stronger linking of science and national goals in the postwar period. For instance, in models propagated by international development agencies such as the Organization for Economic Cooperation and Development (OECD) and the United Nations Education, Scientific, and Cultural Organization (UNESCO), the role of science changed from a general world

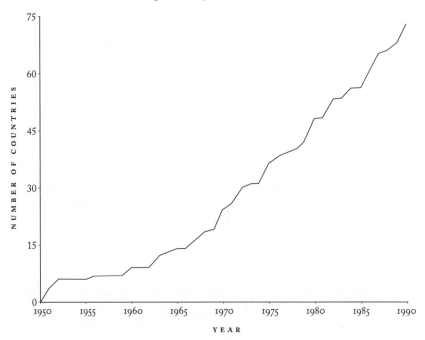

*Figure 5.1.* Cumulative Number of Countries with Science Ministries, 1950–1990. The total number of countries with science ministries as of 1990 was 73. Source: *Political Handbook of the World* (Banks 1950–1990).

good to a specific instrument for national development (Finnemore 1996; Chapter 4), associated with national organizational control (Meyer 1999). As the rationalized myth of "science for development" has evolved on a global level, a large proportion of nation-states has accepted it and established related organizational forms in their central governments, such as science policy bodies and ministries at rates that have accelerated markedly from the late 1960s and the early 1970s to the present (Finnemore 1993, 1996).

On the basis of this general picture of the changing notion of science and its link to national development, I posit the following propositions to investigate how external and internal factors of nation-states influence the establishment of science ministries. I then introduce a shift in explanatory importance from internal and domestic factors to external and world factors over time, from realist, resource-, and interest-related factors to factors increasingly cultural and professional in nature.

MODELING THE INSTITUTIONALIZATION OF SCIENCE MINISTRIES

*Institutional versus Realist Explanations*

What factors shape the global diffusion of ministries of science affairs? What accounts for the expansion of central governments to take formal charge of science issues? Institutional and realist approaches emphasize unique factors as propelling this worldwide trend.

### Institutional Approaches

Institutional approaches emphasize rationalized models and discourse, which lead to isomorphic structures and behaviors of social actors. Institutional effects involve processes that make such rules seem taken for granted. At the world level, two particular producers, or carriers, of such cultural rules or discourse are of importance: first, communities of the sciences and professions, establishing and formalizing a discourse for the proliferation of global models, and second, an organizational network—of international organizations and nation-states—to carry the themes and imprint local structures.

First, world society is made up of not only actors (that is, nation-states) but also "others," nonstate actors such as the sciences and professions, who advise actors on what to do.[1] In the modern world, both actors and "others" are rationalized. Moreover, rationalized others produce normlike talk about what actors should look like and how they should behave (Meyer 1994b). Therefore, nation-states depend on these "others" to become better and more effective actors. For example, economists give advice on how the market economy works, and natural scientists identify the myriad problems of the environment. In other words, the communities of the sciences and professions, sometimes only loosely organized, are one form of the generalized others in the world society, whose jobs are to theorize world discourse in scientific ways (that is, develop and specify abstract categories and formulate patterned relationships such as chains of cause and effect; Strang and Meyer 1993).

Second, international organizations define and provide recipes for proper nation-state activities; they also produce "talk" and thus provide world standards for nation-states. The number of international nongovernmental organizations (INGOs) and international governmental organizations (IGOs) concerned with science issues has grown exponentially over time since World War II (Chapter 3), and these organizations have created stronger and more consensual world cultural discourse and normative pressures in the domains of science and development (Chapter 4). As part of this complex process of influence of international organizations on nation-states in the field of science, international "talk" affected national science policy. For instance, starting in the early 1960s, OECD provided extended discussion and detailed advice on science policy structures to its member countries. Its report, *Science and the Policies of Governments* (1963), was the first of several OECD documents to recommend the creation of a national Science and Policy Office. Similarly, through its regular Ministerial Meeting on Science (started in 1964), OECD gave instruction to its members on the necessity of high-level science bureaucracies to coordinate and direct science activities (Finnemore 1991). Although the science policy strategies of OECD differ dramatically from those of UNESCO,[2] both bodies advocate the establishment of national bodies—governmental or nongovernmental—for science affairs. Illustrating the prevailing doctrines of development and the related need to incorporate science, Spaey (1971:11) writes, "The exchange of views carried out in international organizations, more particularly in UNESCO and OECD, on the science policy of governments, made it possible to apprehend many ways in which science and technology contribute to development of human society."

In sum, the pronouncements and ideas promulgated since the late 1960s by UNESCO, OECD, and other influential international organizations on the importance of systematic management of national science policies have provided the justifications for many nation-states to establish national science ministries and other national science policy bodies. My first hypothesis argues that the world cultural discourse generated by professional associations and international organizations have led nation-states to adopt science ministries.

HYPOTHESIS 1.1.    *World Effect of Professional Associations, INGOs, and IGOs. The amount of global organizational discourse on science and national development, and the strength of state involvement in the discourse, positively affect the founding of science ministries.*

As the globe becomes increasingly more tightly connected, both politically and economically, there occurs a contagion effect among its different "units." Contagion refers to the positive effect of the occurrence of one event on the likelihood that another event will occur, given that one event has just taken place elsewhere. Contagion effects are particularly expected to occur in the founding of national governmental structures such as science ministries because nation-states are themselves the relevant "others," or comparison groups for each other. Under the myths of ultimate similarity of identity

(Strang and Meyer 1993) in which nation-states are formed and legitimated, it is rational to propose copying policies and structures that seem to be successful in dominant or exemplar nation-states. Although nation-states obviously try to influence one another through mechanisms of exchange and dominance, nation-states with *perceived* success (in terms of their economy, military, politics, or other social aspects) also occupy a higher stratum in the world system and exert greater influence on other nation-states by providing global models and examples. For instance, because of their hegemony, Western European countries have provided various economic and political models for other countries in different historical periods. In the same way, Japanese industrial success in the past several decades has led to a wave of copying of Japanese policies, organizational structures, and educational systems (Cole 1989). Although this process of influence, or contagion, affects also the dominant or hegemonic,[3] it is nation-states outside of the core that tend to be most strongly affected by world models (Meyer 1999).

Consequently, as more states—whether dominant, hegemonic, or peripheral—adopt a science ministry, other nation-states *copy* the science ministry to quickly legitimate their own governmental structures. Thus, the cumulative number of nation-states with a science ministry signals to later adopters the degree to which the new model of science ministry is prevailing and legitimate. Moreover, a rapid diffusion within the world system is based on the fact that contemporary nation-states are culturally constructed as formally equivalent (Meyer and Rowan 1977; DiMaggio and Powell 1983; Strang and Meyer 1993).

HYPOTHESIS I.2. *Density Effect, or Perceived Similarity. Prior adoptions of science ministries generate an accessible and legitimate model that accelerates the establishment of science ministries by remaining nation-states.*

### Realist and Resource Control Approaches

We have discussed external world-level environmental factors that affect the founding of science ministries, following the line of institutional theories. However, functional and resource control stories suggest the alternative impact of internal characteristics of nation-states on the founding of science ministries.

Functionalist arguments stress that the structural forms of contemporary governments become more complex to satisfy increasingly necessary functions of nation-states (Harris 1946; Aikin and Koenig 1949; Pemberton

1979). As societies become more complex, the state is expected to expand to meet the demands of a differentiated social system. Emergence of new domestic demands forces each state to create new subunits of the central government to handle them (see Jones 1981; Poggi 1978). The state also expands to manage internal development such as economic growth (Galbraith 1973) and technological expansion (Ellul 1964). At the same time, increasingly complex international interactions require each nation-state to expand its structure to adjust itself to the demands of various environments (see Rosenau 1970; Krippendorff 1975).

According to this functionalist line of argument, developed countries that attain a considerable infrastructure in science and technology are more likely to adopt a ministry of science and technology. It is a simple functional idea that the systematic management of science policies becomes more necessary for nation-states that have more science. This "functional" argument needs to be narrowly qualified in the sense that it refers not to "encouraging and developing" science and technology of nation-states with low scientific development, but specifically to "sustaining and managing" science and technology of states with sizable internal scientific activity.[4]

HYPOTHESIS 2.1.    *Functional Argument. More highly developed nation-states with greater science and technology activity are more likely to adopt the science ministry.*

In contrast to functionalist arguments, resource control theorists such as Tilly (1985) argue that nation-states are shaped mainly by the need to prepare for internal and external competition both within and among societies. Nation-states establish a new ministry to extract additional resources from the population and to construct efficient regulatory systems for effective resource control during internal and external competition (for example, war).

Internal resource control theorists argue that the contemporary expansion of the central government into many social sectors formerly considered as private or professional (for example, education, labor, welfare, science, technology) results from the state's desire to aggrandize resources and establish durable instruments of surveillance over various emerging and munificent social sectors (that is, "state-making," to use Tilly's term). Hence, it is expected that expanded states (in terms of scope, size, and ambition of government) establish the ministry of science and technology at a faster rate, as part of an overall strategy for more effective internal resource control.

HYPOTHESIS 2.2.    *Internal Resource Control Argument. Expanded states are more likely to adopt the ministry of science and technology as part of an overall control strategy.*

Another group of researchers sees the nation-states as actors located in a larger worldwide system of political power, military relations, and economic competition (Wallerstein 1974; Collins 1986; Chase-Dunn 1989). World system theorists, stressing the prominence of external resources, argue that nation-states expand their governmental structures to cope with international military and economic pressure more effectively ("war-making" in Tilly's terms). For example, because the direct or indirect applications of science toward advanced military technology could be helpful to occupy more advantageous positions in world competition, states would foster science and technology as important resources in such external competition. Thus, nation-states that are more concerned with or oriented to military pressures from outside are more likely to adopt science ministries as an effort to manage science and technology toward the strengthening of military capabilities.

HYPOTHESIS 2.3.    *External Resource Control Argument. Nation-states that place greater emphasis on military concerns are more likely to adopt science ministries to mobilize and promote their science resources more effectively for external competition.*

### Changing Trend over Time: From Internal Domestic to External World Factors

This chapter synthesizes both approaches—institutional and realist—in predicting that functional and resource control effects decline over time relative to institutional effects (see Tolbert and Zucker 1983; Ramirez et al. 1997). The success of countries first adopting this new ministry helped to legitimate new world standards and models with regard to managing science ministries and the linkage between science and national development. As the ministry of science became an expected and legitimate governmental structure through influential organizations (such as OECD and UNESCO) and scientific and professional communities since the late 1960s, countries with even low internal pressures for the coordination of science and technology adopted the new ministry. Therefore, the impact of external and institutional factors on the founding of science ministries, as opposed to the internal and functional factors, will increase over time.

HYPOTHESIS 3.    *Changing Trends over Time. The positive effects of internal factors (functional and resource control effects) on the founding of science ministries will decrease relative to the effects of external sources (institutional effects) over time.*

TESTING THE PROPOSITIONS

*Variables*

The dependent variable is the date of establishment (or adoption of model) of ministry of science and technology. Such a date is determined by the first instance of a ministry that includes the world "science" or "technology" in its title, as listed in the *Political Handbook of the World* (Banks 1950–1992) or the *Statesman's Yearbook* (Steinberg 1950–1992).

The independent variables are chosen to represent various theoretical approaches and their various emphases. First, a total of five indicators address the propositions set by realist theories. Real gross domestic product (GDP) per capita (in constant U.S. dollars), which measures the general level of national development, and the percentage of the population enrolled in natural and engineering sciences at the tertiary-education level, which measures the size and complexity of the local science field, serve as indicators of functional propositions (hypotheses 2.1 and 3). Two indicators capture the general picture of state strength, which is the degree to which the state has expanded (hypotheses 2.2 and 3): The proportion of governmental consumption[5] to GDP measures the fiscal power of the central government, whereas the number of ministries for a country reflects the structural complexity of the central government. Last, the proportion of military to the total population measures the degree to which the state is concerned with external military competition and captures the effects of the external resource control argument (hypotheses 2.3 and 3).

Second, three indicators get at the heart of the institutional propositions. The number of annual founding events of world-level INGOs related to science (a global-level variable) was counted to measure the strength of this world discourse regarding science at the world level (hypotheses 1.1 and 3). As discussed above, international governmental or nongovernmental organizations related to science whose activities are devoted to the theorization of world discourse are important advisors in world society. Next, the number of memberships of each country in the International Council of Science Unions (a country-level variable) was employed to measure how strongly each nation-state adheres to the world discourse regarding science and development (hypotheses 1.1 and 3). Membership refers to how many individual unions within the International Council for Science (ICSU) a nation-state is represented in. Last, the cumulative number of nation-states having previously adopted a science ministry was used to measure the degree of legitimacy of science ministries as a governmental structure, or density effects (hypotheses 1.2 and 3).

Finally, period effects for two eras, 1950–1970 and 1971–1990, indicate the baseline founding rates of science ministries over time, when controlling for the influence of the above factors (hypotheses 1.1 and 3). These period effects reflect the roles of prevailing discourse and international organizations such as OECD and UNESCO, specifically the intensification of the "science for development" themes and the related international action around 1970.

Virtually all the economic and social indicators discussed above were collected from the *World Tables of Economic and Social Indicators* (World Bank 1950–1988, 1992), the *National Capability Data: Annual Series, 1950–1988* (Taylor and Amm 1992), and the *Yearbook of International Organizations* (UIA 1950–1992).

*Sample and Research Design*

Because the dependent variable is an event date, event history is the preferred estimation method. In this research, event histories mean quantitative descriptions of the life path of each nation-state as it develops and changes over time, especially with regard to the adoption of a new ministry. In other words, this study explores the rate of change of governmental structure from "not having" to "having" a science ministry (in terms of probability) under various internal and external conditions of nation-states over time. This chapter employs a piecewise exponential model with period specific effects. The basic idea of the *piecewise* exponential model is to split the time axis into periods and to assume that transition rates are constant in each of these intervals but can change between them. The piecewise exponential model with *period-specific* effects also assumes that the effects of covariates—that is, their associated parameters—can vary across time periods. Because this chapter is investigating changing trends over time, with the hypothesis of decreasing positive effects of internal functional factors and increasing effects of external institutional sources over time, it is appropriate to employ the piecewise exponential models with period specific effects.

Given time periods, the transition rate from origin state $j$ (not having a science ministry) to destination state $k$ (having a science ministry) of the piecewise exponential model with period specific effects is

$$r_{jk}(t) = \exp \{\alpha_l^{(jk)} + A^{(jk)} \beta_l^{(jk)}\} \; if \; t \in I_l.$$

For each transition $(j,k)$, $\alpha_l^{(jk)}$ is a constant coefficient associated with the $l$th time period. $A^{(j,k)}$ is a row vector of covariates, and $\beta_l^{(j,k)}$ is an associated vector of coefficients, also associated with the $l$th time period (Rohwer 1994; Chapter 2).

Moreover, the dynamic approach of event history analysis permits the inclusion of additional cases to the "risk set," as new countries gain their independence during this 1950–1990 period. Overall, one hundred forty-six sovereign nation-states during the observation period of this study are included. All colonies and politically restricted territories are excluded because of their limited discretion in adopting new departments or governmental structures.

*Analyses and Results*

Models I to IV in Table 5.1 indicate that while controlling for the covariates, the baseline founding rate has increased from the earlier period to the later period. For instance, in model IV, the baseline founding rates presented as "period effects" have changed from 0.0027 ($e^{-5.9047}$) to 0.0379 ($e^{-3.2707}$). That is, the baseline founding rate of science ministries during the later period is significantly higher than that of the earlier period. This result implies that after much discourse had been produced by these influential international organizations during the 1960s through the early 1970s, the founding rate of science ministries increased at a faster rate.

In line with functional propositions, models I and II show that during the earlier period, more highly developed countries with greater amounts of science activity are more likely to establish a science ministry. In contrast, during the later period, less developed countries with less resources in science are more likely to adopt science ministries. In particular, the gross domestic product per capital (GDPPC) of the later period is negatively related to the adoption rate of science ministries in all models at the 0.05 or 0.1 significance level. In terms of the magnitude, the coefficient for the later period in model IV estimates that one unit increase in logged GDPPC decreases the founding rate of science ministries by 30 percent $[(1 - 0.7) \times 100]$.

TABLE 5.1

Maximum Likelihood Estimates of Founding Rates of Science Ministries

(Piecewise Exponential Models with Period Specific Effects, 1950–1970 and 1971–1990)[a]

| Variable | Indicators | Model I | | Model II | | Model III | | Model IV | |
|---|---|---|---|---|---|---|---|---|---|
| | | 1950–1970 | 1971–1990 | 1950–1970 | 1971–1990 | 1950–1970 | 1971–1990 | 1950–1970 | 1971–1990 |
| Independent variables | GDP per capita (logged) | 0.53** (1.70) | −0.28** (0.76) | 0.02 | −0.35** (0.71) | 0.22 | −0.29* (0.75) | −0.18 | −0.36** (0.70) |
| Functional variables | Population enrolled in natural and engineering science at the tertiary level (%) | | | 10.74*** (46,166) | 0.75 | 9.67*** (15,835) | 0.57 | 9.84*** (18,769) | −0.39 |
| Number of ministries | | | | | | 0.10*** (1.11) | 0.16*** (1.17) | 0.09*** (1.10) | 0.16*** (1.17) |
| Resource control variables | Governmental consumption (% of GDP) | | | | | −0.01 | −0.03 | −0.02 | −0.02 |
| | Military size/Population | | | | | 14.76 | −5.21 | 15.56 | −1.04 |
| | Science NGO foundings (logged) | | | | | | | −0.52 | 1.10** (3.00) |
| Institutional variables | ICSU memberships | | | | | | | 0.10* (1.11) | 0.06** (1.06) |
| | Density of science ministries (logged) | | | | | | | 0.06 | −0.63 |
| | Period Effect | −8.68** (0.0020) | −1.43 | −5.95*** (0.0027) | −1.05 | −9.16*** (0.0010) | −3.83*** (0.2170) | −5.90** (0.0027) | −3.27* (0.0379) |
| Log likelihood, no. of events | | −349.15 | | −337.43 | | −306.28 | | −298.90 | |
| | | 24 | 49 | 24 | 49 | 24 | 49 | 24 | 49 |

[a]Numbers in parentheses are antilogged estimates; total number of spells = 3,848; total number of nation-states in the sample = 146.
*p < 0.1, **p < 0.05, ***p < 0.01.

With regard to the effects of national scientific development on founding of science ministries, in the earlier period, there are strong, positive, and significant results, whereas in the later period, we observe no significant effects. Model II clearly shows these trends, and the fuller models present consistent results. For example, model IV indicates that an increase of 0.0001 in the percentage of population enrolled in engineering and natural science at the tertiary-education level is estimated to increase the founding rate of science ministries by nineteenfold during the earlier period, but no significant effects are detected during the later one. These results support hypothesis 2.1: more highly developed countries that have considerable development in science sectors are more likely to adopt the ministry of science and technology, but only in the earlier stage of the overall diffusion process. As time goes by, however, less developed countries establish science ministries at a faster rate. The negative or no effects of functional variables in the later period partly indicate that the peripheral nation-states are more sensitive to "world fashion," as the cultural model regarding science ministry becomes more globally institutionalized.[6]

In models III and IV, resource control variables are added to the previous models. As expected in hypothesis 2.2, structurally expanded states are more likely to adopt science ministries during both time periods. Yet contrary to expectations, the rate does not decline over time, as expected. The total number of ministries within a government has a positive and significant effect (at the 0.01 level) on the founding rate of science ministries in both periods. Model IV indicates that each additional ministry at the cabinet level increases the founding rate of science ministries by 10 percent [$(1.10 - 1) \times 100$] during the earlier period, and by 17 percent during the later period. These results for the structural expansion of central governments are quite consistent and robust throughout the models. In contrast, the financial expansion of central governments, measured by the proportion of governmental consumption to GDP, has no significant impact on the adoption rate of science ministries in either period in any model. These effects of resource control variables do not significantly change the effects of the functional variables presented in models I to IV.

Models III and IV also include an external resource control variable capturing the impact of state "war-making" efforts on the adoption of a new ministry that might be helpful in mobilizing resources for military competency. Contrary to the arguments set in hypothesis 2.3, this variable does not have a significant effect on the founding rate of science ministries throughout any period.

In the final model, model IV, three institutional variables are added. The total number of annual foundings of international science organizations (logged), which reflects the amount of science-related discourse at the global level, has a positive effect (at the 0.05 level) in the later period, but not in the earlier period. Evidently, then, the intensity of global level discourse itself exerts a positive impact, particularly on the later adopters.

In addition, the number of memberships in the ICSU, which measures the extent to which each nation-state adheres to world-level discourse on science, has a positive effect at the 0.1 level during the earlier period, and at the 0.05 significance level during the later period. Each additional membership in the ICSU increases the founding rate by 11 percent in the earlier period and by 6 percent in the later period. These results indicate that a state with stronger linkages to world discourse on science and development is more apt to establish a science ministry, which is valid during both time periods. As set in hypothesis 1.1, the intensity of global discourse on science and development, and the extent to which the state is involved in the discourse, positively affect the founding of science ministries.

Hypothesis 1.2, on the other hand, is not supported, as the cumulative number of states that already have science ministries does not influence the founding rate. I expected that mimetic pressures from other countries would increase the founding rate of science ministries (DiMaggio and Powell 1983). However, model IV implies that the normative and direct professional pressures from global society—for instance, the amount of world discourse on science and the direct linkage of each state to it—have stronger effects on science ministry foundings than do mimetic pressures from earlier adopters.[7]

The results presented above also demonstrate the changing trend over time, as set in hypothesis 3. During the earlier period, developed countries with considerable levels of science activity are more likely to establish a science ministry. In contrast, during the later period, less developed countries began to adopt science ministries at a more rapid rate. The effects of economic development present a dramatic change from the earlier period to the later period. The amount of science also has no significant effect in the later period, whereas it has strong positive effects on the founding rates of science ministries during the earlier period. These results represent strong evidence for the hypothesized nature of institutional effects.

Are these time-varying effects related to the age (or newness) of the country? Taking into account the establishment of new countries as changing the composition of the world system at any time point, the findings imply that new countries do act in a unique pattern. Evidently, although their internal societies are less functionally complex than older ones, new nation-states on average have larger central governments, and although they adopt new emerging ministries (for example, science ministries) later, they do so at a faster rate than old nation-states. The prevailing modern world ideology legitimates states within the nation-state system on the bases of the organizational forms and rhetoric of central governments to pursue modern progress rationally. States have greatly expanded their governmental structures as well as their formal powers, thereby acquiring constitutional authority and responsibility over many aspects of social life such as economic activity, family, organization, education, and welfare (Meyer 1987). This is especially true for new peripheral states, as they are more dependent on external world-level legitimacy for support and survival within the world society (Jackson and Rosberg 1982). As a consequence, they are more eager to adopt global standards of international organizations and copy exemplary models of developed countries.

In sum, although the functional variables exerted positive influence on the adoption rate of science ministries during the earlier period (1950–1970), the institutional effects have become more influential over time. In particular, the direct linkage to world-level science organizations strongly affects the founding of science ministries during both periods, and even more strongly in the later period, as the model of science ministry becomes more globalized. Structural complexity of a central government also has a strong positive effect on the founding of a science ministry in both periods. That is, state-making efforts to control greater resources in effective ways positively influence the adoption rate of at least this one ministry throughout the periods. However, other indicators of state strength capturing the financial capacity of a central government, and state war-making efforts do not have any significant effects on the founding of science ministries in any of the periods.

IMPLICATIONS

Internal functional conditions predict the adoption of a science ministry by nation-states at the beginning of the spread of science ministries but do not predict adoption very well once the process is well under way (Tolbert and Zucker 1983; Ramirez et al. 1997). Throughout the examined periods of this study, we observe that states with more expanded structural aspects are more likely to adopt science ministries. In addition, we also observe that world-level discourse regarding science and national development, produced by science IGOs and INGOs, has "taught" nation-states the appropriate and necessary role of science policy structure, leading nation-states to adopt the new ministry. Direct linkage to the general world discourse on science (as measured by the ICSU memberships) has especially exerted great influence on the founding of science ministries throughout the periods.

These findings permit a partial integration between functional and resource control approaches focusing on internal factors and on institutional approaches focusing on external sources. Nation-state actors continuously interact with each other and other actors or consultants, such as the UN, UNESCO, OECD, and other international organizations. Through such interaction, they are constituted and constructed as ultimately similar actors under universalistic and rationalized world standards and models produced by these actors and consultants (Meyer 1983, 1987; Meyer et al. 1997a). Although functional conditions or competitive pressures may have initiated the phenomenon, the science ministry as a legitimated model has become by now firmly infused in world political culture. As a consequence, nation-states are adopting the new model regardless of functional conditions or external military and economic pressures. The development and change in the model and standard are produced by the expansion in world-level social roles played by "others" (such as professional associations, INGOs, and IGOs), as well as nation-states that perceive peer countries as "relevant" others. Last, these analyses reveal that specific to the institutionalization of science ministries, the turning point for the globalization of the "science for development" model is circa 1970. This timing is most probably caused by the change in UNESCO to address science issues (and to add an "S" to its acronym) and the corresponding 1970s intensification of the "science for development" discourse (see Chapter 4).

## Chapter 6

# Elements of a Contemporary Primary School Science

ELIZABETH H. MCENEANEY

Primary school science, as encapsulated in pedagogical theories, official curricula, and curricular materials such as textbooks, is a caricature of professional science. It is at once a distilled and watered-down version. Obviously, much specialized technique and sophisticated nuance is omitted from the corpus of primary school science. Yet primary school curricular materials present us with breathtakingly clear statements and pictures about the essence of science: its prescribed ways of thinking and acting, its range of pursuable questions, who is mandated to question and pursue, and who benefits from science. This chapter presents evidence about the transformation of the world's primary school science, focusing precisely on these claims about science as an institution. What are the common elements of primary school science as it is currently constituted around the world? To what extent do they reflect and reinforce cultural models and changes in world society? And which elements of primary school science have become delegitimated over time?

Specifically, I offer evidence of change in three areas. First, school science has become dramatically more participatory over time. This has involved not only a huge increase in the inclusion of depictions of people in curricular materials, but also a shift toward greater personal relevance and emotional accessibility. Second, the iconic image of the expert has changed in recent decades. Children are now imbued with expert status, but apparently not at the expense of professionalized experts. Finally, greater participation in all areas of science and widened claims to expertise both occur in a natural world that is increasingly portrayed, in primary science, as ordered and manageable.

I argue that these changes clearly reflect an embeddedness in a world cultural system that emphasizes such principles as individualism and the authority of rational voluntarism (Boli and Thomas 1997). Chapter 1 in this volume reviews these principles of world culture that shape both science and primary school science, particularly as they relate to the worldwide construction of the modern actor. Yet because these changes in primary science are implemented in mass educational systems around the world, they also reinforce elements of world culture. That is, the changes in primary science I demonstrate here not only presume the actorhood status of individuals, but they also vigorously reinforce and elaborate human agentic actorhood.

The analysis is primarily based on comprehensive coding of 265 science textbooks from sixty countries, published from around the turn of the century through 1995. The textbooks were housed in collections at eight different libraries, including collections at Teachers College, Michigan State, the National Textbook Library in Tokyo, the Institute for Science Education in Kiel, Germany, the German Foundation for Development in Bonn, and the Library for Educational History Research in Berlin. The sample targets science as presented in the equivalent of U.S. grades 4 and 5, but textbooks written for grades 3 to 7 are also represented. Organized by decade of publication, this convenience sample permits us to take a snapshot of the way in which the world is doing school science over time. These snapshots are blurry, particularly at earlier points in time, because not all (or even most) nations are represented and because there may be more heterogeneity within nations than can be assessed with this sample. Yet the trends described below are clear enough to suggest that they would hold even with a more complete sample of textbooks. I supplement textbooks with other data sources. These include information from official statements about curricula as well as articulations of pedagogical theories.

LEVELS OF PARTICIPATION IN SCIENCE

Primary school science has become vastly more participatory since the beginning of the century. The major shift has been from a science that is a delineated set of facts to a science consisting of a range of activities that people do and a range of attitudes that people have. In modern primary school science,

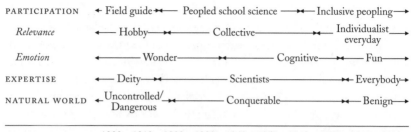

PARTICIPATION     ← Field guide ►◄— Peopled school science —►◄— Inclusive peopling—→

*Relevance*          ←— Hobby—►◄———— Collective————————►◄— Individualist___→
                                                                          everyday

*Emotion*            ←———— Wonder————————►◄———— Cognitive—►◄— Fun—→

EXPERTISE           ←— Deity—►◄————————— Scientists——————————►◄— Everybody→

NATURAL WORLD    ← Uncontrolled/ ►◄———— Conquerable——————————►◄— Benign—→
                        Dangerous

———————————————————————————————————————————————

1900s 1910s 1920s 1930s 1940s 1950s 1960s 1970s 1980s 1990s

*Figure 6.1.* Review of Key Global Trends in Primary School Science

students are to learn that science is a human endeavor. One corollary to this change in emphasis is that primary school science around the world now assumes personal relevance and for this reason is more involving of the child and the child's environment. Another corollary is that contemporary primary school science has become more emotionally appealing to children, with new emphases on humor and fun. An overview of these and other key trends is presented in Figure 6.1.

This set of changes in school science is linked to the cultural expansion of science as an institution, the notion that we denizens of the contemporary world turn to science as a source of answers for a widening array of problems in the natural and social world. As emphasized in the opening chapters of this book, this process of the "scientization of society" suggests that its cultural potency comes primarily from a rationalist image of science as a central con-tributor to national development (particularly economic development) through the building of a skillful and more efficient labor force. Also, non-rational consequences of scientization are legion, including effects on civil and political practices, and in a similar way, government agencies all over the world abandon cavalier treatment of policy-related data (see Chapters 12 and 13; Drori 1997). Thus, scientization necessarily involves sciencelike practices permeating various spheres of social life. Politicians who fail to seek advice from scientists, for example, do so at their own peril.[1] In this changing cul-tural context, we would expect more participatory science textbooks, includ-ing more frequent depictions of personally relevant, everyday applications.

An inspection of textbooks from various countries indicates quite clearly that before World War I, science was depicted as fact-oriented and taxo-nomic rather than broadly processual. Textbooks of this time period were fairly uniform in structure. Commonly, the textbooks march through a panoply of local species of flora and fauna, with fairly regular sketches and an occasional photograph of the species under study. Topics such as simple machines are covered in a similar way. Very few people are pictured or given voice in the text. Thus, textbooks impart early primary school science through a kind of field guide in which the knowledge seems to have dropped from the sky, or is channeled through a very few luminous thinkers such as Galileo.

The modern primary school science curriculum, in contrast, vigorously emphasizes the process of science, especially the idea that the process is people-driven. *Project 2061: Science Benchmarks*, sponsored by the American Academy for the Advancement of Science (AAAS), also illustrates this point. By the end of the second grade, one benchmark states, students should know that "everybody can do science and invent things and ideas." Moreover, there is a mandated responsibility to act. Another benchmark advises that by the second grade, students should "raise questions about the world around them and be willing to seek answers." The motto "science for all" invoked by both the AAAS (1989) and United Nations Education, Scientific, and Cultural Organization (1991) implies instilling a sense of agency in all students—not only science in the service of all, but science *by* all.[2]

Contemporary textbooks from all over the world are filled with pictures and drawings of people not only manipulating and observing the natural, and increasingly, the social world (that is, doing science), but also using technology and benefiting from the products of scientific activity. Between 1900 and 1950, only about 10 percent of the textbooks I analyzed offered depictions of people as a regular feature. By the 1980s, approximately 60 percent of the textbooks in the sample regularly depicted people, rising to nearly 80 percent in the 1990s. At the beginning of this transformation, introductions and prefaces to the textbooks often explicitly exhorted students to participate actively in their science study.

The preface to an early U.S. textbook proved to be a harbinger: "The study of science offers a great many chances to make things happen. . . . Some men and women spend most of their lives performing scientific experiments, because they are so keenly interested in the changes which they themselves bring about" (Nichols 1934:xiii). Concluding comments in a textbook on simple machines published in what is now Ghana were a bit more tentative, although the emphasis on participation by all is similar: "There is no need to think of [machines] as wonderful things which Europeans have made but which other people cannot understand" (Joselin 1948:60). The introductory chapter of a West German textbook has sections entitled "People Understand the Forces in Nature" and "People Investigate and Invent" (Knoll and Knoll 1965:2–5). This approach virtually equates participation in science with personhood. That is, it suggests that understanding, investigating, and inventing are inherent characteristics of people. A Botswanan textbook places the notion of "experimenting" in the following historical context, with implications for students in the present: "[The African people who first made fire] did things. They noticed things. They tried things out. They learned how to make fire by thinking and doing. You can learn to be a scientist. You will learn by thinking and doing" (Nicholson 1989:216). More stridently, a contemporary U.S. textbook offers regular sections simply entitled "Take action!" (Heil et al. 1994). The understanding of science as fundamentally participatory has become so taken for granted in most countries that explicit, detailed statements about the centrality of participation are far less frequent than they used to be.

A key aspect of this "peopling" of primary school science is that a widening array of various social groups are included. This push toward at least nominal inclusiveness is especially evident after around 1970 (see Figure 6.1). Science is no longer constructed as the sole province of white (or majority race), middle-class men and boys—at least, not in primary school textbooks. One of the earliest manifestations of this trend is the disappearance of science textbooks written exclusively for girls, which often emphasized the applications of science principles in homemaking. This kind of textbook was common before the 1930s in various countries, including Great Britain, the United States, and Germany. A British text from 1923 by Evelyn Jardine of this type, for example, was called *Practical Science for Girls as Applied to Domestic Subjects*. Since then, this extreme gender ghettoization of school science has been essentially abolished.

Another manifestation of a widening construction of participation in primary school science is the range of groups represented in pictures in the books. Although depictions in contemporary textbooks are far from an equitable fifty–fifty split, it has become de rigueur to include some images of women and girls since the end of the 1970s. Only two textbooks from the 1980s sample of twenty-two textbooks (9 percent) failed to include female subjects regularly. More dramatically, only one textbook out of the thirty-one sampled textbooks published in the 1990s excluded women and girls. In a similar, although less dramatic, way, textbooks in many countries increasingly show science participants from racial and ethnic minorities, the working class, and the disabled. Before 1970, there was little evidence of this expansion of the roster of appropriate actors in science.

Increasingly, then, people are the prime subjects of depictions in science textbooks from virtually everywhere. Yet there are other manifestations of greater emphasis on participation and actorhood. Subtle shifts in language also suggest a more fundamentally participatory construction. Readability of the language in the text is increased, thus heightening children's access. The passive and objective voices no longer dominate the writing. The choice of pronouns also reveals a trend toward a more participatory science. A South Korean textbook (Ministry of Education [South Korea] 1991) introduces activities with "let's"—as in, "let's think about why a lot of birds lead to preventing a lot of bugs." A Mexican textbook has chapters entitled "How We Study Things," "How We See," and "How We Hear" (Secretary of Public Education [Mexico] 1990). By the 1990s, textbooks virtually ubiquitously pose questions in the second person: "What do you think?" or "What did you find out?" This linguistic device thus brings students into the center of the material in yet another way. Moreover, in languages that distinguish between formal and familiar pronouns, the familiar becomes the standard, thereby narrowing the social distance between student as participant and the textbook as authority.

*Personal Relevance*

Deriving from this tremendous increase in the participatory nature of science is an emphasis on personal relevance. For most of the century and in most countries, science has been constructed in the school curriculum as relevant.[3] The vast majority of recent textbooks, however, present material as personally and immediately relevant to individual students in their everyday world. Textbooks published between the 1920s and the 1960s tended to portray science as relevant, but at a collective level. Often, this took the form of nationalistic demonstrations of scientific and technological achievements and progress. At the beginning of the century, science was depicted mostly as a worthy hobby for men of financial means (Figure 6.1). Early constructions of school science, then, were also personally relevant, but with more emphasis on science's potential for moral uplift than everyday utility.

Relevance took the form of overt nationalism, and so was more collective in its orientation, in many countries' textbooks starting in the late 1920s. A textbook from Italy boasts about the nation's agricultural and industrial production, as well as the country's "perfect" climate (Piccoli 1928). Various generals and politicians adorn the inside cover, and the book discusses the health benefits to Italian youth of Mussolini's opening of the new National Opera building. Particularly in the 1950s and 1960s, textbooks from nations in the first, second, and third worlds celebrated national research facilities and military capabilities. An Indian physical science textbook shows a man using an electron microscope; the caption specifies that the equipment is located at the National Physical Laboratory in New Delhi (National Council of Educational Research and Training [India] 1969). During this time, countries flaunt what they have in terms of research capability, which is often defense related. A Romanian botany textbook from 1969 by Emil Sanielevici and Alexandru Dabija traces the history of the field generally but also includes a section on the "development of botany in our land," including pictures of leading Romanian botanists (8–10). Other Romanian textbooks, as well as some from East Germany, Hungary, and Cuba, pitch pan-socialism rather than nationalism, praising, for example, the scientific and technological accomplishments of "our Soviet brothers" (Meusel and Meyendorf 1960:5). In the late 1960s and 1970s, pictures of astronaut Neil Armstrong were followed in many countries' textbooks by a picture of an astronaut of local origin (for example, Novy et al. 1979). U.S. textbooks of the time never failed to depict the moon landing without noting that it was an American achievement (for example, by showing the U.S. flag alongside the astronauts). One exception, however, is that U.S. textbooks gingerly broached American achievements in the development of atomic energy. In general, though, accomplishments in scientific realms are portrayed in this time period as collective. Honor and, more mundanely, utility accrue to nation-states.

Contemporary primary science textbooks tend not to rely on nationalism to convey the relevance of the material. By the early 1970s, nationalistic portrayals fade in importance, such as in the Israeli textbook that shows a cartoon of an astronaut of no discernible nationality in a space capsule above the statement, "Man has been able to reach the moon" (Curriculum Centre, Ministry of Education and Culture [Israel] 1969a:55). Similarly, a Mexican textbook discusses the "international effort" in space exploration including achievements by "Soviets, North Americans, and diverse nations" (Secretary of Public Education [Mexico] 1990:170). Instead, the main pedagogical thrust has been toward "everyday science," and at higher grade levels, science, technology, and society reforms (for example, Roth 1997; Levine and Johnstone 1995; Zeitler 1981; Solomon and Aikenhead 1994).

Concretely in textbooks, we find much more attention to plausible applications of science in the child's world. One example is an entire unit of a Scottish primary science program entitled "Teapots" (School Council Publications [Scotland] 1982). The unit addresses such central concepts in science and technology as heat, convection, and insulation, but applies them to a revered cultural object in the child's everyday world. An Irish environmental studies textbook rounds off a discussion of the functions and structure of human skin with the following activity: "By watching T.V. advertisements and looking through magazines, find the names of various skin care products. List them under the headings: essential or non-essential" (Kelly and Ryan-Enright 1990:40). Mixing critical consumer education with standard primary school science content, this textbook directs students to gather data in the familiar world of television and magazines.

This trend toward personal relevance for children appears to have started earlier and with more vigor in less economically developed countries. One booklet produced by the African Primary Science Association for Ghana in 1970 is entitled *Common Substances Around the Home* and includes chapters highlighting everyday uses of various household chemicals, such as "A Closer Look at Familiar Liquids" and "Painting, Using Flowers and Liquids" (African Primary Science Project [Ghana] 1970). Far from depicting chemistry as an exotic activity conducted in laboratories, the book portrays science as useful and pertinent in everyday life. Likewise, a series of paperback books sponsored by the Science Education Center from the Philippines in 1979 covers such homespun topics as *From Rice Washing to Delicious Nata de Arroz* (1979b) and *Money in Rabbits* (1979a). A Fijian text from 1978 made a dramatically direct appeal to personal relevance. Concluding a discussion of the uses of friction is a section entitled "How to Avoid Death." Below this eye-catching title is a local newspaper story about a fatal car accident attributed to bald tires (Ministry of Education [Fiji] 1978:37–41). Understanding that friction is what allows car brakes to function properly leads to the personally and highly relevant conclusion that driving with bald tires is dangerous. Thus, although there is a concerted move worldwide toward more personally relevant applications of science knowledge in the textbooks, economically developing countries seem to have led the way.

### Emotional Accessibility

This recent push toward personal relevance can be viewed as part of a broader trend toward making all kinds of public cultural offerings (for example, museums, public radio, television) more accessible and therefore more subject to mass participation. Trends toward greater emotional appeal of primary school science can be seen in this context. Turn-of-the-century science textbooks allowed some degree of emotional content, although it was not necessarily appealing from a child's perspective. For example, aesthetic appreciation of nature is the emphasis in this early British text: "The person who goes through life knowing nothing of the beautiful workings of nature is like a blind man. . . . One who knows how to look at nature's works, and the order of her workings, sees much more: to him the 'yellow primrose' is so many sunbeams bottled up, as it were, in the plant to cheer him again . . . as a thing of beauty" (Tyndall 1883:1).

A textbook used in various parts of East Africa, published in 1928, has a more sectarian emphasis, as apparent in this discussion of geology: "Most of [the geological changes] are very, very slow; but now and again some tremendous outburst takes place to teach us to keep humble before God, who made the world, and rules over all these mighty forces which He put into it" (Rowling 1928:8–9). Thus, the pedagogical imperative of the early textbooks was to develop in students an almost visceral appreciation for nature. They aimed to cultivate awe and wonder, and very often, humility in the face of god's handiwork.

There was a change in this emotional backdrop after World War II, and an even more pronounced shift in the wake of the *Sputnik* launch. Given the demonstrated destructive potential of science, and the fact that science came to be seen as a crucial weapon in the cold war, there was no cultural space for emotional content in primary school science. The attempted appeal was at a cognitive level, rather than underscoring wonderment. This shift is shown in Figure 6.1. An Indian text introduces physics as "this most absorbing game" (National Council of Educational Research and Training [India] 1969:xv). With depictions of focused, expressionless faces toiling in emotionally remote laboratories on relatively abstract problems, the reader comes to understand that "the game" referred to is more like chess than Frisbee.

The cognitive emphasis during this time is also reflected in massive attention to techniques of precise measurement. The first 57 pages of the 165-page Indian textbook mentioned above cover measurement techniques. A Czechoslovakian textbook shows a similar level of attention to measurement technique with virtually no consideration of the phenomena to be measured— that discussion comes later (Novy 1970). In the United States, a radically abstract treatment of science was still evident in the late 1970s. A series of physical science booklets included one entitled *Relative Position and Motion* (Thier et al. 1978). Here, a character named "Mr. O" guides students through paper-and-pencil exercises on the position and motion of abstract shapes from the mathematical perspective of symmetry and transformation. From a contemporary perspective, one doubts that students develop any affinity for "Mr. O" merely on the basis of his radial symmetry.

Starting in the 1970s, a worldwide rise in direct emotional appeals to children became discernible in primary school science textbooks. In some cases, science simply becomes a venue for creative expression. In its preface, a Ghanaian text from 1988 by Saka et al. sets up this possibility: "Do not write anything in this book. Keep a special exercise book for science. Call it your 'Science Expression Book' " (ii–iii). There is an even more pronounced tendency, however, for textbooks to incorporate devices and activities that highlight humor and fun. Among the textbooks in the sample, 10 percent of textbooks published in the 1960s featured elements that emphasized humorous situations or that science is fun. That proportion increased to one-third for sampled textbooks published in the 1980s and 42 percent for 1990s textbooks in the sample. The curricular ideals laid out in AAAS's *Science Benchmarks* (1999) also epitomize this shift. One prescription advises, "By the end of the fifth grade, students should know that . . . science is an *adventure* that people everywhere can take part in" (emphasis added).

This notion of science as an adventure echoes through most contemporary textbooks, and not only in liberal Western countries. The cover of a combined science and mathematics textbook from the Dominican Republic shows a colorful drawing of two children scuba diving for a treasure chest (Garcia 1988). The implicit message is that scientific/mathematical knowledge is valuable and that the pursuit of it is fun. A device used by some textbooks in the 1990s is the introduction of a cartoon character that appears regularly in the text. A Colombian textbook sports a cuddly pink rabbit every few pages (Pardo Miller et al. 1993), and a winged fairy appears recurrently in a Japanese book from 1995 (Ministry of Education [Japan] 1995). Such characters, in these books and others, traipse through the pages, aping the activities described, miming a kind of spin: science is enjoyable, and sometimes it makes for a good gag. And it's clear that the primary function of these characters is emotional accessibility for children. In both the Colombian and Japanese cases, the creatures depicted do not resemble any naturally occurring life forms, so in themselves, they offer no actual science content. They are, however, downright charismatic compared with the 1970's Mr. O.

In the world's contemporary primary science textbooks, ignorance of basic scientific principles can make for silliness: clothes hanging on the line won't dry when it's raining; a sail positioned the wrong way won't catch the wind, etc. The humor has a slapstick quality, and the consequences of such ignorance are individual, personal, and usually a matter of inconvenience; the potential alternative would be to portray collective, life-threatening, and livelihood-threatening disasters. This alternative is in fact not available in contemporary primary school science. The portrayal is of an ordered and lawful natural and social world that is manageable by individuals who understand and apply scientific principles. I will return to this point in more detail later.

To summarize, primary school science (as embodied in textbooks) has transformed from an inert body of facts to principles for action and participation by individuals. Textbooks increasingly highlight the capacity of a wide range of individuals to understand nature and to act on this knowledge. Science is a tool of immediate utility, in a world imaginable to children, both materially and emotionally. In short, the main pedagogical imperative of contemporary primary science is both to cultivate agency in students *and* to construct a cultural field in which students believe they can exercise this agency. Highlighting the destructive, unfathomable, and uncontrollable forces of nature detracts from this goal. Notably, agency now accrues to individuals, the process bolstered by the emphasis on personally relevant applications and personally experienced emotions. In contrast, earlier, post-*Sputnik* constructions of primary school science attributed agency to collectivities (typically, the nation-state). And textbooks published before World War II do not appear to have had the construction of agency—individual or collective—on their agenda at all.

IMAGES OF THE EXPERT

In a related but primarily distinct trend, there has been an evolution in the depictions of experts and expertise in primary school science curricular materials. This can be traced along several dimensions: the range of people who can be experts, the kinds of questions experts ask, and the kinds of procedures experts use to secure answers to their questions. For our purposes, an "expert" in science is a person who has a mandate to systematically pose questions about the natural and social world and the knowledge and skills to search empirically for answers to those questions. Two partially countervailing pressures inherent in the evolution of science influence the way experts are portrayed in primary school science.

First, the cultural expansion of science—its invocation to explain and tame virtually everything—implies ever greater alignment with the needs, interests, and sensibilities of the lay public, thereby blurring distinctions between scientist and nonscientist, expert and nonexpert. I claimed in the previous section that this trend explains the dramatic shift toward a more participatory school science, but it also influences the construction of expertise. At the same time, the institution of science experiences a sustained trend toward greater specialization. The specialization and subspecialization of science, and the increasing importance of specialized credentials, serve not only to highlight distinctions among scientists, between, for example, biochemists and molecular biologists, but also between scientists and nonscientists.

Gieryn (1983:781) highlights this tension in general terms by focusing on the "boundary work" that scientists as a group perform to maintain their privileged status. Moore (1996), stressing a slightly different perspective, points out that the legitimacy, prestige, and financial support of professional science rely not only on the public perception that science is useful to a broad constituency, but also that science is a privileged kind of knowledge and method. Curricula and textbooks are tools to shape this public perception. Although the cultural expansion of science both produces and is sustained by the enlistment of a broad lay constituency, the specialization of science serves to distinguish it (and elevate it) from the lay public. The contemporary view of society as increasingly specialized and technocratic suggests that the castelike status of scientific experts will be maintained. Yet to the extent that there is a cultural trend toward the expansion of science, we should see a softening of claims of distinctiveness made on behalf of scientists. A consideration of the image of experts in the world's primary school textbooks is one way to assess the symbolic status of scientists.

Are depictions of experts requisite in textbooks? When they are, what kinds of people have textbooks generally portrayed as worthy of expert status? Figure 6.1 sketches the trend in how primary school science locates expertise over time. As noted previously, before World War I, school science generally portrayed science as a hobby. Its curatorial quality did not require the posing of questions, although the techniques of collecting, preserving, and documenting specimens were specialized and not part of everyday life. Essentially, then, virtually no people are portrayed as experts in most early textbooks because the depicted science involved a kind of filling in the blanks in a god-given taxonomy. In this sense, expertise is located in the deity, and his work is merely revealed to human practitioners of science.

Thus, it was common in textbooks before World War II to link content coverage to the seasons of the year. In some climes, this might involve beginning the book with a discussion of why some trees shed their leaves in the fall, followed by reasons for animal hibernation in the winter, and so on (for example, Reinders 1920). This heavenly ordering of science knowledge as presented in the textbooks parallels the frequent presentation of a classification scheme by order, genus, and species as divinely inspired. If the biologist Linnaeus is mentioned at all in connection with this classification system, the implication is that he *discovered* rather than *invented* the approach. There is no hint, for example, that living things might exist that do not neatly fit into a category. The march through the various zoological and botanical branches of the system was foundational for textbooks from many countries through the 1920s, but persisted in some places quite a bit longer (for example, Gibitz and Kern 1945; De Oliveira Faria 1973).

Starting after World War I, and particularly after World War II, primary school textbooks began to address the notion of expertise much more directly. An expert is a scientist. An expert works in a special location with equipment that looks unfamiliar to nonexperts. Some questions about nature are more important than others, and the expert in science knows which questions are most important. In fact, some questions are "classic." The existence of such a hierarchy of questions and problems to solve is communicated in many of the textbooks of this time. Scientists have a specialized means of finding answers to questions which is known as the "scientific method." This approach, conscientiously applied, in time yields truth.

Between 1900 and 1919, none of the textbooks assessed in the sample regularly depicted people conducting science for wages, reflecting both the state of science as a profession and the general lack of participatory images in the books. This changes abruptly in the 1920s, when 35 percent of the sampled textbooks featured professional scientists. By the 1950s, that percentage had peaked at 43 percent. Most of these depictions are of scientists in specialized locations such as laboratories. Depictions of biologists and geologists working in the field—that is, less specialized sites accessible to nonscientists—are not nearly as frequent. The questions conceived by scientists are the questions the books portray as worthy of investigation. This is clear in an Israeli textbook called *The Structure of Matter*: "The differences between structures of matter have been known to mankind for hundreds of years. . . . What hypotheses have scientists made over the years? In our future lessons, we shall get to know the hypotheses made by the scientists" (Curriculum Centre, Ministry of Education and Culture [Israel] 1969b:48). This quote from the preface of a textbook in the renowned British Nuffield Science series is in a similar vein: "In this course . . . we want you to understand why scientists wish to know about life and how they go about finding biological truths" (Nuffield Foundation 1966:ix).

In societies as diverse as East and West Germany, India, and Hong Kong, textbooks featured the kind of abstract questions of interest to academic scientists. One section in the Indian physical science textbook queries: "What is motion?" (National Council of Educational Research and Training [India] 1969:69). At around the same time, a West German textbook challenges students with these questions: "Can several bodies occupy the same space simultaneously? . . . How are substances constructed and what are they composed of?" (Baumann 1967:14). An extreme example of the level of abstraction of questions posed, as well as their lack of everyday context, is a sketch of an ethnic Chinese boy in a Hong Kong textbook with the questions "How? Why? What? Which?" floating around his head (Hayworth 1975:5).

In liberal democratic countries in particular, the procedure advocated during this time as a means to seek answers to these questions was "the scientific method." The prominent and widely used Nuffield Science series from Great Britain featured a description of the scientific method, listing the specific steps as (1) problem, (2) hypothesis, (3) experiment, and (4) result (Nuffield Foundation 1966:31–32). U.S. textbooks of the time offered a similar approach, as did textbooks from other countries in Western Europe.

In the last two decades, the tremendous push for more participatory school science began to break down the boundary between expert and nonexpert. Most notably, children themselves become privileged to pose questions as their own curiosity dictates. A book targeted to primary school teachers entitled *Talking Their Way into Science* (Gallas 1995) illustrates this point in its subtitle: *Hearing Children's Questions and Theories, Responding with Curricula.* Contemporary primary school science curricular materials *follow* rather than *lead* students' questions.

This pedagogical stance is manifest in textbooks in a variety of ways. A Swedish textbook prefaces its treatment of biology as follows: "This book is not so structured. It is the basis for the course. The teacher and the students together will decide what kind of knowledge is important in terms of today's society" (Linnman et al. 1981:i).

Here, the idea that professional scientists are the experts charged with developing a hierarchy of important questions (and their answers) is completely uprooted. Similarly, a German textbook regularly depicts children posing substantive questions. In the section "Water—Supply and Processing," there is a research and writing activity entitled "Where does our drinking water come from?" (Herbert et al. 1991:5). A boy asks, "Can people drink water from the Elbe?", and a girl asks, "And why does water cost money?" The rest of the page is blank, leaving students room to write about what their group has learned through their inquiry into these questions. Children in this textbook ask the key questions, which is not quite the same as children in the classroom being charged with that important task. Yet the questions posed in this book are quite plausibly questions that children might ask, grounded in a social, everyday context.

This same book uses another device to confer expert status upon students. The title page of *My Discovery Book 4* lists its six adult authors and then shows an approximately 2- by 2-inch square with the words "and by," so the student can write his or her name in the box, thereby claiming authorship of the school science presented. Textbooks from other countries use similar, although less radical, approaches. South African students are asked to keep a weather diary. The book advises, "Use words to describe the temperature of the air. You simply use your own sense of warmth" (Earle et al. 1993:8). Absent in this example is any imperative to use conventional scientific measures of temperature. Students become observers and arbiters of natural phenomena.

As was the case in the move toward greater personal relevance, there are relatively early examples of this trend in textbooks from economically developing countries. A Kenyan textbook, for example, shows a strong orientation in favor of this reframing of expertise. It features an activity about animal behavior entitled "Ask the Ant Lion." (An ant lion is a small insect common in East Africa.) The section begins, "A little creature, often ignored, succeeded in keeping a class of children happily busy for many days." It goes on to advise students, "Ask the ant lion; it will always give you an answer" (African Primary Science Project [Kenya] 1966:1–5). Significantly, the book does not specify *which* animal-behavior questions the students should pose to the ant lion. Instead, it focuses on principles of systematic observation and leaves it to the students to settle on a question of interest. Indeed, the ant lion is humble fodder for science investigations. It is not endangered, central to agricultural production, or a theoretical missing link in debates over evolution or genetics. The key point is that the insect is part of the child's world, and like every other aspect of that world, subject to children's investigations.

Contemporary textbooks are more likely to expand the number of legitimate sources of expert knowledge in another way. Although not a strong trend currently, texts from a few countries are beginning to give some play to traditional forms of knowledge encoded, for example, in local myths and folktales. A New Zealand textbook notes, "Maori myths tell how Maui lay in wait for the sun, captured it in a net, and beat it severely to make it move more slowly across the sky. We cannot control the sun, but we can make more use of the energy it provides so freely" (Sweeny et al. 1991:88).

Note that Maori traditions are not discounted here. Rather, this story is used as a metaphorical frame for humankind's continuing pursuit to harness the power of the sun. The appeal to traditional knowledge systems can cross national boundaries. Written just before reunification, a textbook intended for East German students complemented a discussion of historical developments in chemistry with traditional Chinese folktales (Meyendorf et al. 1991). Likewise, a Botswanan textbook depicts an old man telling a group of children a story. Next to the picture, the book explains, "Old people have many beliefs about weather. Ask old people in your community to tell you about these beliefs. Which animals are connected with these beliefs? Do people in your community look at the position of some stars? . . . Which of the beliefs are connected with the wind?" (Nicholson 1989:9).

Earlier textbooks would typically have stated explicitly that these folk beliefs do not constitute scientific thinking. No such disavowal is made here, suggesting that "old people" too can possess expertise.

Thus, there is ample evidence that children themselves are now granted expert status in modern primary school science. Recognized sources of expertise are also beginning to encompass traditional knowledge systems in myth and legend. In addition, this shift occurs not only in liberal Western democracies, but also economically less developed countries. Does this imply that professional scientists have a weaker presence in contemporary textbooks? The answer is no. Although recent textbooks widen the circle of expertise, they continue to portray professional scientists at rates comparable to earlier decades. Although one-third of the sampled textbooks published in the 1960s featured depictions of professional scientists, that proportion had slipped only slightly to about 25 percent in the 1980s and 1990s. Thus, the professional scientist retains a position as expert in contemporary primary school textbooks.

One difference is that the current portrayals of professional scientists are more likely to focus on average people as scientists, whereas earlier textbooks tended to highlight luminaries: Newton, Galileo, Watt. Contemporary textbooks often show average people on the job as scientists. A Swedish text from 1981 presents several pictures of "How Biologists Work" in both laboratory and nonlaboratory settings (Linnman et al. 1981). Judging from setting and clothing, all appear to be salaried workers, rather than mere hobbyists. The U.S. focus on explicit recruiting into science careers is unusual but in keeping with the trend to include average scientists. In one textbook, a recurring feature called "Exploring Careers" photographs scientists at work and interviews them about their jobs (Heil et al. 1994).

Hence, we see careful "boundary work" being done in the textbooks of the kind described by Gieryn (1983) and Moore (1996). Given the undeniably greater specialization and technical demands of science as a global institution, no version of primary school science can afford to dismiss professional scientists altogether. Yet the cultural expansion of science necessitates an expansion of expertise.

ORDER AND THE NATURAL WORLD

Finally, there has been a notable tendency of textbooks since World War II to portray the natural world as ordered and manageable through basic scientific principles. Textbooks from the first half of the century were much more likely than later textbooks to portray the uncontrolled, destructive force of nature. Early books showed the aftermath of storms, disfiguring disease and malnutrition, and dangerous animals. A U.S. text cited earlier from 1934, for example, had entire chapters on "injurious caterpillars" and "other injurious insects" (Nichols 1934:xi–xii). Certain species of flora and fauna are indicted as "weeds" or "pests." Textbooks published later decline to make these kinds of distinctions, dwelling instead on ecosystems that are sometimes "out of balance." This suggests that knowledgeable tinkering will restore balance to the system.

A kind of hopelessness is palpable in an Argentinian textbook as it deals with the topic of alcohol use. A diagram entitled "the four roads from alcohol" sketches a tavern in the middle. One "road" is labeled "economic misery" and leads to the poorhouse. "Physical misery" leads to a graveyard and "premature death," whereas the road of mere "sickness" leads to a hospital. Finally, the road of "serious crime" leads to prison (Fesquet 1947:66). This book, published in 1947, is a bit behind the trend in its emphasis on uncontrollable aspects of the natural and social world, yet this example captures the feel of early depictions. The contemporary style, on the other hand, would point out that alcohol is an accepted and welcome part of many cultures, but that should a problem arise, clinical intervention is an option. In other words, there would be a few alternative roads leading from the tavern!

As Figure 6.1 indicates, textbooks published during most of the century emphasize the lawful and therefore conquerable aspects of nature. Many textbooks from the 1980s and 1990s went further to highlight the benign quality of nature. A Portuguese textbook by Francisco Mateus from 1960 is typical in its stance on atomic energy as subject to control by human beings. Nuclear energy's destructive force is implicitly acknowledged with a picture of an atomic bomb blast, but the discussion and a subsequent picture of a nuclear power plant emphasize the possibilities for taming and utilizing this natural force for the good of all. Likewise, a Soviet textbook shows a cartoon of three smiling men standing behind walls of various thicknesses. Each wall blocks a different type of radiation (alpha, beta, and gamma), so each man is ostensibly protected from any danger (Ministry of Education [Soviet Union] 1970:163).

Contemporary textbooks go beyond a depiction of nature as conquerable. Essentially, the line is that nature has been tamed, and the former struggles now scarcely bear mention. A sequence of pictures from a previously mentioned Ghanaian textbook in the sample depicts a woman bringing a baby to a female health worker. The health worker weighs the baby and records the data on a graph. The caption reads, "The road to health" (Saka et al. 1988:80). In short, the message is that there are accessible means to mitigate the potential for disaster. In this depiction, this potential for disaster is left unelaborated.

In a similar vein, a Botswanan textbook cited earlier in this chapter depicts a boy verifying that a drink is not spoiled by tasting a bit of it from his finger (Nicholson 1989:17). Hence, the destructive potential of nature and natural processes is a remote and, in any event, a manageable threat. A Chinese textbook highlights the benign and lawful qualities of nature: "Everything in nature has a close relationship to humans. Everything in nature has scientific reasoning. Only when we understand natural phenomena and the scientific reasoning underlying them can we correctly use them to construct our beautiful lives" (Ministry of Education [China] 1989:1). Earlier, heroic national efforts and "miracle" drugs were a focus in textbooks from many places. In contrast, contemporary textbooks suggest that prudent, sensible lifestyles based on understanding of scientific principles readily solve problems. This stance is common even in textbooks from poor countries—places where nature is in fact quite threatening as a result of the lack of material infrastructure to respond to disease and natural disaster. Globalized, homogeneous understandings about science and nature therefore prevail in textbooks, often despite local material conditions.

CONCLUSION

This chapter has offered direct evidence to support the claim in Chapter 1 of the central place of actorhood in global science. In particular, primary school science bolsters the ontological status of individual actorhood by conveying a consistent, and by now virtually ubiquitous, message that humans are capable of understanding the workings of the natural world, capable of doing science, capable of being experts. Textbooks rely on a number of devices to accomplish this. Primary school science is literally "peopled," with textbooks devoting much space to depictions of people doing science, benefiting from science, and enjoying nature. School science everywhere is constructed as a fun activity, and in general, depictions have sought to make science more emotionally accessible to children. Great care is now taken to situate science within students' everyday world, including emphasizing the personal utility of science and technology. Children are assumed to have questions, and they are expected to pose and actively address them. In contemporary primary school science, "classic" questions are no more worthy of investigation than other questions. And all of this is portrayed as occurring in a benign and ordered natural world.

The student products that emerge from this contemporary primary school science regime need not be god-fearing in the old-fashioned sense, nor should they passively revere science. In constructing young, rational agents, primary school science now primes students to deploy science in every direction. It remains to be seen whether this prescribed deployment of imparted school science will ever be something other than uncritical.

## Chapter 7

# Loose Coupling
# in National Science

*Policy versus Practice*

Previously in this book, we demonstrated that the authority of science has increasingly been expanded in scope and intensified in legitimacy. The authority of science is much evident in the rationalization of society where broad goals of progress and justice are often linked to the development of scientific outlooks, polices, infrastructures, and practices. Moreover, a further and secondary rationalization postulates the desirability, if not the necessity, of linkages among the preferred scientific outlooks, policies, infrastructures, and practices. As demonstrated in previous chapters, multiple science indicators have indeed spread worldwide, a process carried by the putative universalistic character of science that links the scientific enterprise to its presumed applicability. These rationalizations add up to an overall science for progress and justice perspective, which is reflected in ongoing debates among policy makers, as well as debates within public and academic arenas. Such debates focus on such hot issues as (1) the quality of science and mathematics instruction in the schools; (2) the expansion of science and engineering programs in higher education and the access of women and other disadvantaged groups to these fields of study; (3) the availability and management of human resources in science and technology in the labor force; and (4) the national science priorities and expenditures, and others like these.

These debates—and the resolutions and reforms that follow—often reveal the gap between policy intentions on the one hand, and the execution of such policies into action on the other. Regarding girls' and women's access to science and math education, for example, current debates expose the still flagrant imbalance between the sexes in terms of participation and achievement (see Chapter 8), whereas policies encouraging the access of women to science and education employment have been in place for years. Such policy reassessments—which essentially review the development of science—often focus on the gap between the formation of national science policies and the establishment of science infrastructures and practices. This gap between policies and practices is especially evident in case studies of less developed countries. What accounts for this gap between policy and action in the field of science? And why is this gap more evident in less developed countries?

Some of the literature interprets this gap as a technical snafu brought about by the tendency of elites and professionals from less developed countries to imitate scientific developments in the more developed countries. This mimetic process results in national science parodies far removed from local needs; for instance, the creation of national science think tanks in countries where less than 1 percent of the population proceeds to higher education. A more political interpretation of the gap is also found in the literature; according to this power dependency version, national policy makers are coerced into adopting irrelevant policies to satisfy the interests, or demands, of the dominant national powers or international organizations. There are indeed specific examples of both mimetic and coercive processes at work, but these explanations are limited by their failure to come to terms with the widespread triumph of science as institutional authority. Science indeed enjoys hegemonic status in the world today; alternatives to the authority of science—reliance on either traditional worldviews or charismatic leadership and movements—increasingly appear to be problematic, wrongheaded, or *passé*. However, as has been argued throughout this book, the authority and influence of science cannot be reduced to the instrumental interests or needs of national or transnational economic or scientific elites. The latter operate within both the opportunities and constraints of a world characterized by the hegemony of science without themselves becoming hegemons. By this cultural account, it is the hegemonic status of science and of the scientific worldview—rather than the hegemon itself and the mimetic and coercive processes that center around it—that account for the gap between policy initiatives and actual practice. It is also this hegemonic, or authoritative, status of science that accounts for the related phenomenon of institutional similarities, or isomorphism.

In prior cross-national studies, the growing institutional similarities across nation-states have been accounted for by focusing on world culture and its organizational carriers and the extent to which nation-states, organizations, and individuals enacted their appropriate identities. This is the tone set in numerous studies of worldwide diffusion of institutional entities: from Boli's analysis of the worldwide diffusion of state constitutions as an enactment of nation-statehood, to Meyer's and Ramirez's numerous studies of education systems as enacting corporate scripts, to Frank and McEneaney's study of gay and lesbian identities as enactments of personal identity scripts. A core and underlying feature of all these entities is that they are expected to be oriented toward progress and justice. This orientation is often activated and dramatized by invoking and utilizing the authority of science. Not surprisingly, science itself diffuses in similar ways across nation-states, as demonstrated in the accompanying analyses of science ministries (Chapter 5), science and math textbook content (Chapter 6), and women's access to science and engineering education (Chapter 8). This chapter, though, uses the world culture perspective to explain the gap between more common science policies and more variant science infrastructures, a gap we refer to as *loose coupling* (Weick 1976). Instead of mimetic or coercive processes being the initiators of the loose coupling phenomenon, we emphasize normative ritualistic processes of identity enactment that are played by progress-oriented nation-state entities. We argue, therefore, that because the worldwide incorporation of science is a part of the enactment of *global* prescriptions of nation-statehood, science is divorced from its *local*, or national, context. Moreover, because science diffusion is encouraged by various global entities, each enacting a particular agenda, the local institutionalization of science is partitioned along these agenda lines, and its various activities (or dimensions) are only loosely coupled. Loose coupling thus is a result of identity enactment—of the nation-state and of the profession of science.

In this chapter, we first further elaborate this perspective on science globalization and loose coupling. We next turn to an empirical examination of the degree to which various indicators of scientific development covary with each in two distinct groups of nations—namely, more and less developed countries. The main hypothesis is that loose coupling is more likely to characterize less developed countries. Last, we consider further implications of the loose coupling thesis.

LOOSE COUPLING AND SCIENCE GLOBALIZATION

*Conditions Induced by Globalization*

Organizations, of various types, operate as if their most important decisions involve adapting the right policies, combining the correct procedures, and recruiting the appropriately certified personnel. Whether these policies, procedures, and personnel are demonstrably efficacious in attaining the organizational goals often appears to be of secondary importance. This seems to be especially the case when two conditions are met: first, a lack of clarity as to whether there are efficacious technologies for attaining organizational goals; and second, a "thick" organizational field in which expertise is expanded on the elaboration of what constitutes right policies, correct procedures, and appropriately certified personnel. Under these conditions—uncertainty as to the specific steps required, yet confidence in the general path—one should expect to find greater evidence of loose coupling. This outcome is especially evident in resource-poor organizations. Such organizations are more likely to afford and to embrace the less costly demands of the scripts; they are also less likely to adopt the costly implementation of such scripts, thus creating a disconnect between policy scripts and policy execution. Thus, for example, they would choose to adopt mission statements instead of investing money in the execution of the mission itself. These organizations are also more likely to indulge in pick-and-paste strategies, instead of more coherent goals-versus-means programs.

National development is a diffuse and multifaceted goal for which there are less than well-established efficacious technologies. Different organizations encode their unique definition of "development" into their programs, thus emphasizing such diverse paths as economic liberalization, state sponsorship, democratization, women's incorporation, or any combinations of such goals. Moreover, such goals reflect the changing of fashions in the field of development expertise. As is evident in the field of science during the past several decades, science is evoked in various development programs, regardless of their definition of development. Also, the call for science changed during this period with the changing developmentalist fashions, from the 1950s focus on basic science and human capital to the 1990s redirection toward skill development for a globally integrated economy. Overall, therefore, international policies offer no single, coherent prescription for development, or for science's role in achieving development. Rather, in spite of the cumulative experience in science and development programs for the past thirty or more years, there still is no specification of the surefire recipe or for specific steps to achieve the goal of national science-based development.

Nation-states and international organizations act, however, within a fully developed and dense organizational field, in which there is much consensus about the importance of "development" as a social goal. Moreover, within this consensus, there is a general agreement about the significance of science in achieving development (see Chapter 4). Although this broad consensus has its critics, such criticism is couched in language that rather explicitly recognizes the legitimacy and pervasiveness of the reigning models of development. Within this consensus over "science for development" and the related assumptions that science and technology have an immediate impact on national welfare, nation-states are compelled to monitor trends in their science and technology activity. Displacing the goal from connecting science to its social context to showing some evidence of progress on the indicators leads to gaps between intended consequences and actual performance. Such gaps are best conceptualized as loose coupling.

*Defining Loose Coupling*

Coupling describes the degree of connectedness among subsets of organizations or organizational fields. It thus "may vary in strength along a continuum from very loose or decoupled to tight" (Beekum and Ginn 1993). Loose coupling is hence the structural condition where organizational subsets are weakly coordinated, or independently developed and operated. Decoupled, or loosely coupled, organizational units share low levels of responsiveness, directedness, and immediacy of relations. Such discrepancies may occur between various levels or activities within an organizational field[1] and the intensity of such discrepancies varies by the location within the field. Most importantly, however, loose coupling is masked by a high degree of rationalization. In other words, the lack of internal coordination occurs while the organization appears to be highly rational. Organizations therefore maintain ritualized conformity despite conflicting "demands" or conditions.

Applying these terms to science globalization, one may expect a high degree of loose coupling between the institutionalization of scientific activities. Both conditions for loose coupling to occur are evident in the field of science: a general belief that science is a tool for achieving development, yet a lack of a coherent and concrete program by which science is to shape development. This inconsistency is further complicated by the organizational size and complexity of the fields of science and of development. As a result of the expanded nature of these fields, a variety of "solutions" are prescribed simultaneously by different organization sponsors. As a result, although all countries wish to harness science to their development goals, they adhere to the guidelines of specific sponsors. Therefore, for example, some countries follow International Monetary Fund "prescriptions" and establish technology parks to connect their scientific labor-force capability and their industry, and other countries follow United Nations Education, Scientific, and Cultural Organization guidelines and expand their school science education programs to establish a solid basis for science in early age.

Because the scientization of various nation-states is not a uniform process, it results in differential institutionalization of science. Simultaneously, there is a rapid and pervasive institutionalization of one dimension of scientific activity and a slow and sparse institutionalization of another dimension of scientific activity. Adherence to global models defines which of science's dimensions is emphasized and which scientific activity is neglected instead. Thus, for example, most nations coded their "science for development" goals into national policy declarations, yet few of them created the necessary infrastructure to link research and industry. Likewise, most countries established institutions of higher science education, yet few of them link such training programs with national occupational structures. Overall, then, loose coupling is a pervasive condition in the field of science—loose coupling between science policy and practice, as well as loose coupling among different spheres of scientific activity.

## The Predicament of Less Developed Countries

These situations are more acute in developing countries. As mentioned earlier, a higher degree of loose coupling is evident in weaker and newer nation-states, namely in noncore countries. Conflicting demands—such as the pressure to exhibit evidence of progress while little attention is given to the depth of such progress and its effects—pose a special challenge for less developed and more resource-restricted nation-states. This challenge is further exacerbated by the specter of economic globalization, encouraging countries to hastily adopt international practices in fear of being left behind. Developing countries, we argue, scramble to adopt the "science for development" model by institutionalizing as many of its features as they feel they can afford. These countries scramble to satisfy the expectations imposed on them in the name of nation-statehood. They thus attempt to present a coherent and rational plan while being confined by their limited resources.

Developing countries' pick-and-paste strategies mean choosing the elements from the global models that are cheaper to implement or those that seem to resonate more with local themes while neglecting all other dimensions. This is not merely a strategic move on the part of developing countries to gain legitimacy or resources. Rather, it is an outcome of a rationalized culture that calls for affirming core components of nation-statehood, including a national plan for achieving economic development through scientific and technological progress. Kenneth King (1989:100) describes the gap between science education textbooks and the abilities of local African schools to teach such material. He attributes this gap to the imbalanced relationship between local authorities and international aid organizations. Such well-intentioned international organizations, in their attempt to learn lessons from their success stories, transfer their programs from one context to another without adapting it to its new environment. Thus, their overextended generalization and their powerful intervention is met with little resistance from the overwhelmed, grateful, and impoverished receiving countries; these are the roots of loose coupling in science education in African schools. King's (1989) description of loose coupling provides evidence for the workings of the "irrationality of rationality"[2] in the process of emulating global models of science. It is such mechanisms, similar in nature to the "law of unintended consequences," that distinguish between countries unthinkingly enacting global scripts and countries strategically enacting such scripts. The predicament of developing countries is that this ailment of loose coupling is caused by taking the path of rational decision making under the circumstances unique to these countries.

The predicament of loose coupling in developing countries is further aggravated by the state of the global organizational fields of science and of development. First, loose coupling is produced by the multitude of different organizations in each field. This means that various global institutions emphasize the importance of different programs, or elements, to achieve the desired goal. Second, the global model itself is not cohesive but rather is a hodgepodge of themes, goals, and rationalized plans for their implementation. Clearly, great differences in themes and style are evident between programs originating from different institutions or sponsoring nations.[3] The newly established countries are most prone to buying into international programs, regardless of the unique conditions under which such programs were conceived. Science education textbooks were integrated into the new African countries during the 1960s, regardless of their unique logic of pedagogy that did not suit local conditions but rather was "more dependent upon equipment, improvisation and therefore good teaching than other versions of science that emphasized content" (King 1989:100). Overall, therefore, the combination of, first, such a variety of themes and programs, and second, inconsistencies within the global model are detrimental to the planning efforts of developing nations and to the cohesion of actions. Within developing countries, such gaps or inconsistencies reflect the ritualized nature of science globalization and hence further differentiates between core and noncore countries.

## LOOSELY COUPLED SCIENCE: EVIDENCE
## FROM DEVELOPING COUNTRIES

Numerous case studies document evidence of how inconsistent science is, especially in developing countries. They point out that science is often compartmentalized into different functions, different organizational settings, or different social networks. Most often, studies of science in less developed countries focus on the disjunction in scientific communication. Rossum and Hicks (1997) show that during the 1990s, scientists in sub-Saharan Africa hardly interacted among themselves. Therefore, although science thrives on social interaction among scientists, and although all sub-Saharan scientists are involved in a scientific network in the north, south–south relations among social scientists are weak. Similar results are offered by Velho (1986) in regards to Latin American scientists. He shows that only 2 percent of all citations in scientific papers written by scientists from Latin American countries are of scientific literature written by their compatriot scientists.

Disjunction in science exists also in the organizational sense. On the organizational level, action is taken to establish some science institutional settings while neglecting to establish others. For example, most countries establish such science institutions as universities and ministries (see Chapter 5), yet fewer nations establish local research and development laboratories, science-industry sites, or "big science" operations. It seems that resource-poor countries establish science institutions that are either at the core of our notions of what science is (such as universities, being both a training site and a source for national pride) or are offering an appearance of rational and legitimate national operation (such as agencies for policy work and basic research). Satisfying these "subliminal" goals, though, is far removed from the formal goals of such science institutions. Thus, although the formal policies call for such internationally recommended action as the establishment of technology parks, interfacing between science and industry, or placing scientists and engineers in production positions, the execution of such goals is an unattainable burden. The establishment of these policy and "high science" institutions in resource-poor countries, where their policies cannot be implemented, further contributes to their weak coupling with their social context. In general, less developed countries rarely establish the full array of institutional arrangements that the policies call for and that are probably required in order for science to have its intended consequences.

Finally, the most prevalent form of loose coupling is between policy and practice. Most nation-states create a national science policy, coded into texts and declarations and organized into formal agencies, but fail to follow such declarations with action. Practicing science thus continues at its own pace, facing its own challenges, without the guidance of policy makers, who still uphold the unaltered image of the institution of science as a whole. For example, studying the institutionalization of "science for development" themes in Catalonia during the 1980s, Bellavista and Renobell (1998) report the gap between policy and practice. Great achievements in establishing legislative and political foundations,[4] they note, were not accompanied by actual build-up of applied science within Catalan corporations, nor with establishing actual links between research and industry. The thrust of "science for development" institutionalization remained at the policy level: the drafting of several Research Plans and the establishment of the Centro de Información y Desenvolvimiento Empresarial (Center for Technology and Industrial Development), CIRIT, and CICYT.

Overall, then, case studies of developing countries document numerous examples of loose coupling in science. They report on loose coupling both within the institution of science and between science and its local context. They also provide evidence of loose coupling in social interactions, in organizational build-up, and along the policy–practice axis. Yet although such exemplary evidence is plentiful, cross-national comparative analyses of loose coupling in science are sparse. We next provide such a comparative empirical examination of loose coupling in science by focusing on the inconsistencies among the various dimensions of scientific work in developing countries.

INTERNAL INCONSISTENCY IN SCIENCE: A COMPARATIVE
EVALUATION OF LOOSE COUPLING

Because science is institutionalized worldwide in an inconsistent manner, there
is a high degree of loose coupling between various components, or dimensions,
of science in various countries. As described earlier, in developing countries, science institutionalization occurs mainly in the policy sphere while science practice is rather neglected. Similarly, countries institutionalize locally a different
combination of science activities, depending on their global referentials. Bluntly
said, it is easier for developing countries to incorporate policy words than to
direct resources toward their implementation. It is similarly easier for them to
incorporate some scientific activities that are visible worldwide than to assimi-
late complex infrastructure organizations. In a general way, such tendencies are
the result of the ritualized nature of science globalization. Adherence to global
models defines which dimension of science is emphasized and which scientific
activity is rather neglected. In either case, the initiative for emphasizing or
neglecting a specific dimension of science comes from attention given to global
models and from the normative pressure exerted by professional organizations.

Although prevalent to a varying degree in all countries, tendencies toward
inconsistencies demark core countries from noncore countries. Because
weaker and newer nation-states—namely noncore countries—are more sus-
ceptible to accepting global models in a ritualized manner, higher levels of
loose coupling are prevalent in their institutions. In core countries, on the
other hand, where science is a long-established institution and where there is
a weaker, or slower, penetration of global models, we expect to find relatively
low levels of loose coupling among science's dimensions.

Our empirical investigation therefore compares the degree of loose coupling between these two groups of nation-states. We also focus our comparison of loose coupling that is internal to the field of national science—that is,
evidence of inconsistencies among the various dimensions of the national
field of science. This analysis is guided by a simple proposition: the less developed a country is, the greater the degree of loose coupling among its different
dimensions of local science activity. Science policies, procedures, and practices
are more likely to be better meshed in the more resource-rich countries.

*Data and Model*

Acknowledging that science is a multifaceted social institution and that its
multidimensionality is further complicated by national traits, we shy away from

TABLE 7.1

Variables in the Analyses: Descriptive Statistics

| Dimension | Variable | Core Countries | Noncore Countries |
|---|---|---|---|
| Science policy | Science administrative basis by 1995 (0–2 ; scale) Jang, Finnemore | 1.479 (0.11) n = 48 | 0.42 (0.04) n = 273 |
| | National science organizations 1994 | 53.07 (14.04) n = 46 | 3.31 (0.72) n = 100 |
| | Factor: | 0 (0.15) n = 46 | 0 (0.1) n = 100 |
| Science research | Paper publication in hard sciences 1993; SCI | 13,441.85 (5,712.80) n = 46 | 229.96 (89.38) n = 158 |
| | Science publications 1995; SCI | 9588.00 (2,348.75) n = 45 | 366.53(132.68) n = 157 |
| | Factor: | 0 (0.15) n = 45 | 0 (0.8) n = 154 |
| Scientific labor force | Technicians, scientists and engineers 1990 (logged); UNESCO | −0.74 (0.11) n = 37 | −0.56 (0.19) n = 47 |
| | Total tertiary students (logged); UNESCO | 11.62 (0.31) n = 37 | 8.141 (0.53) n = 68 |
| | Factor: | 0 (0.18) n = 31 | 0 (0.18) n = 30 |

Data in table are presented as mean (standard deviation).

a single indicator methodology. Rather, to overcome the problem of establishing a one-to-one conceptual link between an indicator and the concept it represents, we employ a multivariate methodology. On the basis of confirmatory factor analyses we group various cross-national indicators of science into three indexes. Each index thus represents a unique dimension of science activity while being composed of several indicators to affirm its face validity.

The three dimensions of science investigated herein are science policy apparatus, science research activity, and scientific labor force.[5] The science policy depicts the national efforts to form guidelines and coordinating bodies for its scientific ventures. The dimension of science research focuses on actual research performance, thus reflecting elite-type research work that is often guided by international standards of productivity and rewarding. Last, the dimension of scientific labor force represents the scientific capacity (and potential capacity) directed at the local productive sector.[6] These dimensions are empirically constructed by indexing several indicators.[7] Table 7.1 lists the indicators included in these analyses.

After establishing the validity of each such dimension of national science activity, we examine the relationship among these three indexes. We operationalize such relationship as the correlation among the indexes. In other words, we conceive of loose coupling as the association measure—namely, the Pearson correlation score—among the indexes. Previous studies of loose coupling did not quantify such relationships, but rather relied on qualitative methodology. Such studies employed questionnaires (Scheid-Cook 1990) or interviews (Welsh and Pontell 1991; Beekum and Ginn 1993) as their data and interpreted these texts. Our model, however, not only quantifies the phenomenon of loose coupling, but also identifies it as a national trait.

### Findings

Core and noncore countries differ greatly by their levels of cohesion—or loose coupling—among their various scientific activities. Whereas in noncore countries there exists great disjunction among the three dimensions of local science, in core countries, these dimensions are closely associated. More specifically, in developed countries, the correlation scores among policy, research, and labor-force activities in science are all significant. As graphically shown in Figure 7.1A, the Pearson correlation between science policy and research equals 0.69 ($n$ = 44), $r^2$ between policy and labor force equals 0.43 ($n$ = 29), and $r^2$ between research and labor force equals 0.60 ($n$ = 30), whereas all correlations are significant at level higher than 0.2.[8] In comparison, in developing countries (Figure 7.1B), the significant correlations are between science policy and research work ($r^2$ = 0.65, $n$ = 99) and between science policy and scientific labor force ($r^2$ = −0.36, $n$ = 26); the Pearson correlation score between research work and labor force is smaller and nonsignificant ($r^2$ = −0.23, $n$ = 30).[9] Also, it is only in the group of developing countries that some such correlation scores are negative, suggesting inverse relationships: the more intense policy work is in developing countries, the smaller the scientific labor force is. Overall, we find that whereas in core countries all three dimensions of scientific activity in the 1990s are highly correlated, in noncore coun-

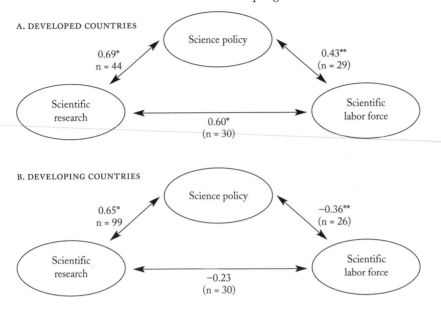

*Figure 7.1.* Loose Coupling Among the Dimensions of Science: Results from Cross-National Empirical Analyses, 1990s. The data in figures are the Pearson correlation score, the significance level, and the number of cases. *$p < 0.01$, **$p < 0.05$, ***$p < 0.10$.

tries, there exist dyadic relationships between science policy and other dimensions of scientific work. Moreover, in developing countries, expanded activity in one sphere of science parallels reduced activity in another sphere, suggesting that in such resource-poor countries, each expansion comes at the expense of another sphere of activity.

In the process of combining the various cross-national measures of scientific activity into meaningful spheres of scientific activity, we came across an interesting conceptual problem. There was relatively weak correspondence between indicators that had great face validity as joint into a single sphere. The Pearson correlation scores between presumably similar indicators, reflecting similar activities, were surprisingly low. For example, although two indicators presumably reflect the national effort to establish a policy coordination infrastructure, in a sample of developing countries ($n = 100$) the Pearson correlation score between the scale of national science affairs agency and the size of the local organizational field of science is only 0.14 (not significant). This is, in our opinion, not merely a measurement problem. Rather, it shows that loose coupling is prevalent even among indicators of a similar sphere, or dimension, of national science. Apparently, loose coupling is an endemic problem worldwide, especially in developing countries.

*Summary*

The models of association among science's dimensions and the comparison between the two groups of nations seem to confirm institutional assumptions in regards to loose coupling in the process of science globalization. These results show that in core countries science policy, research work, and scientific labor-force activity are highly related with each other, and that hence, in these countries, science is a more coherent social institution.[10] In noncore countries, on the other hand, the relationships among these three dimensions are mostly weak and insignificant, suggesting that national scientific activity in noncore countries is disjunctured and compartmentalized. In these noncore countries, the only strong and significant association is between science policy and science research work. What explains the differential cohesiveness of science between core and noncore countries? What also explains the strong association between policy and research work in noncore countries? We argue that the explanation lies in the level of adherence to the global model of science and hence the degree to which a nation-state is immersed in world society.

The strong association between science policy and research work in noncore countries results from the nature of the global model of science, from which noncore countries draw their pattern of national science institutionalization. According to this global model of science, both science policy and scientific research are highly acclaimed and highly legitimate. In other words, the global model of science specifies rather clearly what the traits of science policy are that need to be locally institutionalized—mainly emphasizing the role of science organizations as loci of science coordination and policy making. This global model also specifies what is the highly recognized and rewarded type of scientific work—mainly research work that is visible on international monitors, such as scientific publications, participation in scientific meetings, and other forms of international scientific communication. The global model of science is, however, less specific about the desired nature of the local scientifically trained labor force. International discussions are still inconclusive as for type of scientized labor force that nation-states should emphasize: should nations provide general scientific and technological literacy or should they emphasize more elitist programs for scientific and technological training? Or should nation-states emphasize training in the medical and health sciences or in engineering? Because international organizations and donor countries did not yet resolve these debates and therefore do not yet provide a clear policy path for noncore countries to adopt, noncore countries resort to patterns of pick-and-paste policy making. In other words, they adopt these science training policies in an incoherent manner, which results in greater disassociation with the local context (or "needs"), and hence in greater loose coupling. Overall, it is because of the nature of international science policy, which is specific and clear in regards to some dimensions of science activity and diffuse and ambiguous in regards to other dimensions of scientific activity, that noncore countries have a rather loosely coupled structure of their national science field.

## SCIENCE POLICY AND BEYOND: THE GAP BETWEEN
## POLICY GOALS AND RESULTS

Science policies, both national and international, not only define the various dimensions of science that are desired in each nation-state, but also define the goals for such local science fields. As discussed in Chapter 4, the science policy discourse, which is immersed in developmentalism, defines science as a key to producing national economic growth. In numerous policy texts, the adoption of science is encouraged, with the expectation that scientific activity results in economic prosperity. According to this "science for development" model, economic growth relies on the scientization of the labor force and it in turn relies on scientific and technological literacy. These components of the science policy model are set in a hierarchical and causal manner: scientific and technological literacy shapes the skills of the labor force and sets the foundation for technology transfer, whereas the skillfulness of the workers in turn determines the level of economic development. In this sense, science is causally linked with the results expected from the implementation of science policy. Although the general tenets of this model are clearly stated, and although the causal relations among these factors are routinely assumed in public discussions, there is little research of the validity of these policy assumptions. Building upon the evidence of loose coupling within the field of science, one would also expect some evidence of loose coupling between science and its expected results. In other words, loose coupling is prevalent among science education, science labor-force activity, and economic development, especially in noncore countries.

Following the logic of comparative analyses of science globalization, Annelie Strath (1998) sets the investigation of loose coupling between science and its expected results. She finds that cross-national comparisons provide a general support for the "science for development" policy model: the greater the level of science education, the greater the level of scientization of the labor force; and the greater the scientific skillfulness of the labor force, the greater the level of economic development. However, her analyses reveal that such relationships are not uniform worldwide. Rather, the links between science, local labor force, and the economy are mediated by structural conditions. These relationships, she finds, are stronger in countries that have the relevant infrastructure to facilitate the effective connection between science and its expected results. The models show that since the 1970s, first, the relationship between tertiary-level science education and scientifically trained labor force, and second, the relationship between the skillfulness of the labor-force and economic performance are stronger in countries with an expanded economic activity. For example, the greater the national export market, the stronger the connection between the size of the scientific labor-force and economic growth. Similarly, the effect of the labor force on economic growth is greater in countries enjoying higher levels of capital accumulation (resources in terms of machinery and equipment). Finally, it is in countries that have both a highly scientized tertiary education sector *and* high levels of economic growth—namely core countries rather than noncore countries—that the level of science education positively affects the level of skillfulness of the labor force. Overall, therefore, Strath's results suggest that in developing countries, there is looser coupling between science education and the concentration of scientistists and engineers in the labor force, as well is between the scientization of the labor-force and economic growth.

Country-specific research into the effects of science policy offer ample evidence in support of these conclusions. To continue on the issue of Catalan efforts to create a science-based economic miracle, Bellavista and Renobell (1998) report that in spite of the official formal emphasis on "science for development" agenda, the Catalan government failed to establish a fruitful link between its local science and industry. They cite numerous reasons to explain the disconnect between the extensive efforts to establish local science (as listed before) and their actual effect on local economic performance—

from having too many goals for their science policies and thus spreading their attention too thin,[11] to inertia and tradition in government operations,[12] to goal displacement in science policy agencies.[13] Yet prominent in their analysis of this disjunction between science and its intended economic consequence is the academization of science policy. "Priorities," they write, "are mainly connected with academic scientific disciplines, and the discourse to help industry research it is not applied successfully for the moment. There is no actual promotion of technological innovation and there is no impact within Catalan industry, since the system promoted is mostly academic. Generally speaking, policies for research have been very close to academia and far from industry needs" (Bellavista and Renobell 1998:7).

It is not only the conceptual, problem-setting link with academia that disconnects science from industry; in addition, the reward and resource allocation systems are geared toward academic standards. Thus, for example, in Catalonia, fellowships are the main means of sponsorship during science training, yet the definition of "fellowship goals" is attuned to academic standards, and there are few prospects of getting an industry job after completing work in such a science track (Bellavista and Renobell 1998:4). Overall, then, it is science itself that is biasing the science–industry connection, because science "means" acquiring Ph.D.s or researching academic topics. Academic inertia undercuts science-based innovation.

This analysis of the Catalan experience exemplifies the loose coupling both within the local field of science and between science and its intended consequences. Also, the power of academia to define science, and hence to decouple it from its industry context, reveals something about the global model of science. This global model is clear in regards to some dimensions of scientific work, namely academic elite-type science, and is more ambiguous about science's other dimensions. According to this model, the core of science is still defined in academic terms: working toward academic degrees, emphasis on peer-reviewed publication, and focus on "basic" topics. Also well defined are the policy goals, as codified in the "science for development" policy model. Yet as mentioned earlier, the global model of science is still ambiguous regarding the actual link between science and the economy: is the link between them scientific labor force or transferred technology? Should a country emphasize school-level science education or advanced training in universities—meaning "science for all" or elite science training? This is why emerging nations or regions can easily adopt policy and research flavors but find it enormously more difficult to link between these and local economic

activity. This is also why in these countries there is a strong link between policy action and research work, as shown in the empirical analysis above, yet
both are weakly connected with labor-force activity.

CONCLUDING COMMENTS

As much of this book's research suggests, common science, educational, political, and organizational cross-national patterns can be explained by theories
of world polity and institutional isomorphism. In this chapter, we focus on
the cross-national similarity in structural relationships within science and
between science policy and its expected results, employing the terminology of
structural coupling. Loose coupling is an endemic problem worldwide, and it
is more pervasive the weaker or newer the country is. As much as countries
make efforts to link their scientific ventures to boost local economic performance and as much as the modeling of such goals is elaborated, such
efforts are still mediated by the prominence of global models—of science and
of developmentalism.

   International organizations are widely aware of such problems of lack of
connection between science education, science labor force, academia, and
economic growth. Moreover, they assume the role of linking between such
disjunctured spheres, especially in developing countries, or alternatively,
they encourage local governments to take on this task. For example, in an
effort to repair the extremely weak link between the resource-poor farmers
(about 90 percent of all farmers) and the agricultural research system
in Bangladesh, two international organizations—the Bangladesh Rural
Advancement Committee and Grameen Kristi Foundation—positioned
themselves in this linking role. Grameen Kristi Foundation encourages the
Bangladeshi Agricultural Research Council to develop specific technologies
to assist local farmers and organizes feedback channels from the farmers to
the Council (see Zweekhorst and Bunders 1998). This effort, they hope,
will make local science more relevant for its social "clients." Their goal is to
improve the link among science's dimensions or between science and its
expected results.

   As this organizational "solution" suggests, currently this issue of connecting science with its local context is masked by the language of efficiency and
relevancy. Loose coupling in science is, however, not primarily the result of
lack of competency or effectiveness, although these factors are prominent in

many countries. Rather, it is the institutional features of science globalization—chiefly isomorphism and model enactment—that make loose coupling an inevitable outcome, especially in developing countries. The disjunction in science is a result of the ritualistic affirmation of the authority of science, the value of nation-statehood, and the prominence of developmentalism. So although processes of coercion and imitation do take place and some outcomes of globalization can be interpreted in these terms, other outcomes—such as the cross-national similarity of forms—are best understood in terms of the global institutionalization process involved. Such outcomes appear to be scripted, thus suggesting that the action taken by nation-states is an enactment of global scripts, such as the models of development, progress, and justice.

## Chapter 8

# Women in Science

## For Development, for Human Rights, for Themselves

CHRISTINE MIN WOTIPKA AND
FRANCISCO O. RAMIREZ

In July 1999, the United Nations Educational, Scientific, and Cultural Organization (UNESCO) and the International Council for Science organized the World Conference on Science in Budapest, Hungary. The conference Declaration on Science approvingly highlighted the vast authority of science over a range of global issues, many of which we discuss throughout this book. Not surprisingly, though, the conference also emphasized the disparities in the reach of science and the problems these disparities created at the regional, national, and individual levels. Less access to science and technology (S&T) and less participation in the production and development of S&T are widely discussed throughout the world as inequalities that unquestionably require rectification. This discussion is increasingly a feature of the broader transnational discourse on women's rights. This discourse is dominated by a liberal feminist perspective within which many of the proposed solutions emphasize the importance of equality of access to science in education and in the labor force. Increasing the number of women schooled and employed in science is the overriding aim within this agenda. Some critics, however, have reflected on the institution of science and argued for a more transformative agenda. Their goal is not only to add women to science but also to demasculinize science.

We address both the liberal and transformative perspectives in this chapter. First we provide a rationale for a world society focus on women in science. This rationale emphasizes the ties between science for progress and women in science. Next we examine and assess cross-national patterns of women's participation in science and engineering (S&E) in education and in the labor force. Worldwide trends are identified, regional patterns are compared, and the results of cross-national analyses of the antecedents of women's participation patterns are summarized. Last, we consider the ways in which the notion of women in science has been framed in international conferences and organizations up to and including the World Conference on Science.

## WOMEN IN SCIENCE: LIBERAL AND TRANSFORMATIVE PERSPECTIVES

The globalization of the authority of science exerts tremendous impact on nation-states owing to the legitimacy and structure that it provides these entities. All societies, regardless of their economic, political, or cultural backgrounds, increasingly place immense value on science and seek to affirm their commitment to science for progress (see Chapters 1 and 4). Construed narrowly, progress may be thought of as economic development, and science becomes an instrument for increased production. In fact, what is globally institutionalized is a broader vision of science that includes economic development but also sociopolitical developments. The latter, in turn, involves notions of citizenship, equality, and justice—not just enhanced growth. This broader model of progress and the authority of science raise questions about technology and economic growth and about science geared for economic growth alone. These questions include concerns regarding the putative costs to the environment and to local knowledge as well as the reproduction of national and transnational hierarchies. But, as we repeatedly show throughout this book, the authority of science is also invoked in discussions of sustainable development, local empowerment, and transformations to harness science for the common good.

As applied to women, the narrower issue of "science for progress" has been framed in terms of women as underutilized resources in the process of economic development. The "science for development" vision is contingent on a scientifically trained labor force (Chapter 4). In light of employment trends that predict heavy growth in science- and technology-related jobs, women are recognized as essential elements needed for countries to reach their full potential in the development process. For example, in its 1993 report to the Office of Science and Technology, the United Kingdom's Committee on Women in Science claimed that women were the country's most underutilized resource.

That same year, Canada's report of the Prime Minister's National Advisory Board on Science and Technology recommended that gender equity is necessary in order to position the country for "global competitive readiness" (cited in UNESCO 1996:326). By tapping into women's scientific talents, Approtech (1993) reports that countries may be able to meet their strategic interests. As increasing numbers of women join the labor force around the world at the same time that more and more jobs require backgrounds in science, engineering, or technology, the notion of women in science has become an important component to national economic development.

From this perspective, it thus follows that the notion of women in science for development emphasizes the need for access to those opportunities that make women active participants in this process, such as access to schooling, especially to higher education, and to occupations in S&E. From the liberal perspective, the result has been calls for equal access to science education, part of which can be remedied by making available scientific materials in schools such as equipment, laboratories, gender-sensitive teaching materials, and mentors and role models (UNCSTD 1995).

In addition to the national-level benefits of women's expanded participation in science, individual-level benefits to women exist as well. Seen in this light, the issue of equity is again important, as are links to economic gains. For example, projected demand for workers in S&E on a global basis is expected to grow into the future. Nowhere is this truer than in the United States. According to the U.S. Bureau of Labor Statistics, growth in S&E jobs will advance at three times the rate for other occupations between 1996 and 2006, or 1.36 million new jobs. As can be expected, three-fourths of this increase will take place in computer-related occupations, with the number of these new jobs estimated to double over this time period (U.S. Bureau of Labor Statistics 1997). In short, in addition to the possibility for women to enhance national economic development on an individual level, it is argued that women stand to gain by enhanced workforce participation and higher salaries in scientific fields.

Women in science may also be considered from a broader definition of progress that includes not only development (for example, national economic development and individual betterment) but also justice (for example, human and women's rights). In fact, the two tend to cross boundaries. For example, the above definition of progress that focuses on economic development is often thought of in conjunction with women's rights. As a result, we find arguments in favor of development with equity such as those common to the language of sustainable economic development and women in development. This line of reasoning would also support women's right to jobs in science for the sake of equal opportunities to the relatively high salaries that these jobs offer compared with jobs in other fields. These examples show that the discourse on women in science for the sake of progress emphasizes women's rights with an underlying belief that economic development is key to such rights.

A more transformative approach to women in science for the sake of progress, however, has also been taking place in both theoretical and practical terms. This perspective stresses the role of women qua women in science and focuses not solely on access but on the transformation of science and the creation of scientific knowledge. The ramifications of such an approach are many and include demasculinizing science, challenging the biases of the medical and other science establishments against women, and demanding science created not only by women, but for women and for the issues that they deem to be relevant to their lives. For example, Bix's (1997) analysis of women's activism toward funding and policy making in breast cancer research reveals their success in not only raising public awareness of the disease, but also in making use of statistics to gain research money and to challenge research, testing procedures, and scientific authority. Bix argues that these female crusaders, along with AIDS activists, were the first to actually engage in decisions regarding fiscal spending and to challenge the authority of scientists and medical doctors (see also Rose 1995). Some may even go further, suggesting that women should be in charge of research that concerns them most closely in addition to other types of research that do not, such as primatology (Morell 1993a,b).[1] Although some have argued in favor of role differentiation between men and women in science in order to improve science by and for women, a science that once was assumed to be the natural domain of men has been challenged at policy levels. To reiterate, the transformative perspective calls for more than giving women an equal opportunity to become male scientists; it seeks a science that is more hospitable to women and women's ways of perceiving.

To empirically address the issues raised by these different perspectives, systematic longitudinal cross-national data on multiple dimensions of women in science are much needed. UNESCO's *World Science Report 1996* aptly makes the point, "no data, no visibility, no priority" (UNESCO 1996:307). Comparative data are increasingly available as regards women's enrollment levels but are more limited with respect to labor-force and occupational levels. There are even more severe data limitations as regards the more complex issues raised within the transformative perspective—for example, the gender character of science curricular content and dominant research agendas. The data presented below build on the existing research on cross-national trends in women's access to S&E higher education and on labor-force participation data.

WOMEN'S PARTICIPATION IN S&E EDUCATION

Here, we build on existing comparative research on women's participation in education and the labor force. First we focus on the participation of girls in secondary education. Next we turn to women in higher education. Last, we consider the participation patterns of women in S&E occupations. Much of the literature correctly notes that women's access to secondary levels of science education is greater than their access at higher levels. And in turn, higher education is easier to crack than the occupational sphere (for example, Jacobs 1996). What we seek to ascertain is the direction and magnitude of cross-national trends as regards women's participation patterns.

*Participation of Girls in Secondary-Level Education*

Comparative studies of girls' participation in science education at the secondary level have generally been limited to analyses of gender differences in science achievement as measured by three international studies of science, the latest being the Third International Math and Science Study.[2] Hanson et al. (1996) analyze enrollment rates for twelfth-grade science classes in five industrialized countries (Canada, Finland, Japan, Sweden, and the United States) and in Thailand. They used data from the Second International Science Study[3] to quantify unequal participation in biology, chemistry, and physics. Of the three, biology was the only subject in which more girls than boys participated among the six countries. Enrollment rates for girls were considerably lower in chemistry and particularly in physics for most countries. Thailand was the only country with slightly higher enrollment rates for girls than for boys in all three subjects (54, 55, and 52 percent, respectively).

*Participation of Women in Tertiary-Level Education*

Women's share of enrollment in S&E at the tertiary level relative to total enrollment for both sexes has slowly increased cross-nationally over the past

TABLE 8.1

Women's Share of S&E Higher Education,
1972–1992, by Region[a]

| Site | 1972 | 1992 | n |
|------|------|------|---|
| All countries[b] | 15.4 | 21.4 | 71 |
| Africa | 9.6 | 11.8 | 17 |
| Middle East/North Africa | 19.2 | 27.7 | 10 |
| Asia | 8.7 | 15.8 | 9 |
| Latin America/Caribbean | 15.1 | 26.9 | 8 |
| Eastern Europe | 27.9 | 29.5 | 7 |
| Western Europe[c] | 17.2 | 23.8 | 20 |

[a]Source: *UNESCO Statistical Yearbook,* various years. The women's share of
(S&E) higher education is calculated by dividing the total enrollment of
women in S&E higher education by the total enrollment in (S&E) higher
education.
[b]Includes all countries for which women's enrollment data are available for
1972 and 1992.
[c]Includes Australia, Canada, and New Zealand; data for the United States
were missing.

few decades. Table 8.1 shows that between 1972 and 1992, women's share of
these fields of study has gone from 15.4 percent to more than 21 percent on a
global level. However, stark differences by region are evident. By 1992,
women made up less than 12 percent of students enrolled in S&E in Africa.
In Asia, a region in which large numbers of students are enrolled in these topics relative to other fields of study, less than 16 percent of the students were
women. With women's share at close to 28 and 27 percent, respectively, in the
Middle East/North Africa and Latin America/Caribbean, it may be surprising to learn that these figures surpass that in Western Europe (just under 24
percent) (Ramirez and Wotipka 2001).

Women in Eastern European countries have always participated in S&E
higher education at greater rates than their sisters in other parts of the world.
By 1992, women made up close to 30 percent of S&E students compared
with less than 28 percent in 1972. Although Eastern Europe can boast of
these high numbers, the increase in women's numbers over this same time
period was also the lowest. This suggests that countries starting out with
high numbers are less able to expand. On the other hand, places that start
with relatively low proportions of women in S&E are able to grow at faster

TABLE 8.2
Rates of Change in Women's Enrollment
in S&E Higher Education, 1972–1992, by Region[a]

| Region | Means | (SD) | n |
|---|---|---|---|
| All countries[b] | 11.9 | (39.8) | 92 |
| Africa | 8.5 | (20.9) | 25 |
| Middle East/North Africa | 35.4 | (80.2) | 14 |
| Asia | 23.3 | (52.3) | 14 |
| Latin America/Caribbean | 2.5 | (3.2) | 11 |
| Eastern Europe | −0.03 | (.67) | 11 |
| Western Europe[c] | 1.7 | (.85) | 21 |

[a]Source: *UNESCO Statistical Yearbook,* various years. SD = standard deviation.
[b]Includes all countries for which women's enrollment data are available for 1972 and 1992.
[c]Includes Australia, Canada, and New Zealand; data for the U.S. were missing.

rates in order to essentially catch up with other parts of the world. Alternatively, the slow growth experienced in Eastern Europe may suggest that the region is reaching a ceiling in the proportion of women who will participate in these fields. Below, we return to these questions by considering the region's earlier socialist emphasis on egalitarian ideals and policies.

Another way to examine the data is to consider women in S&E higher education as a proportion of the college-age cohort (in this case, twenty- to twenty-four-year-olds). This method controls for changes in the population "eligible" for pursuing these fields.[4] Rather than looking at raw data for this measure of women's enrollment (which are minuscule in their numbers, given the large number of young adults within this age group, the small proportion of whom are in higher education, and the even smaller numbers in S&E), we instead computed a rate of change between 1972 and 1992, as listed in Table 8.2. Among all young adults of college-going age worldwide, there was roughly a 12 percent increase in the number of women enrolled in S&E higher education during the time period studied. The Middle East/North Africa region experienced women's greatest increase: 35 percent. Although we reported that women still make up a small proportion of S&E students in Asia relative to the college-age cohort, they increased their numbers by 23 percent. Increases for women in Africa, Latin America/Caribbean, and Western Europe were quite small (8.5, 2.5, and less than 2 percent, respectively). We again notice a small decline for Eastern Europe—among the seven countries for which we had data, women's participation in S&E higher education as a proportion of the age cohort decreased by less than 1 percent.

On a country-by-country basis, it is interesting to note exactly where growth and even declines are taking place. By using the same measure of women's enrollment as in the above, we witness declines in ten countries for which data are available (Table 8.3). Of those ten countries, four of them were communist countries during part, if not all, of the time period covered. The Netherlands is the only industrialized country to experience a decline during the time period. Among the other economically advanced nations, none of them make strong gains compared with the poorer nations in the sample. Norway finishes on top at 3.6 percent, followed by Japan and Denmark at 3.2 and 2.9 percent, respectively. Of the ten countries that make gains above 10 percent, only India is considered to have a solid S&E infrastructure. Several of them are Middle Eastern countries, which have traditionally restricted women's access to male-dominated fields of study. The findings clearly show that gains in women's participation in these fields are being made in the majority of countries regardless of economic standing, the strength of the scientific establishment, or the degree of women's rights in society in general.

By using data for the United States on graduates of S&E major fields of study, the National Science Foundation (NSF) reports that American women have made great strides in S&E higher education over the past thirty years or so, particularly in the natural sciences but to a lesser extent in engineering. By 1995, women's share of earned bachelor's degrees in physical science had reached 35 percent; women made up over half of the graduates in biological sciences (NSF 1995). As for engineering, although the rates are far behind those in the natural sciences, American women earned close to 18 percent of the engineering degrees by 1995 (NSF 1995).[5]

It is important to not only consider the situation of women in S&E education relative to their male classmates, but also to consider it in relation to their enrollment in other fields of study. Studies that have examined the growth in the number of degrees obtained by tertiary-level students in S&E have generally failed to consider those numbers relative to other fields of study (Johnson 1994; NSB 1998). It is misleading to rely on reports that indicate an increasing number of students majoring in these fields when we

TABLE 8.3
Changes in Women's Participation in Science and Engineering
Higher Education, 1972–1992[a]

*Countries with increases[b] (n = 83)*

| | | | |
|---|---|---|---|
| Saudi Arabia | Norway | Rwanda | Israel |
| Qatar | Ethiopia | Switzerland | Trinidad and |
| India | Japan | Sweden | Tobago |
| Laos | Mexico | Ghana | Uganda |
| Yemen | Barbados | Turkey | Greece |
| Liberia | Denmark | Malta | Romania |
| Somalia | Malawi | Spain | New Zealand |
| Bahrain | Bangladesh | Niger | Czechoslovakia |
| Morocco | Central African | Honduras | Kuwait |
| Nicaragua |   Republic | Korea, Republic | Syria |
| The Philippines | Malaysia |   of | Paraguay |
| Thailand | Australia | Togo | Bulgaria |
| Mauritius | El Salvador | Belgium | Egypt |
| Mali | France | Iceland | Zambia |
| Jordan | Canada | Senegal | Italy |
| Algeria | Sudan | Finland | Guyana |
| Swaziland | Afghanistan | Gabon | Congo |
| Burundi | Nepal | Lesotho | Tanzania |
| Cameroon | West Germany | Sri Lanka | Papua New |
| Argentina | United Kingdom | Austria |   Guinea |
| Madagascar | Cyprus | Cuba | Iraq |
| Tunisia | Ireland | Hong Kong | |

*Countries with declines (n = 10)*

| | | | |
|---|---|---|---|
| Hungary | Mozambique | Pakistan | Iran |
| Albania | Poland | Yugoslavia | |
| Brazil | Netherlands | Sierra Leone | |

[a]Sources: UNESCO (various years); United Nations (1993). Women's participation in
S&E higher educaiton is measured as women in S&E higher education relative to the age
cohort in 1992 (time 2) minus the value in 1972 (time 1) divided by the value in 1972
(time 1).
[b]Countries are arranged according to magnitude of their declines (or increases), with
those at the top of the list having the greatest declines (or increases)

consider the rapid pace of expansion of higher education in other fields.
When we control for growth in other fields of study, we find that participation in S&E fields has in fact decreased over time for both women and men
in many countries (Table 8.4). For the entire sample of 81 countries, the proportion of all students enrolled in S&E higher education declined from 21.9

TABLE 8.4
## S&E Higher Education Enrollment Relative to Total Enrollment in Higher Education, 1972–1992, by Region, by Gender[a]

| Site | 1972 Mean | SD | 1982 Mean | SD | 1992 Mean | SD |
|---|---|---|---|---|---|---|
| *All Countries*[b] *(n = 81)* | | | | | | |
| Women | 11.4 | 8.7 | 10.6 | 6.6 | 9.9 | 6.3 |
| Men | 27.6 | 14.2 | 26.6 | 11.8 | 25.2 | 11.4 |
| TOTAL | 21.9 | 10.1 | 20.5 | 8.7 | 18.5 | 8.3 |
| *Africa (n= 18)* | | | | | | |
| Women | 12.9 | 13.6 | 6.6 | 5.0 | 7.1 | 5.2 |
| Men | 22.7 | 12.3 | 19.0 | 9.0 | 19.0 | 9.0 |
| TOTAL | 21.0 | 11.9 | 16.2 | 7.8 | 15.9 | 7.7 |
| *Middle East/North Africa (n = 14)* | | | | | | |
| Women | 11.5 | 8.1 | 15.8 | 6.5 | 14.0 | 6.5 |
| Men | 27.1 | 20.1 | 26.6 | 12.5 | 27.0 | 13.8 |
| TOTAL | 20.6 | 10.9 | 22.3 | 9.5 | 20.6 | 9.6 |
| *Asia (n = 11)* | | | | | | |
| Women | 8.0 | 6.1 | 11.2 | 7.1 | 10.0 | 6.2 |
| Men | 24.3 | 14.0 | 27.4 | 12.8 | 26.0 | 9.0 |
| TOTAL | 19.6 | 11.4 | 22.2 | 9.3 | 20.0 | 6.3 |
| *Latin America/Caribbean (n = 11)* | | | | | | |
| Women | 9.1 | 4.2 | 11.2 | 6.0 | 11.1 | 7.0 |
| Men | 25.6 | 5.9 | 32.0 | 7.8 | 27.7 | 11.6 |
| TOTAL | 19.6 | 5.4 | 23.1 | 6.3 | 19.8 | 9.2 |
| *Eastern Europe (n = 7)* | | | | | | |
| Women | 21.0 | 6.6 | 15.8 | 7.9 | 14.1 | 8.8 |
| Men | 47.3 | 11.8 | 37.1 | 9.9 | 33.9 | 12.3 |
| TOTAL | 34.8 | 7.3 | 26.3 | 9.1 | 23.6 | 10.2 |
| *Western Europe*[c] *(n = 22)* | | | | | | |
| Women | 9.6 | 3.9 | 8.4 | 4.3 | 7.9 | 3.9 |
| Men | 28.3 | 9.5 | 26.9 | 11.5 | 24.9 | 10.7 |
| TOTAL | 21.3 | 6.9 | 18.9 | 8.0 | 16.4 | 7.2 |

[a]Source: *UNESCO Statistical Yearbook,* various years. SD = standard deviation. Share of S&E higher education is calculated by dividing the total enrollment of all students (women; men) in S&E higher education by all students' (women's; men's) enrollment in all fields of higher education.
[b]Includes all countries for which women's enrollment data are available for 1972 and 1992.
[c]Includes Australia, Canada, and New Zealand; data for the United States were missing.

percent in 1972 to 18.5 percent in 1992. During this time period, women's S&E participation relative to other fields of study went from 11.4 percent to 9.9 percent. Although 27.6 percent of men were in S&E higher education in 1972, twenty years later, this figure decreased to 25.2 percent. Asia and Latin America/Caribbean were the only two regions to experience slight increases for both women and men. Eastern Europe experienced the greatest decline, although the region also started out with the highest numbers. By 1992, just under 24 percent of all students were enrolled in S&E higher education compared with close to 35 percent in 1972. As a point of clarification, this does not suggest that absolute numbers of women in S&E higher education are declining as a whole. What it does mean is that in relation to other fields of study, fewer women and men are choosing to major in S&E cross-nationally and over time.[6]

WOMEN'S PARTICIPATION IN S&E OCCUPATIONS

Although cross-national and longitudinal enrollment data are available for many countries, there are few equivalent data for women in scientific careers and policy-making positions (UNESCO 1996). In countries where the data are more readily available, the data clearly demonstrate that women are not entering the S&E labor force in proportion to their university training in those fields (Hanson et al. 1996; Alper 1993). For example, in some industrialized countries in which women make up over one-third of the postsecondary students in math and science fields of study, the proportion of women engineers remains close to just 5 percent (Hanson et al. 1996).

Jamison's (1993) analysis of 1990 census data sheds further light on women's employment rates in S&E for several industrialized countries. In Japan, women made up 2 percent of the S&E labor force by that year. Of this number, roughly 60 percent were employed as computer processing technicians, a field that is ordinarily made up of junior-college degree holders. As for other occupational fields, approximately 31 percent of women in S&E were in the more prestigious field of engineering. The same source also reports advances by women in the United Kingdom's scientific and engineering labor force. Twenty percent of employed computer scientists were women by 1990; women had achieved 8 percent of electrical engineer positions and 5 percent in industrial/mechanical engineering. France reports wide gains in the employment of women in natural science (32 percent) and computer science (20 percent), in contrast to just 6 percent of the engineering labor force. In Germany, women's entry into S&E fields has resulted in much of the gains in overall employment in these occupations. As of 1987, 20 percent of those employed in the natural sciences were women. However, German women made up a small percentage of employed engineers, at just a little over 4 percent. The NSF reports that by 1995, American women comprised a little over 22 percent of the S&E workforce. As can be expected, however, their numbers fluctuate depending on the occupational field. For example, although women made up half of the social scientists, they accounted for less than 10 percent of engineers and only 22 percent of those in the physical sciences. Female computer and math scientists comprised less than 30 percent of the total in 1995 (NSB 1998:3–15).

Similar to occupational data for S&E, data on faculty positions are incomplete for all countries. The use of the few data that do exist reveals that the more industrialized countries and those with stronger traditions of strong women's movements actually have fewer female faculty members in physics than do other parts of the world (Barinaga 1994). For example, Hungary, Portugal, and the Philippines lead a group of 31 countries with 30 to 47 percent of their physics departments having female faculty members. On the other end are Japan, Canada, and Switzerland, with less than 5 percent female faculty in this field. In the United Kingdom, less than 2 percent of the professorships in the sciences are held by women (Johnson 1994:11).

Other considerations for employment in S&E include the type of positions women hold within each field, the sector in which their positions are located, and their salaries. Numerous studies have reported on the fact that when women enter male-dominated fields, for whatever reason, they are located in the less prestigious specialties within them (for example, biology within natural science or pediatrics within medicine). In some countries, the distinction between public and private sector occupations translates into vast differences in prestige and wages. In India, for example, public-sector laboratories are the most prestigious and are the largest employers of scientists and engineers. Yet 40 percent of female scientists are employed in the private sector. Among the country's top-tiered research institutes, women fill only 5.4 percent of the jobs. Within all others, women take up less than 4 percent of the scientific jobs (Ramasubban 1996). In the United States, women generally tend to be found in less prestigious sectors such as private nonprofit organizations and nontertiary-level educational institutions. It is interesting to note that of self-employed scientists and engineers, close to a third are women (NSB 1998).

In addition to the above, a gender gap in salaries also exists between women and men (BOSTID 1994; Australia 1994). With median salaries of $42,000 in 1995, American women made 20 percent less than their male colleagues, who brought in $52,000. Part of this difference can be explained by differences in sector of employment and experience[7] as well as occupational field. Although women tend to be relatively well represented in the life sciences, this field also offered the lowest median salary among all S&E fields in 1995 in the United States (NSB 1998:3–15).

Although our review is not meant to be exhaustive, the findings presented here do reveal the nature of women's marginal status in S&E education, occupational pursuits, job status, and salaries. They also point to the need for improved data on women in the sciences at the nation-state level and over time so that more meaningful cross-national comparisons may be conducted. In our discussion of the findings for women in S&E higher education, we raised several questions that required further consideration. We now return to those questions.

IMPLICATIONS FOR WOMEN IN S&E HIGHER EDUCATION

The twenty-first century brings with it much anticipation concerning the continuing global expansion and reliance on science in ways never believed imaginable. For a large majority of the world's population, including women, full access to this process continues to be hampered, as we discussed above. Our discussion of the trends in women's participation in S&E education and employment cross-nationally has raised a number of questions that highlight the complexity of the issue at hand. Now we focus on two of the most interesting findings stemming from our analysis. The first relates to the rate of growth in women's share of S&E higher education and the deceleration in growth taking place in Eastern Europe over the twenty-year time period. We then consider explanations that have developed out of the literature and from our own statistical analyses (Ramirez and Wotipka 2001) that address why certain unexpected countries have experienced considerable advances in women's participation in S&E higher education.

As reported earlier, the Eastern European countries in our sample experienced the smallest increase in women's share of S&E higher education over the twenty-year time period compared with other regions. As a proportion of the college-age cohort, we find actual decreases in women enrolled in S&E

for this region. This raises a number of important questions. First, do the findings for Eastern Europe foretell changes (that is, slowdowns) to come for other places in the world, or is this phenomenon particular to this region? Is it the case that other countries will find their numbers leveling out at the same 30 percent level? If so, why is it the case that gender parity has been attained for enrollments in all fields of higher education in several dozen countries (Bradley and Ramirez 1996; Windolf 1997), but such parity remains elusive in S&E higher education?

Regarding the first question, it is unlikely that the slowdown in Eastern Europe will be mirrored throughout the world. Eastern European science for development combined elements of state-managed economy, gender egalitarianism, and an almost solely instrumental view of science. The result was that women as human resources were mobilized into S&E in education and in the work force in greater numbers than elsewhere in the world. All of this closely adhered to a Soviet model of development. The disintegration of this model has resulted in a weaker state hand in both educational and occupational allocations. With greater choice, more women (and more men) may be opting for fields of study on less than instrumental grounds (Bradley 2000). It goes beyond the scope for this study to determine the extent to which there has been a major shift in the educational decision-making criteria used by Eastern European students. Where gender egalitarianism is a relatively novel phenomenon, however, greater choice may lead to more women opting for S&E majors. This indeed seems to be the pattern in Asian and in the Latin American regions.

In order to examine this problem further, we compared the rates of change for groups of countries at the earlier time point with those twenty years later. Such a method allows us to compare Eastern European countries with other countries that started out with similarly high proportions of women in S&E higher education in 1972. It is expected that countries having more room to grow in order to reach greater parity between men and women will experience greater rates of change. The results are listed in Table 8.5. All countries were sorted into four quartiles on the basis of women's share of S&E higher education in 1972. We computed the averages for both 1972 and 1992 and then calculated the percentage increase for each quartile.

The results for the quartiles that began with the lowest and highest proportions of women are most striking. The twenty-one countries that started out with an average of just over 5 percent women in these fields of study in 1972 jumped to more than 15 percent—an increase of 185 percent. However, among

TABLE 8.5
Growth in Women's Share of S&E Higher Education
1972–1992, by Quartile

| Quartile | Average Women's Share | | Percentage Increase |
| --- | --- | --- | --- |
| | 1972 | 1992 | |
| First (n = 21) | 5.4 (2.3) | 15.5 (10.7) | 185.0 |
| Second (n = 21) | 11.9 (1.6) | 16.9 (7.3) | 42.0 |
| Third (n = 21) | 18.0 (2.1) | 27.3 (5.6) | 52.0 |
| Fourth (n = 21) | 30.0 (7.1) | 30.1 (8.6) | 0.2 |

Source: *UNESCO Statistical Yearbook,* various years. Includes all countries for which women's enrollment data are available for 1972 and 1992. Data are expressed as mean (standard deviation).

those starting out with the highest proportions, women's share increased by less than a percentage point to around 30 percent. Included in this quartile were a number of Eastern European countries as well as others from Western Europe (Belgium, Finland, France, Italy), Latin America/Caribbean (Barbados, Paraguay, and Trinidad and Tobago), as well as Africa (Lesotho, Madagascar, Mozambique, Swaziland) and the Middle East (Israel, Kuwait). In addition, Cuba, another communist country although located outside of Eastern Europe, was similarly included in this group. These results demonstrate that certain other countries in addition to those in Eastern Europe started out with relatively high proportions of women in their S&E educational programs and consequently experienced minimal increases. This suggests that a ceiling on women's participation in S&E higher education may indeed exist in places other than Eastern Europe. For countries in the middle two quartiles, the results do not follow the same trajectory. The third quartile started out with a greater proportion of women in the earlier time period than the second quartile, and yet it was able to maintain a stronger growth rate between 1972 and 1992.

Previously, we also provided evidence that demonstrated the way in which changes, particularly growth, in women's global participation in S&E higher education have taken place in the majority of countries, even those with limited economic means to nurture scientific establishments or relatively weaker attention to women's rights at the national level. In other words, increases have taken place where they are least expected.

Barinaga's (1994) study of women in scientific and engineering faculty positions provides some possibilities based on the opinions of women in the field.[8] These explanations offer a good starting point for thinking about how national culture influences women in science. Some possibilities relate to features of the scientific establishment within a country. For example, it is suggested that older systems discourage women's entry, whereas newer ones, having been created around the same time that women have gained rights in other domains, are more open to their equal participation. The status of science may also determine the extent of gender parity such that women often make up greater proportions in science teaching or research and development, depending on which one is considered to be less prestigious. Among possible factors related to educational systems, it is believed that those requiring all students—regardless of educational track or future plans—to pursue math and science subjects experience greater numbers of women at the end of the science pipeline. In addition, it has been suggested that single-sex schools foster an appreciation of science that coeducational ones cannot offer female students. Finally, societal factors may also explain women's participation in science. These include societal norms regarding the integration of work and family that may dictate the availability of and financial assistance for child care as well as the degree of flexibility in work schedules.

On the basis of our cross-national study of explanations for changes in women's enrollment in S&E over time (Ramirez and Wotipka 2001), the explanations may be even simpler than the ones suggested in the above. By using panel regression analyses[9] to analyze changes in women's enrollment in S&E higher education relative to the age cohort, our findings suggest that the expansion in women's participation in higher education in general is the best guarantor of women's increased participation in S&E, holding constant economic development, male participation in S&E, state power, and women's participation in S&E higher education at the earlier time period (that is, the lagged dependent variable).[10] This does not suggest that educationalists, policy makers, and other concerned parties need simply watch and wait for women's numbers to go up, as they have during this time period. In many industrialized countries, including the United States, women have surpassed their male counterparts in tertiary-level enrollments. Yet these countries remain far from having achieved gender parity in the S&E domain.

To summarize, we find cross-national trends in the direction of greater women's access to S&E enrollments. Much of this growth seems to be driven by the expansion of higher education in general and by women's greater enrollments in the nonscience/engineering sphere. However, we also find an artificial 30 percent ceiling, with less growth in female S&E enrollments manifested by those countries closest to this ceiling. The centrality of science for progress models and the rise of gender equality as a world issue increasingly make the place of women in science a global agenda. What role are women expected to play in the globalization of science for progress? It is to this question that we now turn our attention.

THE GLOBAL DISCOURSE ON WOMEN IN SCIENCE

Here we reflect on recommendations stemming from major United Nations (UN) conferences that deal specifically with women in science, technology, and engineering.[11] Given our focus on women's science education, we are limiting our analysis to recommendations dealing specifically with education (as documented in Table 8.6). Our aim is to examine the underlying perspectives that shape the recommendations to elucidate their changing nature over time. We demonstrate that over the course of twenty years of global action on the issue of women in science, the liberal perspective remains dominant. More recently, though, the transformative perspective has gained some ground.

It is evident that the early conferences and committees—and even those that took place recently—relied on the liberal perspective and suggestions for women's equal participation in science. One of the first references to women, S&T, and education occurred within the Vienna Program of Action on S&T for Development in 1979. Despite its lofty title, its goals as they pertained to women simply sought the promotion of their full participation in S&T for development. In 1991, the World Women's Congress for a Healthy Planet called for more education and training for women in S&T on a global level. Exactly how this is to be achieved was unspecified.

Another liberal approach to women in science education includes recommendations for the elimination of negative stereotypes of women. It is assumed that changes in curricula and society at large may lead to more positive attitudes toward women and, like the eradication of discrimination, will allow women to participate equally alongside their male counterparts. For example, UN Agenda 21 (1992) mentions the need for the elimination of negative stereotypes of women in society in general and specifically in educational curricula. By 1993, attention was still being directed at liberal

TABLE 8.6
United Nations Conferences
and Discussions of Women in Science Education

| Year | Name of Conference |
|------|--------------------|
| 1979 | Vienna Programme of Action on Science and Technology for Development, UN |
| 1984 | Advisory Committee on Science and Technology for Development, UN |
| 1985 | Nairobi Forward-Looking Strategies for the Advancement of Women, UN |
| 1986 | World Survey on the Role of Women in Development, UN |
| 1988 | Report of Participation of Women in Science and Technology Committee, Canada |
| 1988 | Women's Vocational Education and Training, European Perspectives Conference, Scotland |
| 1989 | World Survey on the Role of Women in Development, UN |
| 1990 | Measures Increasing Participation of Girls and Women in Technical and Vocational Education and Training: A Caribbean Study |
| 1991 | World Women's Congress for a Healthy Planet, WEDO |
| 1992 | Agenda 21, UN |
| 1992 | More than Just Numbers, Canada |
| 1993 | Project 2000 and Forum UNESCO |
| 1993 | The Rising Tide: A Report on Women in Science, Engineering and Technology, United Kingdom |
| 1993 | Winning with Women in Trades, Technology, Science and Engineering, Canada |
| 1994 | Why May Women Science Undergraduates and Graduates not be Seeking to Take Up Careers as Scientists? United Kingdom |
| 1995 | Draft platform for Action, Beijing |
| 1999 | World Conference on Science, UNESCO and ICSU |

Source: Gender Working Group, United Nations Commission on Science and Technology Development (1995).

approaches to women in science. Project 2000 and Forum UNESCO (1993) stressed the need for equality in S&T literacy and education for all along with the needed resources.

Throughout this period, most conferences and committees advanced recommendations that included combinations of liberal as well as alternative goals for women in science education. For example, the UN Advisory Committee on Science and Technology for Development in 1984 called for women's enhanced participation in existing technical training programs while also advocating the encouragement of science clubs for girls, the formation of education and materials to suit women's needs, and the creation of apprenticeship programs specifically for women. The World Survey on the Role of Women in Development in 1989 highlighted the need for women's equal access to S&T education, particularly for rural women, while also calling for the representation of women's interests through their participation in planning and implementing policies. Special programs for women in marginalized groups were also recommended. The Report of Participation of Women in Science and Technology Committee in 1988 endorsed the facilitation of access to S&T occupations for women while also stressing the need for educational changes aimed at women, such as information about the importance of mathematics, jobs and working conditions in the sciences, role models, and scholarships. The conference More than Just Numbers, Canada (1992), included a number of different recommendations along with its appeal to eliminate discrimination in associations of professional engineers. Among its recommendations were the promotion of women's self-esteem and interest in the sciences, the need for role models for girls and women, and parental and public encouragement for women's entry into the sciences. Likewise, *The Rising Tide: A Report on Women in Science, Engineering and Technology* (United Kingdom 1993) included the strongest language in favor of equal opportunities by recommending that such efforts be monitored and reported in annual reports and that goals for public appointments and senior positions be made and met. Along with these changes, the report made calls for child care and other ways to balance career and family responsibilities as well as media coverage of women's contributions to S&T so as to improve public awareness of women in science.

Despite the diversity in approaches recommended by the various conferences and committees, the underlying standpoint of the majority of them rests on the liberal approach and to some extent on other perspectives that do little to challenge science or educational and societal structures. It is interesting to note that several of the conferences and committees have provided attention to the difficulties faced by women in balancing family with education and career responsibilities. Despite the seemingly progressive stance that these recommendations make, in the end, the underlying issue remains that of access. For example, the Advisory Committee on Science and Technology for Development (1984), the Report of Participation of Women in Science and Technology Committee (1988), and the World Survey on the Role of Women in Development (1989) have included language that endorses efforts to ensure that family and marital responsibilities do not impede women's progress in the sciences. These suggestions have been made at a time in which global recognition is being made for women's "double burden" of paid employment outside the home along with unpaid child care and other household duties, all of which combine to render women's workdays longer than men's in nearly all countries of the world (Blau and Ferber 1992; UN 1995). Among its most interesting recommendations, the Nairobi Forward-Looking Strategies for the Advancement of Women suggested the introduction of programs to enable men to "assume responsibility for child upbringing and household maintenance" (UNCSTD 1995:323). Several others specifically mention the need for day care.[12] Once these conditions have been met, women are likely to gain better access to science education and careers and to remain in these realms for longer than they would without them, but nothing about the science into which they have entered has changed.

More recent events dealing with science at the global level, namely the World Conference on Science (1999), reveals both the continuation of the liberal model for women in science as well as possibilities for changes made by nation-states to support a transformative approach to science. This would involve efforts to increase women's participation in science—not only for the sake of justice in the form of economic development and human rights, but also by placing value on women's knowledge and by proposing a transformation in the way in which science is conceptualized and practiced. Next, we analyze the results of the 1999 World Conference on Science in order to better understand the most recent global discourse on women in science.

THE 1999 WORLD CONFERENCE ON SCIENCE

The Declaration on Science and the Use of Scientific Knowledge (UNESCO 1999a) along with the Science Agenda—Framework for Action (UNESCO 1999b) recognize both the benefits and drawbacks that develop out of a world increasingly dependent on science. It offers recommendations for science and scientific knowledge in four broad headings: science for knowledge; science for peace; science for development; and science for society. The Declaration advances the liberal notion that women's participation in science education should occur without discrimination. In the same section, women's contribution to food distribution and health care are both recognized as well (UNESCO 1999a, sec. 3, par. 34). In the next section, women's equal access

to science is again advocated while the discrimination faced by women in science careers and decision-making processes is also recognized (UNESCO
1999a, sec. 4, par. 42). Direct attention to women in science is made in these
few instances as well as in the following paragraph found in the preamble:

> Most of the benefits of science are unevenly distributed, as a result of
> structural asymmetries among countries, regions and social groups, and
> between the sexes. As scientific knowledge has become a crucial factor in
> the production of wealth, so its distribution has become more inequitable.
> What distinguishes the poor (be it people or countries) from the rich is
> not only that they have fewer assets, but also that they are largely excluded
> from the creation and the benefits of scientific knowledge. (UNESCO
> 1999a, para. 5)

The Science Agenda—Framework for Action also calls for both liberal
and transformative changes to advance women's status in science. In terms of
science education, it asks for the elimination of gender bias in science education and proposes the development of new curricula, teaching methodologies,
and resources that take into account gender and cultural diversity (UNESCO
1999b, sec. 4, par. 41 and 43). As for women scientists, it is recommended
that they be incorporated into the science establishment in both southern and
northern countries (UNESCO 1999b, sec. 1, par. 17).

Furthermore, the Science Agenda goes beyond these changes to advance
recommendations that recognize the importance of women's knowledge and
their contribution to science. It suggests that women be actively involved in
decision-making practices regarding science policy (UNESCO 1999b, sec.
2.6, par. 56). The Agenda also calls on government agencies, international
organizations, universities, and research institutions to "ensure the full participation of women in the planning, orientation, conduct and assessment of
research activities" (sec. 3.3, par. 78). In terms of less developed countries,
women should be recognized as sources of traditional knowledge (UNESCO
1999b, sec. 3.4, par. 86). The significance of such a proclamation must be
understood in light of the fact that traditional forms of knowledge in science
originating from either men or women have typically been viewed with sus-
picion in the modern era. As an illustration, the Western medical establishment remains hesitant to embrace "alternative" or Eastern forms of medicine
despite the growing popularity of such treatments among Americans.

From our analysis of the most recent global debate on science as it took
place at the World Conference on Science, it appears that both the Declara-

tion on Science and the Science Agenda indeed support a liberal notion of women's equity in science. The language found in the Declaration and the Agenda include common themes of recommendations for the removal of discrimination against women in science education and careers so as to bolster their numbers in the field. Furthermore, as evidenced in the Science Agenda, efforts have been made in the direction of transformative policies that advocate the introduction of women's perspectives to scientific research, policy, and knowledge. As the collaboration among scientists, policy makers, and grassroots women's organizers increases, we can expect to see an increasing proportion of declarations and recommendations that transcend the liberal approach to women's equitable participation in science to transform the very nature of scientific inquiry and knowledge.

CONCLUSION

As our analysis reports, the situation of women in science around the world has improved over the past twenty years, particularly in the realm of S&E higher education and to a lesser extent in S&E occupations. In light of the focus of the global discourse on liberal approaches to women in science, such as access to education and jobs and the reduction of gender-based discrimination, it is little wonder that what changes have occurred have taken place in these realms. And yet the results of the World Conference on Science suggest the possibility for future changes. A transformative approach to women in science would include some of the suggestions that resulted from the World Conference on Science, such as the inclusion of women in scientific policy and research and the recognition of women's traditional knowledge, while also going beyond those changes to demand that science be created by and for women. With increased collaboration practitioners of science, policy makers, and theorists, a greater proportion of declarations and recommendations will develop that transcend the liberal approach to women's equitable participation in science and transform the very nature of scientific inquiry and knowledge. Rose (1995) is optimistic that their deep commitment to democracy means that feminists both in and outside the sciences will continue to work together to build alliances and to avoid the confrontations that have plagued the science wars. Indeed, as the pace with which S&T now continues to accelerate, we cannot afford to let science pass women by.

## Chapter 9

# "Styles" of Science

## *Variations in Global Science*

Previous chapters emphasize the commonalities of science in different nations, arguing that such similarities follow from science's worldwide triumph as a universalistic cultural authority. We show, for example, that similar national science policy structures emerge worldwide (Chapter 5), that similar themes penetrate science textbooks worldwide (Chapter 6), and that the internal cohesion among science institutions is similarly structured within groups of nation-states (Chapter 7). Our studies also emphasize that similar factors—those exogenous to the nation-state—are driving the process of global science diffusion. The belief is that science is the key to national economic and social progress that drive diffusion of a common model of science.

Despite these global commonalities, however, variation still exists among nation-states in scientific institutions. Nation-states differ from each other in the intensity of scientific work, the level of governmental commitment to science affairs, and the organizational base of science—to mention only a few dimensions of national science. Similarly, groups of countries differ from each other in the pattern of scientific interaction, their disciplinary emphasis, and the relationship between their science and other national institutions. In this chapter, we describe such cross-national variation and the factors that shape it. We argue that there are several variants of the global model, which we refer to as "styles of science." They are not unique forms of science disconnected from the global model, but rather reflect variations in institutionalization due to particular regional or political processes.

To develop this argument, we first discuss the global model of science, the variations in it, and causes for such variation. In this discussion, we weave together our own empirical examination of national disciplinary emphasis with a review of evidence from other case-based studies on national differences in science infrastructure. Second, we reflect on the related discussions in current literature about cross-national differences in science: the debate about "nationalization and denationalization" pressures, the applicability of the term *glocalization*, and the direction of change in science styles. We conclude by arguing that even localized variations in science are influenced by, and even created by, the world polity, rather than by local interests or particular histories. The general and abstract nature of the global model and the inherent looseness of its worldwide institutionalization enable variations of science to be institutionalized. Thus, to describe the particular variations in science, one needs to also conceive of the core components of the global model of science, which serves as the "menu" for its localized variations.

## THE GLOBAL MODEL OF SCIENCE

Science became organized as a world institution in industrialized Western nations. In previous chapters, we focus on some of the central features of this global model of science. We describe the "science for development" frame as the dominant theme in policy circles, delivering a narrow vision of science as technical tool of sorts for solving development problems (Chapter 4). This theme is carried by scientists themselves, as well as governmental ministries of science affairs (Chapter 5), science education programs (Chapter 6; Drori 2000), national science associations and learned societies (Moon and Schofer 1998), and national agencies for science policy (Finnemore 1993, 1996). The research university is a main national locus of institutionalization of this global model. Although this institution precedes the modern form of science (Riddle 1989), its modern version epitomizes the current-day character of science. More than just a site of professional training, universities serve to institutionalize the content of science by sustaining a particular set of disciplines—institutionalized in the form of departments, schools, and faculty. Universities worldwide, as if by definition, arrange their work into similar sets of fields: arts and humanities, natural sciences, and social sciences, with some distinct professional schools (medicine, engineering, law, education, and more recently business). They sustain a fairly standard set of disciplines: archeology, history, literature, philosophy, and the arts are part of the faculty of the humanities; the natural sciences include chemistry, mathematics, geology, and zoology; and the social sciences incorporate anthropology, political science, and sociology. A few universities pepper this recognized format with avant-garde disciplines or indigenous science (such as Islamic math). Overall, however, the organizational structure of universities reflects the general and broad nature of the model of science, supported in a wide range of departments.

A cross-national study of universities reveals that this pattern of disciplinary structure is consistent worldwide.[1] A review of publication patterns by scientific discipline, based on ISI/SCI counts, confirms that countries worldwide publish scientific articles in a wide range of disciplines.[2] Contrary to notions of a "scientific niche,"[3] countries worldwide tend to generate publications in *all* major disciplines. According to our data, only a handful of countries, all of which are poor ones, tend to have no publications (per year) in the core scientific fields. It is only in a refined breakdown of scientific publications per discipline that a substantial number of countries fail to produce work in the full array of scientific disciplines.[4] As presented in Table 9.1, countries of all sorts still list publications in all major disciplines identified and compiled by the ISI/SCI.

The global model of science is clear: a recognized set of core scientific disciplines, organized into a recognized set of core organizational formats, producing a recognized set of scientific outputs, and performed by people carrying recognized titles. In this case, the medium is also the message—meaning that the set of science indicators found consistently worldwide signals not only the industry of indicator production, but also the dimensions, or features, of the global model of science.

STYLES OF SCIENCE

Yet as much as the core features of the global model of science are incorporated worldwide, their incorporation is not wholly uniform. Countries differ in their research priorities, science–industry relations, intensity of scientific interaction, and levels of productivity. This phenomenon is not national at its core; it is related to distinct groups of nations. Clustering countries along scientific dimensions identifies global differentiation into styles of science.

Although the Western model of science is quite broad—including social sciences and conceiving of science within the framework of progress-oriented

TABLE 9.1

Identifying Scientific Research Style: Patterns of Research Output
Concentration by Geopolitical Bloc, 1982[a]

|  | *Physics, Chemistry* | *Biology, Clinical/ Biological Medicine* | *Math* | *Engi- neering* | *Earth Sciences* | *Social Sciences* |
|---|---|---|---|---|---|---|
| Developed | 52,824 | 118,507 | 4,416 | 15,550 | 7,765 | 52,277 |
|  | 21% | 48% | 2% | 6% | 3% | 21% |
|  | 68% | 82% | 78% | 80% | 71% | 93% |
| Communist | 14,240 | 7,477 | 507 | 1,443 | 1,354 | 770 |
|  | 55% | 27% | 2% | 6% | 5% | 3% |
|  | 18% | 5% | 9% | 7% | 12% | 1% |
| NICs | 1,007 | 1,424 | 83 | 166 | 118 | 329 |
|  | 33% | 45% | 3% | 5% | 4% | 10% |
|  | 1% | 1% | 1% | 1% | 1% | .5% |
| Other non- west | 9,053 | 16,448 | 654 | 2,262 | 1,636 | 2,973 |
|  | 28% | 51% | 2% | 7% | 5% | 9% |
|  | 12% | 11% | 12% | 12% | 15% | 5% |

[a]Data in table are: total paper publications, row percent, and column percent. Percentage
information does not necessarily add up to 100, as a result of rounding error.

and socially relevant agenda—we find variations in the application of this
model. That is, groups of nations emphasize particular subsets of this broader
model. We refer to these variations as "styles" of science. Science styles are the
scientific structures and practices that take a unique form along national or
geopolitical boundaries. The identification of these patterns refers to the
breadth of science in different groups of countries.

## Constructing an Indicator for Styles of Science

Analysis of ISI/SCI publication data reveals heterogeneity of disciplinary
emphasis of nations in different regional or geopolitical groups (see Frame et
al. 1977 for an earlier analysis of such data). Table 9.1 presents such evidence
of patterns of scientific research output by various national groups based on
data in 1982.[5] We focus on 1982 as a snapshot of the cold war era. Post–cold

war trends are discussed later. Our choice of country groups (for example, industrialized West, communist bloc) has a strong basis in previous research and in exploratory cluster analyses.

Table 9.1 shows subtle but distinct differences in scientific disciplinary emphasis across groups. Developed, (mostly) Western countries disproportionately concentrate their research efforts in the disciplines of biology, biomedicine, and clinical medicine. In contrast, communist countries focus more on the traditional natural sciences: physics, chemistry, and mathematics (see also Kristapsons et al. 1998). Newly industrialized countries (NICs; newly industrialized Asian and Latin American economies) exhibit a mixed pattern, focusing their relative efforts on both the traditional "hard" sciences and the medical sciences. Last, other non-Western countries (such as Middle Eastern or Caribbean nations) follow a pattern similar to both Western countries and the NICs. Western countries are unique in their relative emphasis on the social sciences (economics, psychology, and a general category of social sciences), consistent with the findings of Lange (1985). Although countries worldwide produce publications in the full set of scientific disciplines, groups of countries still vary by their emphasis on particular scientific disciplines.

We emphasize two important points. First, scientific styles differ by sets of nations delineated by political affiliation, region, trade relations, and developmental state. Second, to a large extent, science styles reflect variation along one particular dimension—between a focus on "hard" versus "soft" science disciplines. Styles do not refer to a wholly different package of science. Rather, the full range of scientific disciplines is represented in each and every group of nations, but with slightly different emphases.

*The Factors That Shape National Science Styles: A Regression Model*

The main dimension of national variation in science, clearly evident in factor analyses, runs from national emphasis on hard science (for example, physics, chemistry) to national emphasis on soft science (for example, social science, biology.) We examine a number of factors that determine a country's scientific emphasis on both hard and soft science. The dependent variables, national emphasis on hard sciences (mathematics, physics) and soft sciences (clinical medicine, psychology, economics),[6] are calculated as the relative share of the publications in these disciplines out of total publications. In the following series of ordinary least squares regression analyses (Table 9.2), we

TABLE 9.2
Factors Determining Science Style:
Results of OLS Regressions, Circa 1980[a]

| Model | Independent Variables | Emphasis on Hard Sciences | Emphasis on Soft Sciences |
|---|---|---|---|
| | | *Dependent Variables* | |
| 1 | GDP | .02 | .01 |
| | Membership in INGOs | .02 | .05** |
| | | Constant = −.13, $R^2$ = .06, n = 75 | Const = −.10, $R^2$ = .07, n = 75 |
| 2 | GDP | .04 (.01) *** | .01 (.01) |
| | Democracy scale | −.01 (.01) *** | .01 (.00) *** |
| | | Constant = −.15, $R^2$ = .13, n = 74 | Constant = .19, $R^2$ = .11, n = 74 |
| 3 | GDP | .03 (.01) ** | .01 (.01) |
| | Protestant West (dummy) | −.03 (.03) | .10 (.04) *** |
| | Socialist Europe (dummy) | .09 (.03) *** | −.13 (.04) *** |
| | NICs (dummy) | .06 (.03) ** | −.07 (.04) ** |
| | | Constant = −.11, $R^2$ = .23, n = 75 | Constant = .20, $R^2$ = .30, n = 75 |

[a]Data in table include coefficient (standard error) and significance level.
Summary statistics: constant (two-tailed test), adjusted $R^2$, and number of cases.
*$p < 0.10$, **$p < 0.05$, ***$p < 0.01$

evaluate the effects of various national features: cultural heritage (dummy variable indicating affiliation with protestant Europe), political affiliation (dummy variable for communist Europe), economic restructuring (dummy variables for the NICs), political system (democracy scale; see Gurr 1988), and embededdness in the world polity (membership in international organizations). Finally, we add an indicator of economic development (gross domestic product per capita) to all models as a control variable because national economic resources affect scientific capacity (Cole and Phelan 1999) and might also affect disciplinary emphasis.

National emphasis on the hard or soft sciences is consistently influenced by national characteristics in an either/or fashion.[7] In other words, national characteristics that are positively associated with hard science tend to be negatively associated with soft, and vice versa. For example, democracy has a positive effect on soft sciences and a negative effect on prevalence of the hard sciences; similarly, being among the socialist European countries or NICs increases the prevalence of the hard sciences style and lowers the prevalence of the soft sciences style.

Results show that an emphasis on the hard sciences is prevalent in the previously communist countries, newly industrializing countries, and in countries that are less democratic.[8] On the other hand, emphasis on soft sciences is prevalent in countries that are more democratic, members of the protestant West, and nations that are strongly embedded in networks of international associations. National wealth is associated with increased emphasis in the hard sciences but has no significant effect on the soft sciences.

## Summary

We highlight the existence of science styles through regression analyses of national research emphasis on certain scientific disciplines rather than others. Moreover, we show that the preference of a particular science style is related to regional characteristics, levels of democratization, cultural heritage (religion), embededness in world society, and affiliation with geopolitical alliances (that is, communism).

In additional analyses (not presented here), we found little support for competing arguments about national differences in science. Other factors, commonly mentioned in the literature, include isolation (due to either autocracy or resource breakdown), local inertia specific to science (due to strong forms of indigenous science, such as Muslim science), early institutionalization of science (unique forms due to colonial history or the specific colonizer), or the organization of local science to occupy a niche position (for instance, developmentalism might encourage scientific uniqueness in third world, as in the case of soil sciences). Neither our descriptive analyses nor our regression models found support for these claims.

Interestingly, however, we find that *all* countries publish articles in the *full* range of scientific disciplines, even if to varying degrees. This pattern is repeated even when a more refined breakdown of disciplinary activity is used. This pattern suggests that all countries are drawing from a taken-for-granted model of science, which describes the range of legitimate scientific disciplines. These disciplines are among the core components of the general global model of science.

## CONTEXTUALIZING SCIENTIFIC STYLES

Our analyses find four distinct groups of nations: developing countries, communist countries, the NICs, and the Western core. Now we turn to a discussion of the particular national groups and reasons for their distinctive styles of science. Specifically, what is the local context for the variants, or science styles, that we identified?

### Developing Countries

Most often, studies focus on the science gap between developed countries and developing countries. Traditionally labeled "core versus periphery" by such notable comparative science and technology studies (STS) scholars as Joseph Ben-David and Edward Shils, these two blocs of countries were conceived as two archetypes within the international science community. The scientific core, located in the developed countries, is the global center of scientific training, basic research, and funding. On the other hand, the scientific periphery—much like in economic and political terms—is resource poor, dependent on technology transfer, and looks to the core for funding, training, and new knowledge (Gaillard 1991; Eisemon and Davis 1992; Gaillard and Waast 1992). This asymmetry between core and periphery in global science also influences the pace of science diffusion. As described in Chapter 5, the process of institutionalization of ministries for science affairs differs between core and periphery countries. Whereas in the core countries the institutionalization of science ministries occurs earlier and this process is shaped by the characteristics of the local country, in developing countries, the process occurs later and is determined by their connectedness with the international community.

The core/periphery divide is made greater first by the policy guidelines offered by development-oriented international organizations, and second by the tendency toward loose coupling in science in developing countries. First, the agenda of international organizations powerfully shapes science in countries with weak states and limited resources. The United Nations Education, Scientific, and Cultural Organization effectively dictates their national science and technology policies (Finnemore 1993, 1996) and frequently suggests the development of niche strategies for the sciences. As a result, much of the science of developing countries is biased toward their imagined relative advantage, such as soil sciences or zoology (see Arvanitis and Chatelin 1988), generating variation from the global model of science. Moreover, science and technology policies suggested by the United Nations Commission on Science and Technology for Development are predominantly oriented toward the public sector (see Tiffin and Osotimehin 1992:106); for example, it is government representatives who participate in their conferences and workshops, and the funding review process favors sponsorship of public science and technology projects. This policy direction is unshakable, in spite of more recent efforts to implement Triple Helix[9] policies in developing countries (Etzkowitz and Leydesdorff 1997). A major obstacle to the implementation, or bridging, of such policies into the African context is the lack of infrastructure. In the absence of developed infrastructure science–research–industry linkage is hardly imaginable—and unlikely to be fruitful (Strath 1998).

Second, although the build-up of science and technology in Africa seems impressive at first glance (showing several pan-African scientific associations, national universities, and scientific publications), there is little action behind this façade (Forje 1988:242, 245). This is a classic example of loose coupling: policy is highly developed and institutionalized, but little action follows (see Chapter 7). Such decoupling persists with surprisingly little conflict. The introduction of the global regime of "climate change" to Senegal serves as a nice illustration of this point. Senegal was traditionally concerned with forestry, but this practice was redefined in the 1990s in terms of global environmentalism. As a result, Senegal is currently invoking the language of "climate change,"[10] but the brunt of its work is still executed by geologists concerned with coastal preservation (Engels 2000).

One explanation of this pattern of loose coupling is lack of resources. It is cheap for developing countries to produce policy guidelines, yet it is costly for them to sponsor actual research, technological innovation, or industrial infrastructures. Resources also explain the general gap between developed and developing countries in level of scientific output (Cole and Phelan 1999), thus defining resource-poor and resource-rich countries as distinct blocs (Schubert et al. 1983). This is supported by our empirical analyses: the consistently significant effect of gross domestic product on hard sciences emphasis is probably due to the high expenses of hard sciences activities (laboratories and long experimentation periods) and of the related preference for "big science" activities that are also costly.

Alternative explanations stress the importance of colonial legacies in establishing the backward conditions of science in developing countries (Sagasti 1973; Mazuri 1975). Yet Kenya, Nigeria, and Senegal, for example, are the most scientifically advanced sub-Saharan countries, with relatively large research communities, relatively well sponsored scientific organizations, and a history of substantive research contributions. The different colonial legacies in both these countries—Britain and France as colonizers, respectively—suggest that the cause is not the colonizer per se. Rather, colonial ties link colonies to the world polity. In other words, it is the general level of connectedness with the world polity that shapes the scientific trajectory of peripheral countries. Again, our analyses confirm that such connectedness increases the probability of a country to adhere to the science style that emphasizes soft sciences.

*Communist Countries*

Another historically important demarcation of scientific styles is between the liberal West and the communist world. The communists developed an alternative variant of science, which—because of their isolation from the Western capitalist world polity—remained distinct. Communist countries are marked by a bounded network of scientific interaction (for example, participation in scientific conferences and travel); relatively low intensity of scientific work (Schubert et al. 1983); fewer international awards for scientific achievements (for example, citations and Noble Prize laureates); dramatically less exposure of their science work (for example, publication in foreign journals; Schott 1992a); and dramatically greater proportions of scientists and engineers, both in training and in the labor force[11] (see Lubrano 1993). Also distinct is the Soviet concentration on the traditional natural sciences and on engineering. This disciplinary concentration, also identified by Frame et al. (1977), is confirmed in our analyses: communist countries distinctly emphasize the hard sciences rather than the social sciences in terms of publication rates, higher education enrollment and degrees, size of the scientifically and technically trained labor force, and so on. To this description of disciplinary emphasis, Lange (1985) adds that substantial differences between the West and the Soviet bloc in terms of the *nature* of their social sciences: Soviet bloc countries publish less in these disciplines, and the published articles carry a propaganda-like tone. The area of least chasm between these two blocs is citation: probably as a result of the intense competition, there is a high rate of cross-citation across the political iron curtain (Schott 1992a:428). Organizationally, communist-style science is highly centralized, concentrated in the hands of the Soviet Academy of Sciences. The Academy centrally controlled both funding and promotion, and it also shaped scientific production, mainly by censoring its dissemination and by mixing formal criteria for evaluation with political judgment.

The collapse of the Soviet system in 1989–1991 brought dramatic changes to communist science, reducing the distinctiveness of science in the region. First and foremost, the breakdown of the centralized system meant some relative freedom of the system. As a result, previously communist countries dismantled the soviet-style science academies in an effort to return basic research to the universities and permitted an expansion of a private science system. This massive expansion of tertiary-level education is marked by the incorporation of previously nonexistent fields (such as management) and the dramatic expansion of previously marginalized fields (law, economics, psychology, and other social sciences); "The best of these new institutions do not emphasize natural sciences," report Dezhina and Graham (1999:1303). This shift in disciplinary concentration is also reflected in a new publication pattern. The Baltic states of Latvia, Lithuania, and Estonia, for example, dramatically increased life science publications in a shift away from a Soviet-era concentration on the hard sciences. Also, there is a post-1990 decline in articles authored by Baltic scientists in Soviet scientific journals and a rise in Baltic publications in Western, mainly American, journals (Kristapsons et al. 1998). Second, structural and economic changes resulted in a sharp decline in job opportunities for scientists, in increased reliance on foreign—mainly private—funding of research projects, and in massive efforts to convert Soviet military technology into civil and marketable uses. These dramatic changes are still ongoing.

### The NICs

Many countries in Asia and several in Latin America share a development-focused, authoritarian polity. Such polities emphasize the technically applicable sciences while keeping science from expanding much into the socially relevant domains. As a result, the NICs' style of science is marked by disciplinary emphasis on engineering and the hard sciences, as demonstrated in our analyses. These countries rely heavily on technology transfer and orient their local research toward application, mainly industrial production. Science–industry relations, although put forward mainly by the state, are increasingly reliant on private and corporate sponsorship. These countries are also known for strong student performance in math and science, as measured by international tests.

This is thought to be the result of strong "science for development" ideology coordinated centrally by the state and built directly into education and scientific institutions. This leads to massive investment in scientific human capital but little basic research—particularly in fields such as psychology and medicine (see Chapter 10).

## The West

The Western core is unchallenged in its status as the center of world science. Basic research, citations, and elite scientific education remain concentrated in the core—mainly the United States. Furthermore, the corporate structure of multinationals mirrors and enhances this divide: "Despite various transnational flows, only 10% of U.S. corporate R&D funds are spent overseas. Motorola, for example, derives ½ its revenue from international sales and stations 40% of its workplace abroad, but only 20–25% of its product development and 5% of its basic research are conducted outside the U.S." (Andrew Pollack, "Technology without borders raises big questions for the U.S.," *New York Times*, January 1, 1992, sect. 1, p. 1; also see Mittleman 1997:19). The Western style is characterized by focusing on basic research, potent science–industry relations, and intense commodification of academic work. In terms of disciplinary emphasis, our analyses confirm the findings of Frame et al. (1977): liberal Western countries emphasize research in clinical medicine and biomedical research; they also expand their science domains to include the social sciences, thus showing marked emphasis on such disciplines as economics and psychology. This disciplinary emphasis relates Western science to the legacies of the Western polity: both domains reflect a notion of an agentic actor (see Chapter 1), focusing their attention on the social unit, its care, and its characteristics.

## Summary

Table 9.3 summarizes the major differences in science styles across groups of nations. Both qualitative and quantitative research support these general distinctions. To be sure, exceptions exist. Nevertheless, these four groups of science styles powerfully summarize global variation. These archetypical styles of science also clearly map onto the distinct polity types: the liberal core, the statist postcommunist bloc, the authoritarian NICs, and the impoverished and dependent developing countries.

TABLE 9.3

Identifying Scientific Styles

| Country Characteristic | Exemplar Cases | Science Style |
|---|---|---|
| • Developed/ Core<br>• Liberal<br>• Old | West | 1. Disciplinary emphasis on clinical medicine and bio-medical research<br>2. Science production: intense, varied, formal criteria for evaluation<br>3. Science interaction: intense, organized around common language or common problems, highly competitive<br>4. Research priorities: basic research, commercial R&D<br>5. Science–industry relations: intense, commodification of academic work<br>6. Local science field: highly institutionalized, coherent, mostly decentralized |
| • Statist<br>• Centralized | Communist bloc | 1. Disciplinary emphasis: natural sciences (chemistry, physics, math)<br>2. Scientific interaction: low levels, only within member nations, centralized<br>3. Scientific production: high, sometimes censored and secretive, formal criteria for evaluation mixed with political judgment<br>4. Local science field: highly centralized in the hands of the Soviet Academy of Sciences<br>5. Dramatic changes during post-Soviet era: massive expansion of higher education; decline in hard science emphasis and rise in share of social science fields; sharp decline in work opportunities for scientists; reliance on funding from foreign, mostly private, sources; reorientation of applicability from military uses to industry |
| • Authoritarian | Asian Tigers, Latin America | 1. Disciplinary emphasis: engineering<br>2. Scientific interaction: consolidated around common problems or regional location<br>3. Research priorities: focus on applications<br>4. Science–industry relations: emerging private sponsorship<br>5. Local science field: set as workshops of the West<br>6. Tensions due to dual embeddedness |
| • Less developed<br>• Peripheral | Africa, weak Latin American countries | 1. Scientific production: low, no formal criteria for evaluation<br>2. Scientific interaction: weak to non-existent; all contacts are with, or through, the core<br>3. Science–industry relations: weak, lacking mediating structures<br>4. Local science field: established institutional arrangements, yet lacking "real" work; intense loose-coupling |

SOME OPEN QUESTIONS

*Denationalized versus Global Science*

Current discussions of variations in science describe the global/local tension as opposing trends of nationalization and denationalization of science (Crawford et al. 1993; Elzinga and Lanstrom 1996). To emphasize the trends of science nationalization, they cite the reliance of science on public and local funding, its allocation according to national priorities, and the fascination with global ranking of national scientific performances as evidence that science is perceived as a national institution and a national resource. Such perception motivates the recent shift from regional science institutions to national ones and the action taken by nationalist and multiculturalist social movements to encourage the institutionalization of localized (national, ethnic, tribal) sciences, thus calling further attention to the fusion of science with local themes. As a result, attempts to "nationalize" science produce such action as the adaptation of science textbooks to local themes: illustrations in Ghanian textbooks show black-skinned people working, and Hungarian physics textbooks include a chapter about renowned Hungarian physicists (Chapter 6).

On the other hand, comparative studies of science emphasize that international scientific collaboration expands and intensifies, effectively denationalizing science. Big science ventures (such as space research) are mostly international because of their high cost; science policy is increasingly coordinated, especially in Europe, because its implications are cross-border, and transnational coauthorship intensifies because of intensifying international collaboration. In addition, the dominant professional norms still claim science to be an international venture (Merton 1942/1973).

However, to describe the intensifying trends of international cooperation as "denationalization" and to set them as an opposition to "nationalization" is rather deceiving. Empirical evidence supports the existence of both global pressures and local variations in science. We argue that these processes are complementary rather than contradictory.

*Does a Style Constitute a Form of Glocalization?*

Roland Robertson's (1990, 1992) intriguing term *glocalization* was coined to describe observed forms that are often a mixture of the global and the local—when local forms are infused with global themes and global models are altered to incorporate local flavors. In this sense, a style of science *is* a glocalized form of the global model of science: it is a particular combination of components, selected from the variety of the "package of science." The NICs' style of science, for example, is an adoption of those components of the global model of science that pertain to its technical applicability (the traditional natural sciences and engineering) while curbing the expansion toward the more socially relevant science domains.

Yet the literature about glocalization implies the added notion that local variations involve mobilization on behalf of uniqueness and localism; in other words, it is implied that action is to be taken to align the global model with local traditions. That element is not apparent in our research. Regional styles of science do not build on truly localized models of science and knowledge. Truly "local" or "indigenous" forms of knowledge certainly exist—but they are not the ones diffusing rapidly around the world. Rather, the successful "indigenous" forms of science (for example, ethnobotany) are coordinated by credentialed Western scientists and international agencies and thus infused with the global model of science. For example, Allwood (1998) describes the attempts to advance local forms of psychology and to forbear Western psychology, detailing the assistance given to local organizers by the International Association of Applied Psychology and the International Union of Psychological Science; even this description of local resistance efforts clearly demonstrates the influence of the global machinery in the internationalization of indigenousness. In general, glocalized institutions, whose institutionalization involves the mobilization around issues of uniqueness, are heavily coordinated by international bodies that propagate the global model of science.

The term *glocalization* usefully characterizes the hybrid nature of global models as they are modified within local contexts. But it is important to recognize how little the "local" is uniquely local in most instances. In the case of the diffusion of modern science, the global similarities in "science styles" vastly outweigh the differences.

## The Dynamics of Styles

Our analyses identify distinctive patterns in regional styles of science based on a snapshot of the 1980s. But styles of science clearly evolve over time with changing conditions—for example, changes of the world hegemon and break of colonial ties. Home and Kohlstedt (1991) describe Australian science, changing its reference point from Britain to the United States. In the process,

Australia changed its conception of itself from being a laboratory and research field for British science to gaining prominence and visibility on its own, from focusing on marginalized taxonomic work to an integration into the international web of professional science with some notable achievements in certain disciplines. Similarly, the collapse of the Soviet bloc caused the reorientation of science in the previously communist countries toward the forms of science practiced in the Western core (Kristapsons et al. 1998; Dezhina and Graham 1999).

Although evidently countries change their style of science and reorient their scientific system toward the current-day hegemon, such changes are rarely comparatively investigated. Most recent studies acknowledge the intensifying centrality of American science, but they fail to comparatively describe the trends. Also, it is still unclear whether countries turn to American science for information, training, and recognition or whether they also emulate American science in their own countries. Clearly, American science is highly attractive in terms of training and publications, yet in organizing their national science fields, most countries apparently do not emulate the United States. Whereas in the United States such matters are extragovernmental affairs left in the hands of both public and private associations, most countries regard science policy as a governmental responsibility (see Chapter 5), thus not fully emulating the form of American science.

The question of changes in a country's science orientation, or style, is clearly an empirical issue. A related empirical question is the direction of such change in orientation: does it move toward global uniformity or in diverging directions? Evaluating the possibility of global convergence toward a single style of science (or, alternatively, the persistence of national, or geopolitical, differences) can be studied comparatively. A quick look at measures of heterogeneity over time (see the coefficient of variation for various science indicators, 1950–1989, Table I.1) reveals contradictory results. Evidently, there is a general trend toward convergence in the cross-national rates of membership in international science organizations and in science enrollment ratios in tertiary-level education, yet persistent cross-national differences (if not divergence) in scientific publication rates and in the labor-force share of scientists and engineers. This preliminary evidence may suggest that any trend—convergence or divergence—is highly sensitive to the choice of indicator of science style. Further research is needed to resolve the issue and current trends in styles of science.

CONCLUDING COMMENTS

In referring to the tensions between global and local, much of the current literature emphasizes the dilemmas facing individual scientists. Kenyan physicist S. Kionga-Kamanu, reflecting on his own work, notes, "There is little glory or recognition for improving a 'jiko,'[12] for instance, compared to work in the area of nuclear physics. It is obvious, however, which of these efforts has more direct applicability and potential use for the people in Kenya" (cited in Bruchhaus 1985:26).

Erroneously, the struggle of scientists with the issue of orientation is conceived as a matter of choice, either by national governments or by scientists themselves. Also erroneously, whereas scientists themselves struggle with issues of dependency (for example, Mazuri 1975), STSers formalize this conflict by searching for criteria for scientific independence (Basalla 1967). They posit "globalization" as opposed by "resistance." The common explanations for diversity emphasize national interests, coercion, or external dependence, thus imagining sets of push/pull forces. This is a dramatization of a much more subtle process of mutual influence.

Globalization pressures do introduce countries to a common global model. This global model is, however, met with different types of local environments—and varying levels of local institutions that may resist (or transform) global models. The metaphor is "punching jelly or punching rock," referring to the stability and strength of the local forms that may slow the process of adaptation of the global model and probably increase the likelihood of glocalized reinterpretation of the model. Such local models can be found in the groups we identified: communist ideology, the development-centric authoritarian regimes of the Asian Tigers. These local models act as filters of the global model of science, creating regional variants.

Glocalization is accompanied by loose coupling, which describes the disconnection between practice and policy or among spheres of activity. It is loose coupling that allows the institutionalization models distinct from the "standard" global form. Senegalese policy makers, for example, promote a change in their nation's approach to environmental problems so as to reflect the new-age concerns and sensitivities, yet local professionals continue to practice forest and coastal preservation methods as they did before the triumph of global environmentalism in the early 1990s (Engels 1999, 2000). This disconnect is sometimes justified in the name of unique local circumstances in countries not (yet) fully capable of emulating the science as institutionalized in Western countries. The use of pick-and-paste strategies (see Chapter 7) is similarly justified, imagining that only some components of the global model of science are transferable or that only some components of the global model are appropriate for the local context.

Second, glocalized forms of science are enabled by the general and diffuse nature of the global model of science. The global model carries enough clout as a general proposition that the institutionalization of any of its components is considered virtuous—whether they be national policy agencies, universities, national professional science associations, science education programs, or labor-force policies for scientists and engineers—even if they are incomplete or if they differ slightly from global models and blur with "local" scientific forms. It is the commitment to the general picture of science that accounts for science's global expansion. But the global model is sufficiently decentralized and diffuse to allow for glocalization in different social context.

Part IV    The Impact of Science
Globalization

# Introducton to Part IV

Part IV shifts the discussion away from the factors that contribute to the accelerated process of science globalization, as is the focus of Parts II and III, toward a description of the consequences of the process of science globalization. Here we consider the societal effects of the global expansion of science, focusing particular attention on the contribution of science to economic growth (Chapter 10), to a growing consciousness of environmental conditions (Chapter 11), to the diffusion of democratization reforms (Chapter 12), and to the standardization and rationalization of national decision making (Chapter 13).

Although current literature on this subject is still immersed in economics when explaining the exponential trends of science expansion or when investigating the effects of science globalization, we offer an alternative sociological explanation for these matters. As we argue in Parts II and III, the worldwide expansion of scientific activity reflects the expanded authority of science and the nature of the current world polity, rather than arising from instrumental functions, powers, or interests within society. This is also true when considering the consequences of science globalization. In Part IV, we show that science affects not only economic growth (Chapter 10), but also broader notions of social progress. We argue that although science is designed to be chiefly a tool for achieving narrowly defined development, it also has a wide range of additional effects on the societies that are subject to its globalization. Such effects extend to a wide spectrum of modernizing institutions: political structures, human rights conditions, social welfare, environmental concerns, and transparent social structures.

*217*

In problematizing the conventional, econocentric notion of development (Chapter 10), and in relating science to the worldwide institutionalization of environmentalism (Chapter 11), rights discourses and democratization (Chapter 12), and rationalization of decision making (Chapter 13), we connect science not to its intended consequences but to its source—namely, the progress- and justice-oriented global model of science. Overall, the effects of science on the nation-states that incorporate it into their institutions are varied. The affected social spheres range from the economy to the environment to the political system to procedures of governance and management. These reflect an even broader notion of what constitutes social well-being: economic prosperity, social welfare, environmental protection, political participation, and accountability/transparency of public organizations. Our findings support recent international initiatives to redefine social progress and redirect it away from the current exclusive pursuit of economic development. Yet such initiatives are new to the fields of science policy and of science, technology, and society. In this sense, our findings show science to be a central feature of the worldwide move toward a new interpretation of social progress.

Two central themes connect among the various social effects of science globalization: ontological constitution and rationalization. First, science globalization contributes to the constitution of scientized entities. Science penetration encourages the construction of a scientized environment (Chapter 11) and a scientized citizen (Chapter 12). The constructed agentic actor is thus a scientized actor. In other words, the definition of "the actor" as a bounded, distinct, or unique entity relies on scientific "evidence." Social actors initiate such scientific confirmation of their distinctiveness, thus drawing on the authority and legitimacy of science to constitute their own. Environmental activists, aiming to encourage support for their social cause, draw on scientific research and bandy about sciencelike apocalyptic visions of environmental degradation. Similarly, ethnic rights activists rely on anthropological evidence to support their claims for ethnic distinctiveness. Overall, these social actors scientize their appearance and definition to constitute their entitlement and solidify their assertions.

Second, science expansion leads to the rationalization of national practices and institutions. Such rationalization is manifested in the appearance adopted for various social practices and accounts. Political leadership, for example, is reflected in polling results; similarly, consumer choices are accounted for by measures of satisfaction with, and usefulness of, the products purchased. Rationalization is thus the use of codified science-based measures to account for social behavior. Rational action is also exemplified in a greater tendency toward standardized action. Rational nation-states follow international law and adopt the institutions of a "proper" state; rational corporations follow the guidelines acceptable by the business community in their strive for success; and rational individuals use the language of costs-versus-benefits calculations when accounting for, or planning, their actions. Rational actors thus take appearance of "proper" actors, and such appearance is most often a standardized appearance. Standardization is thus the mediator between scientization and proper actorhood. Being a proper actor is currently a central component of the definition of social progress.

How is science affecting social progress? What are the mechanisms connecting between science globalization and the worldwide trends toward environmental protection, democratization, and standardization?

Realist theories (modernization and dependency theories) attribute this effect of science on its social environment to the efficiency of the outcomes, thus focusing—and mythologizing—the stand of science as an education and socialization tool. Yet other than agreeing on the nature of the mechanism, these two theories offer opposing predictions as for the direction of such science effect: modernization theory sees science as a meliorist institution, whereas dependency theory focuses on its manipulative and repressive qualities. These theories are currently immersed in a debate over the technicalities of the modeling (see Drori 1998, 2000). Most importantly, the technical nature of this debate overshadows any questioning—and explaining—of the basic conceptual tenets: why science? Why focus on economic benefits? Why such divergent findings?

Essentialist theories—by definition—attribute these effects of science on its environment to the inherent qualities or characteristics of the institution of science. They refer to science as inherently reflecting an actor-based mentality and a rational attitude. Yet if science inherently encourages activism and thus social participation, what, for example, would the cases of Nazi Germany and Soviet science (where extraordinary scientific leaps coincided with total disregard for political rights, environmental protection, or social welfare) teach us? These cases show that the relationship between science and social progress is historically specific, or valid only during the later part of the twentieth century. It also shows that essentialist explanations are not well founded.

Alternative explanations that focus on science as a professional group emphasize the role of scientists as the leaders of social reforms. Scientists holding advisory positions directly shape procedures of governance and social reforms (Jasanoff 1990). In this sense, science—being a highly esteemed professional group—is the source of the notable effect on social progress. This explanation is still ahistorical, yet may be appropriate for the current period, when professionalization processes are most intense. It also does not account for the process that led scientists to be a powerful and authoritative professional group, thus ignoring the embeddedness of science professionalization in the context of the modern world polity.

Last, science may be perceived as a networking institution that "channels" global cultural trends (or policy fashions) into even remote parts of the globe. Practiced as a global venture and hailed as a disinterested and universalistic profession, international scientific interaction paves the way for the penetration of national society by global customs. Science institutionalization is thus the carrier of global models and modes. Yet if the power of science draws only from its international rationalizing qualities, then how is science different than any other internationally praised professional activity that is globally networked, such as law? Science is distinct from other network-based professions because science faces little opposition to its globalization. Although law is assumed to be more directly linked with political agenda, the only opposition to science (as in, for example, the Islamic countries) is in the name of "true science." Nation-states worldwide welcome the bridging of global customs into national practices by science because they assume all such science-related customs to be economic in nature. Yet as is evident in the research offered in the following chapters, science-related networking mechanisms also bridge additional, noneconomic global customs into national practices. Such additional practices—although reflecting the current global liberal agenda—are unintended consequences of science globalization. Science thus carries broadly defined modes of social progress, yet it does so unintentionally as a result of its presumably meliorist networking.

The authority of science as a vehicle of progress, narrowly defined in economic terms, expands its authority as a carrier of social progress, broadly defined. Drawing on its legitimacy as a tool for economic development, science globalization encourages noneconomic changes in nation-states worldwide. Science introduces not only modes of operations, but also an alternative Western-based rationale. In this sense, science holds some of the properties of a religion, offering a secular "sacred canopy" and a rationale for modern social order, positioning humans within their environment, and granting an alternative mode for collective authority.

## Chapter 10

Science and National Economic
             Development

We now turn to the issue of science's impact on national economies. It is taken for granted that expanded national investments in scientific activity have beneficial effects on economic growth, primarily through improved technology and a more efficient labor force.[1] We expand on this simple story by locating it within the context of a globalizing rationalistic world culture. The assumption that science is a "goose that lays golden eggs," which is very much a part of world culture, has accelerated the global spread of science. Policy professionals and major international organizations (for example, the United Nations) routinely advocate expanded national research investments (OECD 1971; Spaey 1971) and expanded educational instruction in science and technology (UNESCO 1979; OECD 1971; Caillods et al. 1996), leading to rapid globalization (see Chapters 4 and 5). Science now expands world-wide as a stylized policy package and may not necessarily be effectively integrated into domestic economies—undermining its effectiveness in practice. Moreover, as part of a more general rationalistic, progressive world culture, science may help foster conceptions of development and progress that affect national economies.

Science has tremendous authority and legitimacy in modern society, and thus scientific knowledge and scientific depictions of the world become important bases for justifying various new social policies and encouraging the political regulation of economic activity. Medical research findings, for example, have supported progressive worker-safety agendas since the nineteenth century, leading to increased regulation of industry. Likewise, environmental science (for example, Rachel Carson's *Silent Spring*) galvanized the environment movement, helping to create an extensive regime of environmental regulations and new reporting requirements that substantially increase the short-term costs of doing business (Caldwell 1982). Many would agree that gains in health, safety, and environmental protection are worthwhile, even at the cost of lower levels of economic expansion. Nevertheless, science plays a role in changing economic activity that is not often discussed. In adopting and institutionalizing the Western science "package," nations around the world may be importing a mechanism that encourages similar tradeoffs, producing some measurable negative effects on levels of national economic growth. We develop these ideas and then turn to an empirical analysis of the effects of national scientific infrastructures on economic growth over the period 1970 to 1990.

THEORETICAL BACKGROUND

In the contemporary period, the dominant image of science is instrumentalist. Science is thought of as a tool for socioeconomic development. Scientific activity can produce improvements in human health and life, as with medical knowledge. It can improve competitive social effectiveness and technical efficiency in firms and in countries. It can improve societal relationships with the natural environment. Such notions are taken for granted in both theory and policy, whatever private doubts modern individuals may have about it.

Two main causal paths are generally envisioned: first, scientific training is thought to be a valuable form of human capital that improves labor-force efficiency and productivity (Schultz 1961, 1963; UNESCO 1979; Psacharopolous 1973, 1984; Jones 1971). Thus, for instance, the weak performance of American students on science tests or dropping science enrollments are thought to forebode weakness in American performance in the world economy. Second, scientific research produces knowledge, innovation, and technical adaptation that, it is understood, will improve socioeconomic efficiency and generate new products (Williams 1964; Matthews 1973; see Drori 1993, 1998, 2000 for a discussion of this view).[2] Again, competitive lapses (for example, in research spending, new patents) are thought to portend future economic decline.

Thus, two central hypotheses to be investigated in this chapter are as follows:

HYPOTHESIS 1.    *Expanded scientific and technical labor-force training and skill increase subsequent national economic growth, with other factors held constant.*

HYPOTHESIS 2.    *Expanded scientific research activity increases subsequent national economic growth, with other factors held constant.*

The validity of these hypotheses is taken for granted in much modern discourse and policy. Yet few comparative empirical studies examine the issues carefully. The literature on education as human capital suggests positive economic effects of expanded mass education but has not identified the influence of science education in particular (see Barro 1991; Fuller and Rubinson 1992; Benavot 1992a; Rubinson and Brown 1994). Aggregate national-level comparisons typically fail to show economic benefits for expanded higher education of any sort (Benavot 1992b). However, one comparative study does show positive effects of expanded tertiary-level scientific and technical training on the economy (Ramirez and Lee 1995).

The effects of expanded levels of scientific research on the economy are also not well understood. Historical case studies have traced the contribution of effective research to national economic success and the rise of particular industries, but they compare only a few countries (for example, Locke 1984). Direct attempts to compare substantial numbers of countries over time have been hampered by the absence, until recently, of comparative and historical data on scientific expansion. Working with such data, Schott (1994) provides some limited evidence that expanded scientific research activity enhances subsequent growth. Williams (1964) and other economists, however, have observed a negative relation between overall research and development (R&D) spending and economic performance among industrialized nations (Jones 1971). More recently, Shenhav and Kamens (1991), looking only at developing countries, also observe that national research output has negative economic effects. The empirical literature has not reached consensus on the issue.[3]

*Critical Reflections*

The instrumentalist conception of science as a tool for economic growth does not capture the nature of the Western (and now world) commitment to science. The instrumentalist view assumes that science is primarily organized in a manner to generate "progress" and that progress comes primarily in the form of economic expansion. But historically, modern Western science arose only loosely and intermittently connected to productive economic activity. Thomas Kuhn (1977:142) puts it bluntly: "technology flourished without significant inputs from the sciences until about one hundred years ago." Rather than being tied to early industrialism, science grew and became institutionalized in the context of Enlightenment philosophies and optimism regarding the possibilities for social reform and "progress" broadly defined (Hall 1983).

In the past century, the link between science and economic activity has been strengthened—for example, through the rise of modern industrial research laboratories. However, science retains much from its earlier history. Moreover, its expansion is mainly linked to the worldwide spread of universities, with their broad notion of sociocultural progress. The link between science and "progress," throughout the modern period, has had broad cultural and ideological dimensions, and has by no means been focused principally on economic and technological development (Hall 1983). Much of the "progress" envisioned by early (that is, Enlightenment era) science consisted of improved understandings of God's Order, which would glorify God and serve as a basis for a better and more moral society (Hankins 1985). Later, this evolved into more explicit ideologies of social improvement, but again, economic and military activity were not central to the picture (Hall 1983). Science, the embodiment of rationalism, was to transform all aspects of social and moral life for the better. The "progress" that is expected to flow from science includes everything from the improvement of individual life (psychologically, medically, politically, morally), to the rationalization of every domain of social organization (including the polity, family life and law, and a wide variety of noneconomic organizational arrangements), to the improved human understanding of and relations to all aspects of nature (for example, astronomy, physical anthropology, environmental studies).

The coevolution of science and modern social institutions solidified the link between science and various progressive agendas in society. Over time, progressive, rationalistic and scientific ideologies became enmeshed with the modern state and conceptions of governance (Foucault 1971; Shapin and Schaffer 1985; Ezrahi 1990), modern bureaucracy and organization (Ellul 1964; Weber 1978), and modern notions of the individual (Frank et al. 1995). This coevolution resulted in greater interpenetration of science and social institutions: science became more "socialized," that is, organized to address particular social concerns; and society became more "scientized" and rationalized (see Chapters 1 and 3; Schofer 1999b).

On one hand, this has led Western societies to devote substantial resources and social attention to forms of science that have social or cultural, rather than technical, relevance. Efforts such as sending a man to the moon (primarily a technical feat, but also a scientific one) are potent symbolic demonstrations of progress, despite being far removed from productive economic activity. Likewise, science addresses issues of cosmology, a cultural concern, not a practical one. Scientists are given resources to study the origins of the universe, the moons of Jupiter, the possibility of life on Mars, the search for extraterrestrial intelligence, the origins and evolution of humankind, and so on. Such understandings count as "progress" of a sort. However, it is difficult to imagine that such investments produce enhanced immediate economic returns, especially relative to other possible investments.

More importantly, the pursuit of science as part of broad conceptions of social progress has resulted in expansion of the breadth of Western scientific activity. At the same time that science has rapidly expanded (Price 1963), it has become more and more "socialized" (see Chapters 1 and 3; Schofer 1999b), addressing an increasing variety of social concerns and issues. Applied technical sciences seem most likely to produce immediate economic effects, as with computer or car industries. Yet as carriers of "progress" broadly defined, scientists legitimately address almost any topic of social relevance. In doing so, new social problems are constantly uncovered. We have learned, for example, that our children suffer from low self-esteem, subtle learning disabilities, and attention deficit hyperactive disorder. Our society now must bear the expense of various psychologists and therapists, special education programs, and the provision of Ritalin to a substantial percentage of children.

The authority of science in society provides bases for enlarged environmental understanding and protections, for all sorts of newly discovered human medical and psychological and welfare needs, and for new dimensions of rationalized social organization. Such scientific activity, however beneficial for human life and progress in the long run, also generates substantial short-term economic costs. For every scientist who produces economic goods, one can find others who discover new and costly environmental problems: endangered species; ozone layer problems; global warming; chemicals in the air, water, and land. These problems are costly to address and often justify regulating economic activity. Still other scientists, working on broad, culturally legitimated progressive issues, discover and trace new problems for human individuals: medical scientists identify and develop health and safety problems, and social scientists discover and document inequalities arising from the exploitation of labor (Deyo 1989). Finally, there are the scientists discovering and correcting problems in social organization—the broad cultural commitment to science and progress extends here, too. Management has become a "science" and generates prescriptions that alter social organization. Family and welfare arrangements are rationalized; organizational structures expand to control new domains (departments of planning, accounting, information systems, legal control, safety, the environment, personnel, and so on). Most of these developments, however progressive in general, create constraints on short-run economic growth.

A variety of mechanisms mediate between science and new limits on economic activity, each deserving future research. Scientific knowledge can spur and legitimate social movements that, in turn, generate social and political change. The environmental movement is the clearest example. Although environmentalists have existed since the nineteenth century, it was not until strongly validating scientific research appeared that the social movement took off (see Chapter 11). Social movement activity resulted in new legislation and regulation that affected economic activity—for example, by making profitable forms of strip-mining illegal, requiring expensive pollution control devices on cars and factories, and limiting the development of potentially productive land or natural resources. Often scientists themselves join directly in social movements as political actors—for example, against child labor in the nineteenth century or on behalf of many public health initiatives in the twentieth century (for example, antitobacco legislation). However, the most potent mechanism is the least visible. New research findings are simply accepted as truth and implemented as policy by legislators, uncontested. For example, the 1969 U.S. National Environmental Policy Act passed through Congress with little notice, despite having massive ramifications for industry (effectively creating modern Environmental Impact Assessment reporting requirements) (Caldwell 1982; see Chapter 11). This latter process is the strongest testament to the authority of science in modern society.

It is worth considering the counterfactual: a society in which the institution of science is wholly (rather than partially, as is the case in the contemporary world) subordinated or "captured" by powerful social interests. Tied to weapons or widgets, scientists would not legitimately and authoritatively address broad social concerns (for example, environmental or worker safety issues). Scientists would not be likely to address environmental, health, or social concerns, nor would they have diffuse social authority to comment on those domains. Such a system might maximize the positive economic or military benefits of science but yield less "social progress." The closest modern example is perhaps the Soviet Union, where science research infrastructures were tightly linked to particular military or industrial activities. But even there, science was part of a larger cultural vision of creating an orderly society, and scientists had some degree of social authority (Josephson 1996).

In theory, a nation could organize research to solely address issues of immediate technical utility (for example, applied sciences, engineering, agriculture), creating a truly instrumentalized research enterprise, without letting scientists address noneconomic concerns. However, most nations do not do this. Non-Western nations adopt Western systems of science as a wholesale package, rather than creating uniquely adapted local scientific institutions (Strang and Meyer 1993; Finnemore 1996). Most nations incorporate science as they implement broad Western-style university systems (which provide a great deal of scientific autonomy), rather than simply creating applied research institutions solely focused on engendering economic growth. As further evidence of this, we observe that the distribution of national research output across scientific disciplines is quite similar in industrialized and developing nations, suggesting wholesale adoption of Western-style science with little local adaptation.[4]

These critical reflections do not lead us to modify hypothesis 1 above, the idea that expanded scientific labor-force skills enhance economic growth. Improved labor-force quality is perhaps the most direct indicator of instrumentally organized science. An expanded scientific labor force indicates productive capacity fairly directly. Indeed, the largest technical labor forces are typically found in nations without much research output (for example, South Korea).

But the arguments above lead us to consider modifications to hypothesis 2—that expanded scientific research activity increases national economic growth. We propose that expanded scientific research activity imposes substantial costs and constraints on economic activity in the short term:

HYPOTHESIS 2.1. *Expanded national scientific research activity slows short-term economic growth.*

Expanded research activity is one indicator of the expanded autonomy and authority of science, and thus of its potential to generate constraints on economic activity. Empirically, a better indicator may be the expansion of science tied to progressive social agendas, such as medicine, psychology, and environmental science, which we refer to as "socially relevant" science.

Socially relevant scientific research is not meant to denote science that is done with an explicit social agenda. Rather, it refers to science on topics that overlap with important social concerns, such as research on the causes of disease or the nature and functioning of the ecosystem. Such forms of knowledge have more capacity to justify limitations on economic activity compared with fields such as mathematics, where new findings typically have few social implications. In general, nations are isomophically adopting science infrastructures that incorporate socially relevant science (see note 4). However, some exceptional nations (for example, South Korea) attempt to restrict the expansion of science, emphasizing only the "technically relevant" sciences that serve state military or development goals. Without the socially relevant component, science seems more likely to produce economically useful innovations and less likely to generate constraints on the economy. Therefore, we propose the following hypothesis:

> HYPOTHESIS 2.2.    *Expanded national scientific research activity in the more socially relevant sciences (rather than technically relevant applied science) will have a negative effect on short-term economic growth.*[5]

## Unit of Analysis

We have discussed the impact of scientific expansion on economic growth with both variables conceived at the national level. But in fact, both variables are organized at the world level, which complicates the picture. Science has long had an international presence and is increasingly organized as a world institution (Crawford 1992; Chapter 3). It is possible that scientific expansion could have positive economic effects at the world level that are not captured by national-level economic analyses. Many triumphs of science, such as the eradication of smallpox, benefit every nation in the world. The fruits of scientific research can spread rapidly and obviously have had effects beyond their country of origin.[6] For instance, late-nineteenth-century British scientific innovation was poorly captured by the British economy, but it led to economic advance in the United States and elsewhere (Ranis 1977). Likewise, it is commonly argued that American scientific advances in the 1980s were captured in Asian economies (Fransman 1986).[7] The claim, then, is that some countries bear the economic and social costs of scientific research while others reap the benefits. This argument usually applies to scientific research and intellectual property, which diffuses rapidly when published. Scientific human capital also crosses boundaries—moving from the periphery to industrial nations through brain drain and back to the periphery as foreign-educated students return home—what came to be called reverse brain drain (Vas-Zoltan 1976; Adams 1968). The economically constraining effects of science, too, are global. For example, scientific discovery of the Earth's ozone hole led to a treaty ratified worldwide that banned the use of certain refrigerants. The impact of scientific knowledge extended far beyond the countries that conducted research on the issue.

To examine the overall economic impact of scientific expansion requires world-level time-series analyses, which are beyond the scope of the present research. Or one could analyze economic development for a set of countries, each of which was potentially affected by scientific expansion in all (or each of) the others. However, such an analysis is not feasible with available measures.

Globalization processes, if carried to extremes, would eliminate relative economic advantage (or harm) derived from science for a given nation compared with all others (that is, if knowledge, human capital, and world culture penetrated equally and moved effortlessly around the globe). This degree of globalization has not yet occurred, so we expect national-level variance to have domestic impacts. New knowledge and patents, even if they are eventually copied elsewhere, are likely to produce immediate benefits at home. Domestic science training, despite brain drain, increases the resident scientific labor force (UNESCO 1974). And local scientists presumably have direct constraining effects on a nation's economy, either by "discovering" problems themselves (for example, new endangered species), or by creating local awareness of problems discovered elsewhere.[8]

Available data allow us to address a related issue—the question of whether the expansion of science differentially benefits industrialized nations. It is clear that the scientific culture involved is linked to the core and Western world (Schott 1993). Thus, Shenhav and Kamens (1991) argue that the adoption of scientific activities and structures in the non-Western world is likely to be detrimental to non-Western and nonindustrialized societies. Dependency theorists have made similar claims (Cooper 1973). In copying Western scientific forms, such societies incorporate knowledge of little use to their own economic development—and also incorporate many of the economic costs noted above (Drori 1997, 2000; see Chapters 12 and 13). Modern ecological science, which stresses the virtues of rain forests and similar habitats, may slow the rapid expansion of export agriculture in developing countries. Or modern medical and social research on labor conditions may identify harsh and exploitive conditions, reducing the deployment of low-cost industrial labor in the less developed world (Deyo 1989).

This suggests the following interaction effect: science research activity particularly hinders less developed nations. Thus, we consider an additional modification of our second main hypothesis:

HYPOTHESIS 2.3.    *The negative short-term economic effects of expanded scientific research activity are especially pronounced in less developed countries.*

We turn now to an empirical investigation of the effects of scientific expansion on national economic growth.

DATA AND METHODS

*Measuring Scientific Activities*

A first goal is to develop a meaningful typology of national science activities that can be operationalized and used in models of economic growth. Science, as an attribute of a given nation-state, comprises a wide variety of activities, from the training of individuals in the principles of science to the development of science as a profession to the practice of laboratory research. All the various components of science are prevalent in developed nations and thus seem to make up a complete package. Yet some nations, especially outside the West, vary in their propensity to support various kinds of science. Korea, for example, takes a strong manpower-planning view of science, possessing a scientifically trained labor force that is one of the largest in the world in proportion to the size of the economy (UNESCO 1974, 1992). However, Korea rates low as a producer of new scientific research (Institute for Scientific Information 1973, 1982). So although science is often discussed as a single package, it involves multiple dimensions on which nations vary.

Here, we introduce two main dimensions of scientific infrastructure: scientific research and scientific labor force.[9] We derive these from the theoretical arguments discussed above, as well as from the existing literature on science indicators (for example, Morita-Lou 1985). Later, we utilize exploratory factor analysis to determine if this typology is empirically supported.

*Scientific Research*

Scientific research refers to a nation's emphasis on the production of new basic and applied scientific knowledge. Nations vary substantially in their support for and production of scientific research, even controlling for level of development. Industrialized nations typically emphasize scientific research, as do certain poorer nations such as India and Kenya. Indeed, the latter typically produce more scientific publications in a year per dollar of gross domestic product (GDP) per capita than most European nations (Institute for Scientific Information 1973, 1982). In contrast, many Asian and Middle Eastern nations produce low levels of new research in proportion to national resources.

In our quantitative analyses, scientific research is measured with two highly correlated indicators of research output: number of scientific publications per capita and number of scientific citations per capita. First, we will discuss the number of scientific publications per capita (logged). Journal publications are one primary goal of scientific research and thus are direct evidence of ongoing research activity. Publications in a large number of scientific, engineering, and applied science journals are tracked in the *Science Citation Index*, published by the Institute for Scientific Information.[10] On the basis of the institutional addresses provided for the authors (for example, Harvard University; Universidad de Costa Rica), we were able to count the number of publications produced in different nations.[11] Nations range from zero publications total, in a tiny handful of developing nations, to 0.5 or more publications per thousand people in countries such as Canada, Israel, Sweden, and Britain. This variable is divided by population size and logged to reduce skewness.[12]

Publications are listed by scientific field, allowing the disaggregation of research into various fields or groups of fields. This allows the creation of technically relevant and socially relevant science indicators. Technically relevant science is measured as a nation's total number of publications in the fields of engineering, chemistry, and physics; socially relevant science is measured by publications in clinical medical science, biology, and psychology.[13]

A second measure is the number of scientific citations per capita (logged). The *Science Citation Index* can also be used to count citations to authors within a given nation. Such a measure is thought to reflect the centrality and (although there is disagreement on the issue) the quality of scientific publications. As a national-level measure, it is another plausible indicator of amount of research activity in a nation. Indeed, the correlation between national science citations and national scientific publications is in excess of 0.9. This measure is per capita and logged. Again, disaggregation allows the creation of technical and socially relevant scientific research measures.

Other relevant measures include the following: National expenditures on research and development (UNESCO 1974), either per capita or as a proportion of GNP per capita; and national patent grants or patent applications (WIPO 1983). These reflect additional aspects of national research activity and are highly correlated ($r = 0.8$ or more) with citation and publication measures. Ideally, these indicators would be used in the analyses.[14] However, they are available for relatively few nations, and thus their inclusion would substantially restrict the size of our sample.

### Scientific Labor Force

The scientific labor force refers to a nation's emphasis on producing scientifically trained workers that can participate in the labor force. The scientific labor force is reflected by the size of a nation's science education system, as well as by measures of scientifically trained personnel actually in the labor force. This is not to be equated with the scale of scientific research. Indeed, the size of a nation's scientific labor force is not highly correlated with scientific research, if overall national development is held constant. Typically, a large part of a nation's scientifically trained personnel work in capacities other than producing original scientific research. In industry, scientific manpower is often focused on applied technical development, producing little in the way of new research. Also, scientifically trained personnel may work in management, education, or, as is common in many nations, within governmental bureaucracies. Rather than a research focus, the size of the scientific labor force reflects a nation's level of human capital in the area of science, as well as a national commitment to science education. Also, it has been argued that a large scientific labor force is indicative of more general national industrial and educational policies designed to favor economic growth (Hage et al. 1988).

We measure scientific labor-force infrastructure by means of three indicators, all deriving from data collected by the United Nations Education, Science, and Cultural Organization (UNESCO) and published in statistical yearbooks (UNESCO 1974, 1992).

1. Number of scientists and engineers in research and development, per million people (logged). This is a direct measure of R&D scientists and engineers in the labor force in proportion to the size of the overall population (UNESCO 1974, 1979). High-scoring nations, typically in northern or Eastern Europe, include both East and West Germany, Japan, Czech-

oslovakia, Poland, and Norway, with two thousand or more scientists and engineers per million people. On the low end are African nations or peripheral Asian nations such as Kenya, Bangladesh, Mali, and Chad, with fewer than thirty per million.

2. Number of enrolled tertiary-level students in the natural sciences as a proportion of the university-age population (logged).

3. Number of enrolled tertiary-level students in engineering as a proportion of the university-age population (logged). These two educational measures are correlated with each other and with scientists and engineers in R&D at roughly 0.7. Both measures reflect a nation's labor-force potential in scientific and technical areas.[15] Given that many nations have centralized educational systems that channel enrollments, these measures also often reflect broader national commitment to producing scientists and engineers and to manpower planning in general. Nations that score high on these measures include northern and Eastern European nations, as well as, for example, Israel, Japan, Singapore, and Australia. Again, less developed African nations score lowest.

*Modeling Economic Growth: Methods, Measures, and Control Variables*

Both scientific research and the scientific labor force potentially have an impact on the national economy.[16] To determine their effects, we develop quantitative models of national economic growth for a large sample of countries.[17] Independent variables include science indicators and relevant control variables.

National economic performance is most commonly assessed by modeling the annualized growth of real GDP per capita over a period of time (Barro 1991; Barro and Sala-i-Martin 1995). Growth of GDP per capita from period $t1$ to $t2$ can be measured as $(GDPpc_{t2} - GDPpc_{t1})/GDPpc_{t1}$, or more accurately by taking compounding into account as the natural log of $GDPpc_{t2}/GDPpc_{t1}$. We use the latter method. This is then annualized simply by dividing by the number of years between $t1$ and $t2$. Independent predictors are generally measured at or near the starting time, $t1$.

Economic data are obtained from the Penn World Tables (version 5.6; Summers and Heston 1991). Economic growth is typically measured over a substantial period of time—at least ten years (for example, Barro and Sala-i-Martin 1995), and as many as thirty years or more (for example, Levine and Renelt 1992). We opted to use a twenty-year span to allow substantial time for science investments to have an impact.[18] Because many independent variables of interest are not available before 1970, our analyses (and measures of per capita GDP growth) are based on the period 1970 to 1990. Models are estimated via ordinary least squares (OLS) regression.

We also corroborate these models with additional analyses that are based on a broader measure of industrial development. It is composed of three indicators: (1) GDP per capita (logged); (2) energy consumption in kilowatt-hours per capita (logged) (World Bank 1992); and (3) percentage of the labor force outside of the agricultural sector (World Bank 1992). Indicators were combined by use of factor analysis to estimate the underlying variable. Rather than calculating a growth rate from the standardized factor variable, panel models were used (see Meyer and Hannan 1979). This measure was constructed for two reasons. First, multiple indicators are a useful strategy to address the potential for error inherent in cross-national economic indicators. Second, we were interested in whether the findings can be generalized to broader (that is, not just economic) measures of societal development.

To properly specify our models of economic growth, we must control for relevant variables known to affect the economy. A wide array of factors have been proposed at one time or another (see, for example, Barro and Sala-i-Martin 1995). However, Levine and Renelt (1992) find that only a small subset of the variables proposed in the literature have robust and stable effects. They observe consistent effects of investment, the expansion of secondary enrollment, and initial level of GDP per capita. These serve as a starting point. We also include measures of trade and political democracy in our base model.[19] To this we add our science indicators, as well as additional control variables. Variables in our base model are as follows (Table 10.1):

*Gross domestic product per capita in 1970 (logged).* Barro (1991) and others have consistently observed that wealthy nations tend to grow at lower rates than poor ones, producing convergence of GDP per capita among nations. Thus, we control for initial level of GDP per capita, which we expect will negatively affect economic growth.[20]

*Investment.* We also control for national variation in investment in new infrastructure, capital goods, and so on. Economists have shown that such investment has a strong effect on future levels of economic activity (Maddison 1995; Barro and Sala-i-Martin 1995; Levine and Renelt 1992). Even foreign capital investment, long thought to be an indicator of economic dependency, has been shown to have economic benefit (Firebaugh

TABLE 10.1
Descriptive Statistics for Variables Used in Analyses

| Variable | Mean | SD |
|---|---|---|
| Annualized growth of real GDP per capita | 0.02 | 0.02 |
| Log real GDP per capita 1970 | 7.59 | 1.03 |
| Investment rate 1970 | 17.40 | 10.90 |
| Secondary enrollment 1970 | 33.01 | 28.36 |
| Democracy 1970 | 3.24 | 4.17 |
| Trade 1970 | 48.96 | 29.30 |
| Log article publications per capita 1973 | −4.30 | 1.98 |
| Log article citations per capita 1973 | −3.13 | 2.61 |
| Scientific research index 1970 | 0.19 | 1.01 |
| Log scientists and engineers in R&D per capita 1970 | 5.04 | 1.72 |
| Log science students per age group 1970 | 1.42 | 1.04 |
| Log engineering students per age group 1970 | 1.58 | 1.38 |
| Scientific labor force index | 0.04 | 1.03 |
| Log real GDP per capita 1990 | 7.89 | 1.16 |
| Log energy consumption per capita 1990 | 6.13 | 1.91 |
| Percentage nonagricultural labor force 1990 | 60.14 | 29.70 |
| Development 1990 index | −0.05 | 1.04 |
| Log energy consumption per capita 1970 | 5.82 | 1.83 |
| Percentage nonagricultural labor force 1970 | 49.60 | 28.94 |
| Development 1970 index | −0.01 | 1.04 |
| "Technically relevant" science index 1970 | 0.01 | 1.04 |
| "Socially relevant" science index 1970 | 0.01 | 1.00 |

1992, 1996). Investment is measured by the level of investment as a proportion of GDP in 1970 (Summers and Heston 1991).[21]

*Secondary education.* An educated labor force has also been shown to spur economic growth (Psacharopolous 1984; Barro 1991; Levine and Renelt 1992). We use a conventional measure, secondary school enrollment as a proportion of the relevant population age group in 1970 (UNESCO 1974).[22] We expect a positive effect.

*Political democracy.* Several regime characteristics are thought to affect the economy, including democracy, political and economic freedom, and regime stability. Democracy may enhance a variety of social investments, including investments in education (Barro and Sala-i-Martin 1995). Likewise, political freedoms are thought to enhance participation in labor and capital markets (Barro and Sala-i-Martin 1995). We use Gurr's ten-point index of political democracy measured in 1970 to capture these hypothesized positive effects (Jaggers and Gurr 1995).

*Trade.* Economists have argued that trade should benefit a nation's economy, either directly or by enhancing investment (Levine and Renelt 1992).[23] We measure trade as total imports and exports as a proportion of GDP in 1970 (Summers and Heston 1991). Again, a positive effect is expected.

To this base model of control variables, our science indicators are introduced: scientific research in 1973 and scientific labor force in 1970. Conventionally, both are expected to have a positive effect. However, we outlined arguments that predict a negative impact of scientific research. Further, we introduce an interaction term to test the hypothesis that scientific research may particularly constrain economic growth in poor nations. We employ an interaction variable science research in 1970 × less developed nations (dummy).[24]

As an additional test of robustness, we then add (individually, so as not to lose many cases as a result of missing data) the following indicators to our models[25]: (1) Tertiary-level educational enrollment in 1970 as a proportion of the twenty- to twenty-four-year-old age group (UNESCO 1974); (2) percentage of the population under age fifteen in 1970 (World Bank 1992); (3) government consumption share of GDP in 1970 (Summers and Heston 1991); (4) multinational corporation penetration in 1967 (Bornschier and Chase-Dunn 1985); (5) a measure of political rights in 1970 (Gastil 1978); and (6) a dummy variable for oil and mineral rich nations (IBRD 1971). The first four of these indicators are generally expected to have negative effects on economic growth. The latter two are expected to have positive effects.

## RESULTS

We start with an exploratory factor analysis to see if our typology of national science dimensions is empirically supported. Exploratory factor analysis will help us identify the factors or latent variables that underlie our many science indicators. Such an analysis can show if the two types of science we propose (labor force, research) empirically correspond to observable patterns of variation in science infrastructure among different nations. In this particular analysis, the science indicators are residualized by GDP per capita.[26] Thus, the resulting factor loadings are net of the relation between all the indicators

TABLE 10.2
Exploratory Factor Analysis of Eight National Science Indicators:
Rotated Factor Loadings, Circa 1970

| Variable | Factor 1 (scientific research) | Factor 2 (scientific labor force) | Factor 3 (scientific organization) |
|---|---|---|---|
| No. of science publications per capita | *0.94* | 0.15 | 0.19 |
| No. of science citations per capita | *0.95* | 0.05 | 0.20 |
| No. of memberships in the ICSU | 0.20 | 0.34 | *0.85* |
| No. of professional science organizations | 0.12 | 0.16 | *0.76* |
| No. of Memberships in international science organizations | 0.18 | 0.11 | *0.88* |
| No. of scientists and engineers in R&D per capita | 0.47 | *0.68* | 0.17 |
| No. of tertiary engineering students per age group | −0.15 | *0.88* | 0.18 |
| No. of tertiary science students per age group | 0.28 | *0.48* | 0.22 |

All variables are residualized by GDP per capita; principle components extraction; and varimax rotation. Italics indicate factor patterns.

and GDP per capita. This was a reasonable strategy, given the goal of our factor analysis. We are uninterested in the obvious fact that resource-rich nations invest more in science than poor ones. It is much easier to investigate additional patterns among the variables once this confounding pattern (high correlation of scientific activity with development) is purged.[27]

Table 10.2 lists the results of an exploratory factor analysis of eight different science indicators. The exploratory analysis yielded three factors. Factor loadings in Table 10.2 strongly support our science typology. National production of science publications as well as national recognition through academic citation both load at over 0.9 onto one factor. Measures of scientists and engineers in the labor force and in higher education load onto a second factor. Finally, measures of scientific professional organization load primarily onto a third science factor (see note 9). This factor analysis, in combination with the theoretical discussion above, supports our distinction between scientific research and scientific labor force.

We now use scientific research and labor-force indicators in a series of quantitative models of economic development. Presented first are OLS regression analyses of per capita GDP growth over the period 1970 to 1990. Table 10.3 shows results of the base model, with each science measure added individually and then together. Finally, an interaction between science research and less developed countries is added.

Control variables in Table 10.3 have effects consistent with the literature. Initial GDP per capita has a negative, significant effect, highlighting the convergence effect among nations: larger economies tend to grow more slowly than smaller ones. Investment and secondary education both have positive and significant effects. Democracy and trade have a positive effect, each significant in two of the five models.

In model 2, science research is added to the base model. Science research has a negative, significant effect on growth of GDP per capita. In model 3, the science labor-force measure is added to the base model. It has a positive, significant effect on per capita GDP growth. Model 4 contains both science indicators. The respective negative and positive effects remain. Together, these measures improve the explained variance a great deal over the base model, with the adjusted $R^2$ value increasing from 0.21 to 0.36.

The interaction between scientific research and less developed nations, added in model 5, has a nonsignificant negative effect on economic growth and does not improve the model. Hypothesis 2.3 suggested that the potential negative effects of scientific research should particularly affect poor, nonindustrialized countries. Although the effect is in the direction we hypothesized, we cannot conclude that the negative effect of science research affects less developed nations differently from developed nations.[28]

Table 10.4 adds a series of additional variables to the model. The tertiary educational enrollment rate has a nonsignificant negative effect on per capita GDP growth, a nonintuitive but consistently observed effect (see Benavot 1992b). A large child-aged population, presumed to be a drag on the economy, has a negative, insignificant effect on growth. Government consumption share of GDP has a negative, significant effect, consistent with the literature (Barro and Sala-i-Martin 1995). Multinational corporation penetration, a measure of economic dependency, has an insignificant negative effect. Also, a measure of national political rights has a negative, nonsignificant effect. Finally, our measure of oil/mineral wealth has a positive, but nonsignificant, effect on growth of GDP per capita.

Coefficients throughout the model remain quite stable with the addition

TABLE 10.3
OLS Regression: Effects of Science Indicators
on Annualized Growth of Real GDP Per Capita, 1970–1990

| | Model 1 | Model 2 | Model 3 | Model 4 | Model 5 |
|---|---|---|---|---|---|
| Adjusted $R^2$ | 0.21 | 0.23 | 0.28 | 0.36 | 0.36 |
| | 112 | 111 | 87 | 87 | 87 |
| Log real GDP per capita 1970 | −0.0094** (0.0031) | −0.0075** (0.0032) | −0.014*** (0.0038) | −0.011** (0.0037) | −0.013** (0.0046) |
| Investment rate 1970 | 0.00062** (0.00025) | 0.00076** (0.00026) | 0.00060* (0.00028) | 0.00073** (0.00027) | 0.00072** (0.00027) |
| Secondary education 1970 | 0.00035** (0.00013) | 0.00044*** (0.00014) | 0.00019 (0.00017) | 0.00028* (0.00015) | 0.00028* (0.00016) |
| Democracy 1970 | 0.00023 (0.00052) | 0.00057 (0.00056) | 0.00073 (0.00063) | 0.0012* (0.00061) | 0.0012* (0.00061) |
| Trade 1970 | 1.4 e-5 (5.7 e-5) | 7.6 e-6 (5.7 e-5) | 7.3 e-5 (6.8 e-5) | 0.00011* (0.000065) | 0.00011* (0.000065) |
| Scientific research 1973 (log article publications per capita, log article citations per capita) (two-tailed test) | | −0.0072* (0.0036) | | −0.0138*** (0.0039) | −0.0125*** (0.0040) |
| Scientific labor force 1970 (log scientists and engineers per capita, log tertiary science students per age group, log tertiary engineering students per age group) | | | 0.0078* (0.0047) | 0.012** (0.0046) | 0.011** (0.0047) |
| Science research 1973* less developed nation dummy | | | | | −0.0015 (0.0023) |
| Constant (two-tailed test) | 0.062** (0.021) | 0.043 (0.023) | 0.099*** (0.028) | 0.073** (0.028) | 0.088* (0.037) |

*$p < 0.05$, **$p < 0.01$, ***$p < 0.001$ (one-tailed tests, unless indicated). Values in parentheses are standard errors.

TABLE 10.4

OLS Regression: Effects of Science Indicators on Annualized Growth of Real GDP Per Capita, 1970–1990, with Additional Control Variables

| Science Indicator | Model 6 | Model 7 | Model 8 | Model 9 | Model 10 | Model 11 |
|---|---|---|---|---|---|---|
| Adjusted $R^2$ | 0.34 | 0.35 | 0.40 | 0.41 | 0.36 | 0.35 |
| n | 86 | 87 | 87 | 80 | 87 | 87 |
| Log real GDP per capita 1970 | -0.012*** | -0.011** | -0.014*** | -0.0093** | -0.012*** | -0.011** |
| | (0.0037) | (0.0037) | (0.0037) | (0.0041) | (0.0038) | (0.0040) |
| Investment rate 1970 | 0.00070** | 0.00071** | 0.00064** | 0.00068** | 0.00076** | 0.00073** |
| | (0.00027) | (0.00028) | (0.00027) | (0.00028) | (0.00027) | (0.00027) |
| Secondary education 1970 | 0.00032* | 0.00026 | 0.00023 | 0.00026 | 0.00028** | 0.00028* |
| | (0.00017) | (0.00019) | (0.00015) | (0.00016) | (0.00016) | (0.00016) |
| Democracy 1970 | 0.0011* | 0.0012* | 0.0012* | 0.00071 | 0.00044 | 0.0012* |
| | (0.00061) | (0.00066) | (0.00059) | (0.00065) | (0.00091) | (0.00062) |
| Trade 1970 | 0.00009 | 0.00011* | 0.00013* | 0.00006 | 0.00011* | 0.00010 |
| | (0.00007) | (0.00006) | (0.00006) | (0.00006) | (0.00006) | (0.00006) |
| Scientific research 1973 (two-tailed test) | -0.012*** | -0.014*** | -0.009** | -0.010** | -0.013*** | -0.013*** |
| | (0.0040) | (0.0043) | (0.0040) | (0.0038) | (0.0039) | (0.0040) |
| Scientific labor force 1970 | 0.012** | 0.012** | 0.011** | 0.011** | 0.013** | 0.012** |
| | (0.0048) | (0.0047) | (0.0058) | (0.0047) | (0.0047) | (0.0047) |
| University enrollment 1970 per age group | -0.00031 | | | | | |
| | (0.00050) | | | | | |
| Percentage population under age 15 | | -0.00021 | | | | |
| | | (0.00040) | | | | |
| Government consumption share of GDP | | | -0.00062** | | | |
| | | | (0.00025) | | | |

| | | | | | | |
|---|---|---|---|---|---|---|
| Multinational corporation penetration 1967 | | | | −0.00008 (0.00005) | | |
| political rights 1973 | | | | | −0.0019 (0.0017) | 0.0002 (0.0058) |
| Mining/oil dummy | | | | | | |
| Constant (two-tailed test) | 0.076** (0.028) | 0.081** (0.032) | 0.106*** (0.030) | 0.063* (0.030) | 0.090** (0.032) | 0.073* (0.029) |

$*p < 0.05$, $**p < 0.01$, $***p < 0.001$ (one-tailed tests, unless indicated). Values in parentheses are standard errors.

of more controls. In all cases, the negative effect of science research and the positive effect of scientific labor force remain statistically significant.

The results provided in Tables 10.3 and 10.4 suggest the following: the greater a nation's science research activity in 1970, the lower its level of per capita GDP growth from 1970 to 1990. Hypothesis 2 is not supported. Instead, support is found for the converse, hypothesis 2.1. Also, nations with a larger scientific labor force in 1970 experience greater growth of GDP per capita over the period. Hypothesis 1 is supported. The latter result is not particularly surprising or controversial, but the former is. We conduct a variety of explorations to assess the stability of these findings.

In Table 10.5, we test to see whether the findings hold up with a broader measure of industrial development, rather than GDP growth. The same negative, significant effect of science research and positive effect of scientific labor force are found. Additional analyses (not presented here) were estimated by using structural equation models with latent variables (see Bollen 1989 for a discussion of this method). The multiple indicators of development, science research, and scientific labor force allow us to employ such models, addressing concerns about measurement error in our data. Again, findings are consistent.

Another concern is that outliers or influential cases affected results. We explored a variety of diagnostics to identify these.[29] Partial plots noted South Korea as an extreme low outlier on science research, but the case is not distinctive in terms of leverage or other measures of influence. Its removal does not change the results. Zambia is highlighted as influential by a large Cook's $D$, and by a large negative studentized residual. Singapore is flagged as having high leverage and a high Cook's $D$. Finally, Mauritius has a marginally high Cook's $D$ and a large studentized residual. The removal of these cases individually or as a group has no substantial effect on the models.

We also checked for various violations of regression assumptions. Histograms and scatter diagrams of residuals suggested no substantial problems of heteroscedasticity or nonnormality. Multicollinearity is also a potential problem. However, the overall stability of coefficients and standard errors across differently specified models provides some confidence that multicollinearity is not affecting our results. As a further test, we estimated our models by using ridge regression, which employs a more efficient (although potentially more biased) estimator than OLS (Marquardti and Snee 1975). Results of ridge-regression analyses (not presented here) were consistent with the main findings listed in Table 10.3.

TABLE 10.5
OLS Regression Panel Model: Effects of Science
Indicators on National Development, 1970–1990

| Science Indicator | Model 12 |
| --- | --- |
| Adjusted $R^2$ | 0.95 |
| n | 84 |
| Development 1970 | 0.92*** |
| (log GDP per capita, log energy consumption per capita, % nonagricultural labor force) | (0.07) |
| Investment rate 1970 | 0.0091** |
| | (0.0037) |
| Secondary education 1970 | 0.00074 |
| | (0.0022) |
| Democracy 1970 | 0.0051 |
| | (0.0083) |
| Trade 1970 | 0.00042 |
| | (0.00093) |
| Scientific research 1973 (log article publications per capita, log article citations per capita) (two-tailed test) | −0.28*** (0.055) |
| Scientific labor force 1970 (log scientists and engineers per capita, log tertiary science students per age group, log tertiary engineering students per age group) | 0.19** (0.074) |
| Constant | −0.24** |
| | (0.094) |

*p < 0.05, **p < 0.01, ***p < 0.001 (one-tailed tests, unless indicated).
Values in parentheses are standard errors.

Finally, we conduct an exploration of the persistent negative effect of scientific research on economic growth. Although it is beyond the scope of this chapter to thoroughly assess possible explanations, one issue will be examined. We argued (hypothesis 2.2) that many costs of science derive from its cultural centrality, which produces great expansion in the size, scope, and authority of science. One empirical correlate of expanded science is growth in the socially relevant sciences and the social sciences (for example, medical and biological sciences, as opposed to engineering and math). Nations most culturally committed to Western science have not only the largest research infrastructures, but the most diverse (with the most socially relevant science and social science), reflecting scientization of all areas of the natural and social

TABLE 10.6

OLS Regression: Effects of "Technically Relevant" Science and "Socially Relevant" Science Indicators on Annualized Real GDP Per Capita Growth, 1970–1990

| Science Indicator | Model 13 | Model 14 | Model 15 |
|---|---|---|---|
| Adjusted $R^2$ | 0.21 | 0.27 | 0.28 |
| n | 108 | 108 | 108 |
| Log real GDP per capita 1970 | −0.0082** | −0.0073** | −0.0081** |
| | (0.0033) | (0.0031) | (0.0031) |
| Investment rate 1970 | 0.00064** | 0.00061** | 0.00050* |
| | (0.00026) | (0.00025) | (0.00025) |
| Secondary education 1970 | 0.00041** | 0.00042** | 0.00037** |
| | (0.00014) | (0.00013) | (0.00013) |
| Democracy 1970 | 0.00034 | 0.00088 | 0.0012* |
| | (0.00055) | (0.00056) | (0.00058) |
| Trade 1970 | −0.000031 | −0.000087 | 0.00012* |
| | (0.000058) | (0.000060) | (0.000061) |
| "Technically relevant" science research 1970 (two-tailed test) | −0.0046 | | 0.0083 |
| | (0.0029) | | (0.0048) |
| "Socially relevant" science research 1970 (two-tailed test) | | −0.0089*** | −0.016*** |
| | | (0.0028) | (0.0048) |
| Constant | 0.050* | 0.038 | 0.045* |
| | (0.023) | (0.022) | (0.022) |

*$p < 0.05$, **$p < 0.01$, ***$p < 0.001$ (one-tailed tests, unless indicated). Values in parentheses are standard errors.

worlds. Indicators of both technically relevant and socially relevant science research are included individually and then together in a model of economic growth, replacing the general scientific research index.[30]

Results are listed in Table 10.6. Individually, both technically relevant and socially relevant science research have a negative effect on per capita GDP growth over 1970 to 1990, similar to the aggregate scientific research measure. The negative socially relevant science research coefficient is larger and statistically significant (model 13), whereas the technically relevant science research coefficient is smaller and falls just shy of significance (model 14). When combined together in the same model (model 15), the sign of the technically relevant science coefficient reverses, becoming positive. The effect of socially relevant science remains negative and significant.

Although the absolute level of technically relevant science appears not to benefit the economy (as seen in model 13), an emphasis on technically relevant rather than socially relevant science seems to work against this negative effect. Controlling for the size of socially relevant science (in model 15), the relative amount of technical science activity has a positive, although not significant, effect. These findings support hypothesis 2.2. They are consistent with the argument that the negative economic impact of science results from its expanded scope and authority in progressive or socially relevant domains. Nations with high levels of technically relevant science, controlling for socially relevant, are those nations that have limited the authority and expansion of science into socially relevant domains, such as medicine and environmental issues. In contrast, nations that adopt the Western model of expanded scientific authority incorporate types of socially relevant science capable of constraining economic growth.

DISCUSSION

We have shown that national science infrastructure can be characterized by two important dimensions: scientific research and the scientific labor force. In our models, scientific research has a substantial negative and significant effect on economic growth over the period 1970 to 1990. Evidence does not support our supposition that this effect is larger for developing nations than it is for industrialized ones. On the other hand, a nation's skilled scientific labor force has a substantial positive and significant effect on economic growth. These findings are stable after the inclusion of a variety of control variables. Finally, we provide preliminary evidence that the negative effect of scientific research activity derives from the expanded scope and authority of science, as indicated by the expansion of science in more socially relevant disciplines.

The positive impact of the size of a nation's scientific labor force supports standard human capital arguments: nations that educate their populations in scientific and technical skills experience improved economic performance.[31] Although intuitive, this finding is surprising given that expanded tertiary-level education does not produce the same positive economic effects at the national level (Fuller and Rubinson 1992; Benavot 1992a; Table 10.4). Past research has shown that economic returns accrue to primary and secondary education. It is important, therefore, to show that some form of tertiary education does produce clear economic benefits—namely education in the sciences and engineering. However, we should point out an alternative explanation of this finding. Nations with expanded scientific labor forces also tend to have strong state-centered control of the economy and strong macroeconomic development policies (for example, newly industrializing Asian countries). It may be that expansion of the scientific labor force is a byproduct of strong state-centered development policies and economic coordination, and that its apparent direct effect on economic growth is partly spurious. This issue deserves further study (see Ramirez et al. 1998).

The negative impact of expanded scientific research is less intuitive, especially from the commonsense instrumentalist view of science. One explanation is that scientific knowledge flows easily across borders. New scientific knowledge may produce massive benefits worldwide (for example, curing smallpox, introducing computer technology) without providing much advantage to the nation that actually produced the research. However, this better accounts for the absence of a positive effect than a large negative one. Most nations do not spend enough on science to substantially affect their economy simply because they fail to recoup science investment costs.

Scientific research can have negative effects on economic growth because it was historically institutionalized as a highly valued and authoritative activity, linked to broad visions of social progress. The expansive authority of Western science encourages the conferral of constraints on the private sector economy, whereas many benefits of science are noneconomic (for example, improving life expectancy and health at the end of life, reduced loss of biodiversity), or accrue over the long term (for example, avoiding global warming in the twenty-first century). This model of science has diffused worldwide, replicating these processes in economies around the world (Finnemore 1996; Chapter 4).

This view is corroborated by a number of empirical findings. Drori (1997) shows that expanded scientific research infrastructure is correlated with many societal transformations that have long-term or noneconomic returns: improvements in human rights, increased political participation, improved national data collection and computerization, and others (see Chapters 12 and 13). Similarly, domestic science has been linked to the growth of national environmental organization and regulation (Chapter 11; Frank et al. 2000a). Scientific expansion hastens the creation of national environmental ministries, national parks, and environmental legislation and leads to increased participation in international environmental organizations (Frank et al. 2000b; Meyer et al. 1997b). This research, particularly the findings regarding environmental legislation, provides direct evidence that science expansion can lead to regulation of economic activity.

We suspect that national (and world) science activity is associated with various sorts of progressive political action and social movements. Further research is needed to verify that such expansions of science activity lead to new economic regulations and that these regulations actually reduce short-term economic growth.

Although the focus here has been on economic growth, the other outcome is equally sociologically interesting—that science activity also legitimates progressive social agendas and enhances social movement activity in certain domains. Scientific activity may, in fact, lead to societies that are more environmentally protected, healthy, and equitable. The role of science in the environmental movement is particularly evident (Frank et al. 2000a). The extent to which science activity legitimates progressive social change or social movements in other domains of activity (such as public health or the rights of the disabled) deserves further study.

Social science, too, is implicated strongly in these processes. A great deal of sociological research, for instance, is organized around identifying forms of inequality and discrimination, which justify social welfare and reform. Indeed, an explicit agenda of many sociologists is to limit exploitation and reduce inequality. It is not unreasonable to expect that expanded sociological research within a nation may help legitimate social movements or legislation around issues of social inequality. It seems unlikely that this would lead to increased short-term economic growth, but social inequality may be reduced as a result. Further research is needed to identify such processes.

What does the future hold? The broad, Western model of science is still diffusing worldwide. The main historical alternative to Western-style science is a manpower planning vision, embodied in the former Soviet Union and in certain newly industrializing Asian countries. The collapse of the Soviet Union and the recent stumble of the Asian economies have served to further delegitimate those more instrumentalized national science policies. Now even those countries are moving in the direction of Western models, with expansion in the socially relevant sciences and with less focus on engineering and applied science. If nations isomorphically adopt the Western model of science, national variance in science infrastructure will decrease and science will presumably cease to have negative (or positive) effects on national economic growth. However, new innovations may introduce alternate models, creating variance on new dimensions—for example, nations worldwide are now trying to produce their own Silicon Valleys by subsidizing entrepreneurial science in the private sector and expanding academia–industry relations (Gibbons et al.

1994; Nowotny et al. 2000). Such changes may eventually alter the economic impacts of science. However, dramatic changes will not happen easily or quickly, given that Western-style science is well entrenched worldwide.

To return to a central theme of this chapter: Western science is not simply an instrument of social actors (the state, industry, individuals). It is a broad-based institution—indeed, one of the dominant institutions in Western societies. The authority and legitimate status of science allows science to constitute and reconstruct the goals and interests of actors, including expanded environmental protections, expanded health and welfare, and ever more rationalized organization and governance. This yields "progress" by Western definitions, but may reduce economic growth in the short term.

Chapter 11    Science and the Environment

ANN HIRONAKA

Scholars have long been aware of the impact of science on the current environmental movement. Indeed, the publication of the book *Silent Spring* in 1962 by biologist Rachel Carson is often considered the trigger that initially sparked environmental awareness. Her findings showed that pesticides such as DDT were carried down the food chain and increased the fragility of birds' eggs, leading to the near-extinction of several wild bird populations. These findings led directly to environmental policies banning the use of DDT in the United States (Caldwell 1982). There are many other cases in which scientific findings identified biological threats, justifying expanded environmental concern and mobilization.

It is widely recognized, therefore, that science has had a significant impact on the current environmental movement. This impact is usually thought to result from the expansion of human knowledge about the workings of the natural world. As humans learn more about natural processes, environmental policies can be increasingly effective. Science takes the natural world as its domain, examining a larger realm of natural processes than is available to the sensory experiences of nonexperts. As science elucidates these natural processes, environmental policy is assumed to straightforwardly step in to manage or protect those environmental processes. We recognize that environmental policies face social and political obstacles, and scientific findings may be incomplete or flawed. Yet in general, the link between science and environmental policy is seen as straightforward and is rarely questioned. As science learns more about natural processes, environmental policies become correspondingly more sophisticated. To use the terminology from the Introduction, this is the *instrumental* aspect of science.

This perspective on the relation between science and environmental policy overlooks, however, a fundamental aspect of the impact of science on the environmental movement. Science has done more than expand human knowledge about natural processes. In addition, science has created a new framework of nature that provides the fundamental basis for the current environmental movement, supplanting previous frameworks. Science provides the basic template from which environmental problems are defined. This is the *ontological* aspect of science, in the terminology of the Introduction.

One of the most important aspects of the scientific framework for environmental policy is the scientific perception of the natural world as a place in which all beings and processes are interconnected. Nature has been described as a web of life, implying that all natural processes influence each other to greater or lesser degree (Andrews 1999). This conceptualization has allowed science to greatly expand the range of environmental problems. Environmental problems tend to be seen as global in scope, in which local problems are connected with larger-scale ecological dynamics. This allows us to perceive the polluted creek in the backyard as connected to national or continental problems of industrial wastes. Science has also enlarged the depth of environmental problems by creating lengthy chains of environmental causality. Thus, the hair spray in one's bathroom can be connected to the stratospheric hole in Antarctica and the hamburger in the refrigerator relates to environmental practices in Brazil.

One of the consequences of the framework is the expansion of the use of science for studying environmental problems. Before the current environmental movement, scientific attention was typically turned to other areas than the study of potential environmental problems. This meant that even if one wanted to document environmental policies, there were few scientific data available. For instance, little scientific attention was paid to the effects of pesticides before Rachel Carson published her book. As the environmental movement gained momentum, however, all sorts of environmental problems became the legitimate study of scientists. Once science provided the framework, scientific attention started to turn toward investigating environmental problems. Thus, one sees the massive expansion of scientific knowledge that is relevant to environmental policy once the environmental movement is created.

There are two other major consequences of this scientific framework for the environmental movement. The first of these is the globalization of environmental problems to all countries of the world. One might think that environmental problems would be conceptualized as local problems, the result of the specific convergence of particular floral, faunal, and natural dynamics. Instead, the globalization of environmental problems is based on the characterization of the environment as interconnected, allowing the environmental problems of one locale—its pollution, resource scarcity, or overpopulation— to be conceptualized as affecting the environment in distant locales. Thus, the environment has become the target of many international organizations, including the United Nations Environment Programme, and is understood as a global concern. Citizens of all countries are expected to be involved with environmental issues in other countries, over lands and sites they will never lay eyes on and over endangered species that they have never heard of.

This leads to the first proposition:

PROPOSITION 1.    *Science has made environmental problems a global concern, and thus, all countries in the world are seen as responsible for environmental improvement.*

The environmental movement is not limited to Western industrialized countries but instead has become a truly global movement, encompassing all geographic regions and increasingly addressing the environmental problems of the third world as well as the industrialized West.

The second consequence of the scientific framework is the consolidation of environmental problems under one rubric. The scientific conception that natural processes are interconnected has created a broad category under which all environmental problems can be lumped. To us in the early twenty-first century, this seems reasonable: because all natural processes are interconnected, all problems regarding nature ought to be connected also. However, historically, this was not at all the case. Air pollution was historically considered a quite separate issue from water pollution, and in turn, both were distinct from land-based waste disposal; these types of pollution affected different groups of people, and the state responded with different legislation and even different administrative agencies. Various natural resources— forestry, mining, energy—also had their own legislation and regulatory systems and tended to be conceptualized as different issues. However, the adoption of the scientific vision of the environment in the late 1960s united these disparate issues into a single, broad category of environmental issues.

This leads to the second proposition:

PROPOSITION 2.    *Science has caused previously disparate social problems to be consolidated into a single set of environmental issues.*

This has allowed the broadening of today's environmental movement to an expanded range of issues, much more extensive than the narrow and specific nineteenth-century programs against coal smoke or for more efficient management of timber cutting.

These propositions are examined in the following brief history of the environmental movement. The history shows that the conceptions of environmental actors and of nature created by science have greatly altered the tenor of environmental discourse and action. Because the current environmental movement is so fundamentally based on scientific frameworks, we assume that all environmental movements must be similarly founded on science. However, historical environmental movements were not primarily based on science. Science has only become the authoritative voice on environmental issues recently, although environmental movements have existed for centuries. The following histories show that science has reframed environmental debates, encouraging the globalization and consolidation of environmental policies in recent decades.

THE SCIENTIFIC CONSTRUCTION OF NATURE

Science has fundamentally contributed to the current environmental movement through its conceptualization of nature. We are aware of environmental problems only because science tells us what the ideal state of nature should be and defines an environmental problem as a deviation from that scientifically defined ideal. It is only because we believe so deeply in the scientific vision of the environment that we can think of environmental problems as obvious.

The scientific perspective begins with a world without humans. In this world, all natural processes are in balance—the water cycle, oxygen and carbon dioxide cycles, nitrogen cycle, and a host of other cycles. Plant communities move smoothly through successive stages, ending with a stable climax community. Predators regulate the expansion of wildlife, keeping populations within an ecosystem's carrying capacity. Even the occasional disasters—forest fires, volcanic eruptions—turn out to be necessary events that contribute to the harmonious balance of nature.

Into this idealized theoretical world, humans enter. From the scientific perspective, everything that humans do disrupts the balance of nature. Human activity is not ever conceptualized as part of natural processes. The one exception is indigenous peoples, who may be seen as living in a state of nature unless they become too proactive—and still, their activities are often not seen as part of natural cycles in the way that other animals' activities are. When human disruption becomes too severe, negative cycles begin, leading to erosion, ozone depletion, and a host of other disharmonious consequences. The current environmental movement is based on this scientific conception of nature, where pristine nature is in balance and the activities of humans create discord.

This conception also suggests the direction for a solution: humans must limit their impact on nature in order to restore natural cycles. Again, it is science that defines the limits to acceptable human impact. For instance, science defines, or claims the authority to define, limits such as the carrying capacity of the ecosystem, the standards for acceptable amounts of air and water pollution, or the point at which the greenhouse effect becomes irreversible. Even if science is unable to precisely set those limits, the whole concept of the balance of nature is based on science.

Standing behind the Clean Air and Clean Water Acts, strip-mining controls, protection of wilderness and endangered species, and the creation of the Environmental Protection Agency is a vision of what an environmentally desirable United States should be. Similarly, in 1972, when the Club of Rome published its computerized picture of a dismal future by a.d. 2000, which was based on increased pollution, population, and consumption, it also had in mind a corrected and improved world, based again on ecological balance (Opie 1985:24).

Thus, science literally defines environmental problems by conceptualizing the ideal state of the environment. The use of another standard for the ideal environment—aesthetic standards of natural beauty, or standards of human health or convenience—would lead to the definition of a radically different set of environmental problems and suggest a different direction for environmental action. Science provides the template on which environmental problems can be measured.

THE SCIENTIFIC CONSTRUCTION OF ENVIRONMENTAL ACTORS

A second major aspect of the scientific framework has been the conceptualization of humans as environmental actors who are capable of affecting the environment and are charged with doing so responsibly. Today, it seems only reasonable that humans should act in response to the environmental crises that scientists tell us exist. However, this has not historically been the case. Before the twentieth century, human beings were not conceived of as having the ability to protect or destroy nature. Nature was simply too vast to be affected by the actions of humans. Nature impinged on humans and human society rather than the other way around. This understanding began to change in the mid-nineteenth century and of course has been completely reversed in the late twentieth century. The modern environmental movement of the late twentieth century focuses almost wholly on human actions and the ways in which humans unbalance the natural environment. Humans today have become environmental actors, charged with protecting the environment from themselves.

At the beginning of the nineteenth century, it was assumed that nature disrupted human activities, causing disease and bringing natural disasters such as flood or pestilence. This contrasts sharply with today's perspective, in which humans disrupt nature's balances. Two hundred years ago, humans strove to protect themselves from the dangers of nature, not the other way around. For instance, wetlands were drained to protect humans from the theorized dangers of bad air from the swamps (Arnold 1996). Today, wetlands are protected as complex ecosystems by international treaties.

It was scientists who were the first to note the impact of humans on the environment. Although human impact seems obvious to us nowadays, it was controversial a hundred and fifty years ago. The science of geology was the first to conceive of humans as environmental actors. In the early nineteenth century, geologists Buffon and Lyell included humans as minor agents in the list of forces shaping the Earth's surface. Traditionally, Earth's history had been based on the Old Testament periods of before, after, and during the flood (Glacken 1985). After geology had begun to theorize a nonbiblical version of Earth history, incidentally increasing the age of the Earth, it became possible to see humans as environmental actors that influenced the Earth's geology.

The impact of humans on the environment was perceived with increasing frequency over the course of the nineteenth century. The forests, which had been viewed as a nearly infinite wilderness in the frontier lands of the United States, Canada, and Australia, began to be tamed to such an extent that people began calling for their preservation rather than their destruction. During the nineteenth century, humans came to the realization that natural resources needed to be managed and conserved. This realization was based on changed conceptions of humans as environmental actors, not on unprecedented environmental destruction. In previous historical periods, large-scale environmental destruction had occurred, creating the sand dunes of Denmark and the removal of the forests of Britain, for instance. However, because humans were not conceptualized as environmental actors in this period, humans did not respond with efforts to reverse environmental deterioration.

It is science, therefore, that is largely responsible for turning humans into environmental actors. Science has noted the effects that humans have on the environment and declared when those effects have created an environmental problem. Humans are now held responsible for their deleterious effects on nature and are charged with reversing these effects so that human society can coexist harmoniously with the environment. The result is that humans are increasingly empowered (as well as blamed) in their role as environmental agents.

## SCIENCE AND THE RECONSTRUCTION
## OF ENVIRONMENTAL DISCOURSE

The current environmental movement is so deeply rooted in science that it is difficult to imagine a program of environmental protection that doesn't rely on scientific concepts and theories. However, the environmental movements of the late nineteenth century utilized science only minimally (Marx 2000; Nash 1989). The growth of the authoritative voice of science into environmental discourse redefined earlier environmental activities, embedding them ontologically in a scientific worldview. This pattern also emerged outside the Anglo West, in countries such as Germany (Dominick 1992) and Japan (Broadbent 1998). The following histories examine the development of the national parks and the growth of air pollution controls, emphasizing their globalization and consolidation as environmental policies.

### National Parks and Biodiversity

One of the nineteenth-century movements, retrospectively labeled *environmental*, was the movement to create national parks. This movement originally had few ties to science. Instead, it was based on romantic visions of the beauty of nature and a rejection of the growing industrialization and urbanization of the period. By the late twentieth century, however, science had greatly altered the goals and rationales for national parks. The earlier vision of romantic wilderness had been redefined as the preservation of complex ecosystemic habitats and the maintenance of biodiversity.

The rationale for national parks in the late nineteenth and early twentieth centuries was the preservation of the beauties of nature. American naturalists such as John Muir and Aldo Leopold eloquently described the splendor of Yosemite and the loveliness of the redwood forests. Their British counterparts enthused about the English moors and countryside, extolling the virtues of long walks for building one's constitution (Marshall 1998). These naturalists argued for the creation of national parks as a means of protecting the aesthetic aspects of the wilderness. One consequence of this focus on the beauty of nature was the emphasis on the amazing, the splendid, or the exotic. This led to the protection of breathtaking scenery such as the Grand Canyon, the cliffs of Yosemite, and unusual landscapes such as the geysers of Yellowstone. However, land of only everyday beauty was deemed to have little moral or aesthetic value and consequently did not merit protection (Runte 1987; Foreman 1998).

The preservationist movement led to the establishment of national park systems in the United States, Canada, and Australia. The goal of the national parks was to preserve wilderness for future generations. In order to do that, human activity was limited to hiking and camping; activities that interfered with nature on a larger scale such as farming, herding, or lumbering were forbidden. The 1964 Wilderness Act of the United States defined "a wilderness, in contrast with those areas where man and his own works dominate the landscape . . . as an area where the earth and its community of life are untrammeled by man, where man himself is a visitor who does not remain" (Wilderness Act of 1964, sec. 2[c], cited in Callicott 1998:349).

The concept of national parks was based on Anglo-American values and the romantic literary traditions of those countries. However, the concept of national parks was not necessarily appreciated in cultures that did not share those Anglo-American values. To the Western mind, pristine nature can only exist without humans—and thus, the preservation of wilderness requires restrictions on the impact of humans. However, many non-Western cultures do not share this Western dualistic view of humans in opposition to nature (Turner 1998; Kalland and Persoon 1998).

These differences have led to criticism of the ideology of national parks by environmental philosophers in the third world. In developing nations, the land set aside as "wilderness" was often the home and livelihood of indigenous peoples (Guha 1998). Conforming to Western definitions of the national park has resulted at times in the forced relocation of these peoples, often entailing significant hardship (Harmon 1998). Additionally, Western notions of the exotic and splendid are not necessarily the same as indigenous standards. These differences made many third world countries slow to create national parks.

The injection of science into the environmental movement has, however, transformed the meaning of national parks, allowing their widespread globalization. Beginning in the 1970s, ecological scientists began to realize that large numbers of species were becoming extinct (Foreman 1998). These extinctions were due partly to direct killings of previously plentiful animals such as buffalo, gray wolves, blue whales, and rhinoceros, but were due in larger part to the disappearance of animal habitats. Scientists argued that these species were necessary to maintain the complex ecosystems that humans inhabit, and not merely for aesthetic appreciation.

Scientific theories claimed that all species had value, which was attributable to the hidden benefits of biodiversity. Ecological biodiversity is claimed to maintain critical water, air, and biological cycles, and to have the potential to genetically improve domesticated plants and animals and provide the resources for hitherto unknown medicines (McNeely et al. 1990). Furthermore, these benefits are presented as objective benefits for a large proportion of humanity, rather than being limited to the appreciation of a few hikers and campers. In earlier times, people might appreciate a particularly beautiful site or interesting animal, but this appreciation was limited to people who actually observed and had contact with the sites and animals. Today, people are expected to protect animals they will never see or know about, simply on the basis of abstract principles of the value of biodiversity.

National parks are now seen as potential repositories of biologically diverse genes and natural processes. The scientific perspective has also forced preservationists to appreciate the small and mundane species and landscapes as well as the magnificent ones. Indeed, the failure of national parks to slow the extinction of species has been blamed on the tendency to protect the magnificent but infertile, as opposed to lush, fertile, and biologically diverse, lands (Grumbine 1998). In addition, the scientific perspective has encouraged the placement of biodiversity on the international agenda. This has resulted in several international treaties and programs, such as the 1972 Convention on the International Trade of Endangered Species and the 1992 Biodiversity Convention.

Thus, the grounding of the national parks movement in scientific ontology has encouraged its globalization. Environmentalists in the West are still moved by aesthetic considerations, but aesthetics are less appreciated in third world countries, where basic health and economic needs are paramount. Scientific claims to objectivity and value-neutrality imply that the sponsorship of biodiversity is not merely a Western value, but an objective good for all humanity. As a result, nearly every country in the world has signed the 1992 Biodiversity Convention, and the environmental movement is no longer a Western phenomenon but has become fully global (Meyer et al. 1997b; Frank et al. 1999).

## Air Pollution and Climate Change

The second major ontological effect of science on the environmental movement was the creation of the category of "environmental problems." Before the modern environmental movement in the 1960s, this category did not exist. Although the state dealt with problems of air and water pollution, these problems were seen as aspects of urban planning or health policy rather than as aspects of the natural environment. The state dealt separately with these problems, regulating them through different pieces of legislation and administering them via different state agencies.

Thus, early environmental legislation was not conceptualized as environmental protection; rather, it was a response to diverse concerns about public health or the control or management of industrial activities. This early legislation was a result of the broad expansion of the scope of state activity that had been occurring since the nineteenth century (Boli 1987), rather than a concern for the environment itself. Thus, core countries such as Britain and the United States already had a system of air and water pollution legislation in place by the time the modern environmental movement began (Vogel 1986; Andrews 1999).[1] These early environmental efforts utilized science incidentally, if at all. Typically, early environmental legislation was prompted by the direct perception of obvious problems—bad smells in the air or water, the loss of forest as a result of ill-managed cutting. These problems were tangible problems that clearly affected quality of life, and scientific observation was not needed for these problems to be observed or rectified.

By the 1960s, however, science had come to play a much more ontological role in the definition of environmental problems. Air and water pollution began to be seen as instances of environmental degradation, rather than separate phenomena. This led to one of the major innovations in environmental policy, in which a single ministry or agency came to be responsible for the category of "environmental problems." The first was the United States' creation of the National Environmental Policy Act in 1969, which created a single agency, the Environmental Protection Agency, to coordinate all policies relating to the environment. Other countries quickly took up this practice of having a basic umbrella environmental policy and environmental ministry (Janicke and Weidner 1997).

The history of air pollution legislation illustrates these points. Serious complaints of air pollution in England go back at least to 1257, and attempts to control the causes of air pollution go back to 1273 (Brimblecombe 1987; Markham 1994; Vogel 1986). The famous London fog of the nineteenth century was the result of air pollution, chiefly coal smoke mixed with mist and trapped by inversion layers (Grinder 1980). The effects of air pollution in the nineteenth century were widespread. The Victorians found it impossible to wear white suits in London; they tended toward cream-colored clothing and dark furnishings as a response to the omnipresent smoke. Indeed, some speculate that umbrellas are traditionally black because the English rain was so filled with soot that any other color would have been impractical (Brimblecombe 1987). More seriously, days of severe fogs were associated with increases in the death rate. Even as late as 1952, four thousand deaths were attributed to four days of particularly nasty fog in London (Brimblecombe 1987).

Despite these terrible conditions of air pollution, worse than in many less developed countries today, little was done to control air pollution. Although it was obvious that humans were responsible for the pollution, humans were not yet environmental actors charged with environmental health. London residents passively accepted the smoky air, adapting by planting their gardens with increasingly hardy species of plants and wearing duller-colored clothing (Brimblecombe 1987). Not until the late nineteenth century, after centuries of air pollution, did a social movement begin, with the creation of the Coal Smoke Abatement Society. At about the same time, a popular movement against coal smoke was begun in the United States, which succeeded in getting city ordinances against smoke passed in St. Louis, Chicago, and Pittsburgh (Grinder 1980; Tarr 1996). However, these societies were narrowly targeted against coal smoke, not against air pollutants or pollution in general.

Science was not called on in these early air pollution reforms. Even before the nineteenth century, it was fairly clear to everyone that the problem lay in the burning of coal for fuel. Although scientists may have joined societies advocating cleaner air, their word carried no more weight than that of other concerned citizens. Unlike today, when science is necessary to know when an environmental problem exists in many cases, the air pollution of the nineteenth century was obvious to the sensory experiences of nonexperts. This did not mean that the problem was easy to fix but rather that the definition of the problem and the direction of the solution did not require science to comprehend. The British Parliament passed the first Smoke Nuisance Abatement Act in 1853, which did little to improve London air. Britain passed further legislation, the Clean Air Act, in 1956 (Kormondy 1989).

Today, countries routinely enact clean air legislation, frequently modeled on the British or American acts (IUAAPA 1991). Over time, particulates and gases such as sulfur dioxide, carbon monoxide, and hundreds of others have become included as air pollution (Kormondy 1989). These pollutants require scientific instruments to measure and monitor, and frequently emissions standards are set after consultation with scientists. Furthermore, air pollution has become globalized and transboundary effects have become recognized, such as the role of U.S. emissions in the creation of acid rain in Canada.

Science has played an even more dramatic role in the definition of the problem of the depletion of the ozone layer. The chemical compound ozone was first discovered by scientists in 1839 (Benedick 1991). The ozone layer in the stratosphere is only observable by scientific instruments—indeed, the stratosphere itself is beyond the direct experience of the nonexpert. The hypothesized effects of ozone depletion are also obscure, such as a statistical increase in skin cancer and a gradual warming of the Earth's climate (Benedick 1991). Moreover, the area of greatest ozone depletion is over the Antarctic, where few humans will ever visit, much less live. Science has played a critical role in the detection of the problem of ozone depletion and also suggested standards by which to halt the depletion. The suggestions of science quickly became a worldwide issue and were acted on in 1987 with the creation of the Montreal Protocol on Substances That Deplete the Ozone Layer.

Other types of state activity that are now categorized as environmental protection have similar histories, such as the protection of natural resources such as forests and mineral deposits, or control over various types of water, noise, and hazardous waste pollution. Just as with air pollution, nation-states attempted to control these problems as part of their responsibility over public health or land use but did not consider these problems as manifestations of a single category of environmental problems. Instead, each type of problem was treated with separate legislation and regulatory agencies. The ontological role of science in the current environmental movement has redefined these formerly disparate problems into a single category.

QUANTITATIVE ANALYSES

So far, this chapter has argued that science has played a fundamental onto-logical role in the current environmental movement, encouraging both its globalization and its consolidation under a single rubric. Today's environ-mental movement includes programs such as (1) the creation of national parks, (2) the encouragement of biodiversity and the international protection of endangered animals, (3) the creation of environmental ministries and envi-ronmental umbrella legislation, and (4) the international protection of the ozone layer. These programs are among many that are ontologically rooted in the scientific perspective of the environment.

National parks are an example of an environmental movement that did exist before science had much impact on the environmental movement. However, science has allowed the Western form of the national park to diffuse to most countries of the world. In 1920, national parks were limited to the United States, Canada, Australia, South Africa, and New Zealand, along with a couple of Scandinavian countries. However, by 1990, one hundred twenty-nine countries had national parks, a dramatic indicator of globalization.

The ontological impact of science on the environmental movement is also evident in the creation of new mechanisms of environmental protection. The scientific perspective allowed the creation of mechanisms that would not have existed, or would have existed in a radically different form, if not for the authoritative input of science. For instance, the creation of environmental ministries and environmental umbrella legislation began in the late 1960s after science had become the authority on the environment. Only seven countries had environmental ministries in 1975, but by 1993, this number had increased to forty-seven.

Furthermore, the 1972 Convention on International Trade in Endangered Animals, the 1987 Montreal Protocol on Substances That Deplete the Ozone Layer, and the 1992 United Nations Convention on Biodiversity Convention literally did not exist until the scientific perspective had become dominant in the environmental arena. By 1992, nearly every country in the world had signed these treaties, constituting yet another indicator of the globalization of the environmental movement.[2]

The impact of science is further illustrated by the following event-history analyses. These analyses show that domestic science activity is a significant predictor of environmental protection such as the creation of national parks, the building of environmental ministries, and the signing of two environmental treaties. A separate event-history analysis is performed for each of the four dependent variables.[3]

The first analysis examines the rate at which states make national parks. The first national park was Yellowstone Park, created in 1872 in the United States. Other frontier countries, such as Canada, Australia, and South Africa, followed rapidly, eventually followed by European countries. After 1970, however, national parks globalized rapidly as the concept of biodiversity became widespread, spreading to countries all over the world.[4]

The second analysis examines the rate at which states create environmental ministries.[5] The first environmental ministry was created in the United Kingdom in 1971. Other states, such as Canada, the Netherlands, and Norway soon followed. By 1990, states such as the Congo, Mali, and Mauritius also had environmental ministries—and often the corresponding umbrella environmental policy to match.

The third and fourth analyses study the rate at which countries sign two particular environmental treaties. One treaty is the 1972 Convention on the International Trade of Endangered Species. This is the most important international treaty protecting endangered animals, particularly endangered animals that have commercial value, such as the rhinoceros, the gray wolf, and the elephant.[6] Science has played a major role in the globalization of concern for endangered species by transforming the international discourse to one based primarily on the merits of biodiversity rather than on aesthetic or sentimental rationales.

The fourth analysis examines the parties and signatories to the Montreal Protocol on Substances That Deplete the Ozone Layer.[7] The protection of the ozone layer is fundamentally based on scientific evidence of the connection between chlorofluorocarbons (CFCs) and ozone depletion, on scientific measurements of the ozone layer, and on scientific hypotheses on the negative effects of ozone loss. Environmental protection of the ozone—and even proof of the very existence of the ozone layer—depends on science.

Four independent variables are used for each of the four analyses. The first is the level of domestic science in each country, measured by the number of domestic scientific organizations (logged).[8] The next three variables are controls measuring the increase in environmental problems over time. One vari-

TABLE 11.1

Effect of Domestic Scientific Organizations on Nation-Level
Environmental Activities: Results from Event History Analyses

| Variable | Creation of National Parks | Founding of Environmental Ministry | CITES Treaty on Endangered Species | Montreal Treaty on Ozone Protection |
|---|---|---|---|---|
| Science | 0.14 *** | 0.39 ** | 0.51 *** | 0.26 ** |
| | (0.03) | (0.16) | (0.16) | (0.11) |
| Urbanization | 1.94 *** | − 0.31 | − 0.07 | 0.26 |
| | (0.24) | (1.10) | (0.78) | (0.57) |
| Industrialization | 0.05 * | 0.22 ** | − 0.35 *** | 0.08 |
| | (0.03) | (0.11) | (0.13) | (0.09) |
| Population | 0.27 *** | − 0.20 ** | 0.17 ** | 0.02 |
| | (0.02) | (0.10) | (0.08) | (0.06) |
| Constant | − 6.85 *** | − 1.79 | − 5.95 *** | − 2.68 *** |
| | (0.38) | (1.44) | (1.18) | (0.88) |
| $\chi^2$ (df) | 728.92 *** | 24.85 *** | 28.08 *** | 30.31 *** |
| | (4) | (4) | (4) | (4) |
| No. of events | 1182 | 46 | 78 | 121 |
| Years of analysis | 1900–1990 | 1971–1993 | 1972–1988 | 1987–1995 |

$^*p < 0.10$, $^{**}p < 0.05$, $^{***}p < 0.01$. Data are expressed as coefficient (standard error).

able is the level of urbanization, measured as the percentage of the population in cities with more than a hundred thousand population (data from Banks 1990). The second variable is the level of industrialization, indicated by iron and steel production per capita, logged (data from Singer 1990). The third variable is population, logged (data from Banks 1990).

Table 11.1 lists the results of four event-history analyses. In all four analyses, the level of domestic science in a country is a strong positive predictor of environmental protection. This supports the propositions that science has encouraged environmental protection activities. Countries with more domestic science are more likely to do all four environmental activities than countries with less domestic science.

The effects of urbanization, industrialization, and size of population are weaker and less consistent than the effects of domestic science on these environmental activities. Urbanization is positive and significant for the creation of national parks, but nonsignificant for the environmental ministries and the signing of the environmental treaties. Industrialization is positively predictive of air pollution legislation and negatively predictive for endangered species protection. And population is a strongly positive predictor for national parks and signings of the endangered species treaty, while having negative effects on the likelihood of creating an environmental ministry.

Overall, these results suggest that science has had a significant impact on environmental protection activities. The sheer number of countries that participate in these activities indicates that science has encouraged the globalization of environmental protection. Nearly every country in the world participates in some environmental protection actions, and the more domestic science they have, the more likely they are to participate. Also, the creation of environmental ministries is a common occurrence in the world. More than 30 percent of countries in the world had an environmental ministry by 1995, and more than 50 percent of countries had a ministry devoted to some aspect of environmental protection, such as water resources or air pollution (Steinberg 1950-1995). This provides further evidence that the scientifically created category of "environmental problems" has become accepted and is frequently seen as meriting the attention of an entire ministry.

In sum, science has fundamentally shaped current environmental policies. Science has constructed humans as environmental actors, responsible both for the destruction of the environment and enjoined with the improvement of environmental problems. Science has also defined the ideal state of nature, constructing current understandings of environmental problems and implying the form of environmental solutions. This has resulted in the dramatic globalization of environmental problems and the consolidation of previously disparate problems into the single cognitive category of environmental degradation. Thus, science has had a major ontological impact on today's environmental movement.

## Chapter 12

Science, Democracy, and
Governmentality

During the August 1991 uprising against the Soviet Union, the world followed the events as they unfolded. University faculty members in Moscow used E-mail to inform and update the global community, circumventing Soviet censorship. Reports through this medium helped shape this political episode: once foreign powers learned of the political turmoil and its detailed, moment-by-moment unfolding, they exerted pressure on Boris Yeltsin and helped bring about a peaceful conclusion to the political drama. This story and similar ones about other political hot spots around the world are often cited as examples of how scientific activities and linkages, such as the availability of academic E-mail, shape political events. This kind of interpretation of the link between science and political openness is too narrow: it connects the specific actions of scientists to political events, rather than emphasizing the general and cultural influence of science on local polities. The literal stories understate the global authority of science to comment on social priorities, to define and construct social entities, and to establish homogenizing connections among national polities. In this chapter, we apply a broader vision of science and of its global place by investigating the impact of globalized science on democratization and the modern governmental system.

The accelerated globalization of science, carried along by seemingly routine formal policies anticipating that science leads to prosperity (Chapter 4), has additional unrealized consequences. Most dramatically, we show here that the globalization of science results in a cross-national movement toward a liberal and participatory polity. Although science is globalized in the name of advancing national economic development and thus is seen as a tool for achieving this social goal, the overall process also diffuses modernist notions of liberal governmentality (see Foucault 1991; Rose and Miller 1992). In this sense, the effects of expanded modern science are most profound on many more general features of the modern nation-state. Science globalization influences a worldwide constitution stressing the liberal form of nation-statehood, as well as the associated identities and practices. The worldwide institutionalization of science supports the global expansion of democratization, the general empowerment of a wide range of political actors, and growth in many forms of political participation. Overall, this linkage between science globalization and the worldwide diffusion of liberal nation-statehood illustrates the ways globalization processes in different social spheres are interwoven. Globalization processes involve the worldwide diffusion of *packages* of reforms and policies, and democratic foci on human rights certainly make up one of these packages.

To discuss this link between scientization and democratization, this chapter is structured along two main themes. First, we discuss globalized science as having broad political implications for nation-states that incorporate it. On the basis of cross-national evidence, we show how the institutionalization of national science supports the rationalization of national practices and, specifically, democratic and participatory politics. We thus advance the argument that science sets part of the cultural basis for liberal nation-statehood. Second, we argue that globalization processes in various social spheres are interlinked and coconstitutive. Hence, we discuss the general relationship between science and nation-statehood.

SCIENCE AND PARTICIPATORY POLITICS

*Science as Shaping Local Polities*

The instrumentalist expectation that science leads mainly to economic development through the mediating effects of technology and a skilled labor force limits any regard for science as a general and cultural framework (see Chapter 4). But science globalization alters the nature of nation-statehood through setting the basis—normative and networking—for participatory politics.

How does science pave the way for actor-based forms of governance? Both the *actions* of scientists and the incorporation of the *culture* of scientific conceptions of nature fuel changes in political culture. On the action side, drawing on the legitimacy and credibility of science, scientists act to change both the form and direction of governance. They serve as advisors to political elites and their views assist in shaping governance procedures (see Jasanoff 1990). Political action is often mobilized around scientific evidence and expertise, on issues running from human medical rights to the environment to proper forms of rational management. Beyond such internal changes in societies, scientific training and work are based on international cooperation. So science serves as a networking mechanism between each nation-state and global society. Scientists participate in international conferences and routinely communicate and collaborate with scientists from other countries. They bring their countries into intense interaction with the world polity, serving as "receptor sites" for such international fashions as democracy and human rights (Frank et al. 2000a). Overall, then, the behaviors and performances of scientists directly affect local governance.

Second, the *culture* of science permeates modern societies that incorporate the seemingly benign scientific practices into their elite structures. This process of scientization imprints national practices with Western modernist concepts embedded in science. One such Western modernist concept, a part of the paradigm of modern science, is the notion of social actorhood. Here, we argue that this concept of actorhood, introduced to various nation-states partly through the process of scientization, leads polities to become participatory and democratic, as well as more accepting of a variety of human rights. We discuss participatory politics along two dimensions: first, the construction of political actors, and second, the empowerment of such constructed political actors into political engagement. We focus on liberal democratic practices, or practices that are interpreted as participatory and democratic in nature.

*The Culture of Science and Actorhood*

The modernist concept of actorhood sees an actor as a constructed social entity (for example, an individual, a social group, an organization) with attribute agency supporting its proper role in the shaping of social destiny (Meyer and Jepperson 2000). Hence, the modernist understanding of an actor constructs, or defines, the entity or category and also empowers that entity for action by attributing legitimate agency to it.

Scientization plays multiple roles here. First, it constructs entities in nature (social and other) and sets them in relation to each other through causal processes; this general process of cognitive rationalization is at the background of all rational action. Second, among the constructed entities are human social actors, from individuals to societies (and now global society), whose rights and roles are rationalized and articulated; this enables a rationalized polity to be set in place and supports the process of defining its participants. Third, through the attribution of agency to these constructed entities, scientization encourages their political engagement. Overall, the world (including the social world) is a rationally ordered place, and human actors can understand, use, and make the laws involved. In this fashion, scientization is central to the institutionalization of participatory polities in various nation-states (see also Chapter 1).

One element of this is the role of science in the modern project of rationalizing social and natural processes and entities so that coherent public action is possible (Toulmin 1990). Science, as the authoritative voice of modern thinking, codifies rational thinking and rational order. It thus links cause and effect and replaces various forms of authority with the authority of deliberation, calculation, and judgment. Such a cause-and-effect world is one that can and must be managed, and within it, various entities are regarded as the source of action and are thus actors. Science, as the epicenter of the modern rationalization process, both codifies the ground on which actorhood operates and defines actors.

More concretely, the taxonomic nature of science constructs and standardizes knowledge categories (including such categories as political actors). Scientization constructs differences and similarities among these objects or across these groups. For example, zoological taxonomies differentiate insects from arachnids and group house cats, tigers, and leopards together as members of the cat family. Similarly, eugenics differentiates among human racial categories, whereas anthropology links ethnic and racial categories. This scientific, taxonomic attitude is employed for both the natural and the social worlds. Hence, science, by making the social craft of creating knowledge categories thinkable, makes the crafting of other categories thinkable too. Hence, groupings of objects construct both the object and the group. For example, the scientific distinction of women as a unique category supports the construction of womanhood as both potential object and subject of public action. It both defines women as a distinct social and political category, and it labels each member of the group as "a woman," thus deeming them eligible to make political claims that are based on this label.

The construction of social entities relies greatly on scientific reasoning and scientific evidence. Scientific discourse provides the "language tools" and the "evidence" with which one category is differentiated from another. This is obvious in the case of the physical world, in which scientific analyses govern physical resource definitions, technologies, and conceptions of proper (for example, environmental, safety, medical) constraints. But it is striking in the social world, too. For example, gay and lesbian activists cite genetic evidence from the Human Genome Project, which points to the existence of genetic markers of one's sexual orientation, to define themselves as a distinct social group. By drawing on such scientific evidence, they legitimize their definition as social actors. In a similar ontological move, consumer groups rely on theories of the economic market to define their social actorhood. The social category of "the consumer"—defined in contrast with "producers" or the "government"—is constructed from a scientific theory on economic relations. Social groups enjoy greater legitimacy for their claims of distinctiveness, once their claims are grounded in so-called scientific evidence.

The role of scientization in the institutionalization of actorhood goes beyond the construction of social and natural entities. Scientization also empowers these social entities through the conceptualization of their actorhood as competently agentic. Agency is defined in terms of commanding assertiveness over destiny, or taking action. This modernist conceptualization of actorhood is immersed in the Judeo-Christian religious notions of one's control over one's destiny. In science, this attitude is secularized to take the obvious meaning of human control over nature. In other words, "doing" science means to observe, sort, categorize, and analyze—or in general, control nature. As Meyer and Jepperson (2000:103) observe in their description of contemporary analysis, "Nature is tamed and demystified through the extraordinary development, expansion, and authority of science." The approach of modern science, which distinguishes it from premodern science, is that nature is passive and humans are its manipulators. These activist, or agentic, notions translate into, among other things, actor-based norms about the nature of the social environment.[1] In this sense, science-embedded notions of agentic actorhood establish a normative basis for the institutionalization of participatory politics.

*Empirical Support from Cross-National Investigations*

The effects of scientization on the nature of polities worldwide can be examined empirically. Next, we present the results of cross-national longitudinal models investigating these effects during the past three decades. In these models, scientization is operationalized as the extent of national science practice, indicated by cross-national measures of citations of scientific papers from the *Science Citation Index*, of science books published (from the data files of the United Nations Education, Scientific, and Cultural Organization), and membership in International Council of Scientific Unions organizations (for detailed sources, see Drori 1997).[2] Participatory politics is operationalized with a variety of measures of the broadening of social rights in a country (Table 12.1) and of the intensity of political engagement in a country (Table 12.2). In all these analyses, a basic measure of national development is held constant as a control variable[3]: the expansion of science practice is highly correlated with national development, and our aim is to show the impact of science over and above the effects of development.[4]

Most of the analyses reported here take the form of panel models—that is, we look for the impact of early scientific expansion on later social rights with earlier levels of rights held constant. Regarding the modeling of these relations, we control for the influence of the level of national development, and we examine these relations casually (although in some instances, data limitations restrict us to cross-sectional models). Estimation methods vary: we rely on both structural equation models or ordinary least squares regression models (see Chapter 2 for methodological details). And the overall time periods considered fall within 1970 to 1995. Specific details for each analysis (estimation method, design, and time period) are provided in Tables 12.1 and 12.2.

The empirical cross-national analyses reported in Table 12.1 show that in the contemporary period, the expansion of national-level science practice is associated with the extension of political rights to various social categories. Science practice is associated with the cross-national promotion of various discourses of rights (Table 12.1). These define human beings by their gender (women's rights[5]), sexual preference (gay and lesbians' rights[6]), market position (consumers' rights[7]), or generalized "selfhood" as humans (human rights[8]), or by their relation with their natural environment (environmental rights[9]). These findings are consistent and robust in spite of the slight variation in method, and in spite of the variety of indicators used to connote rights.

These rights' discourses cover a wide range of rights issues: humanistic versus market rights, individual versus group rights, and even human versus non-human rights. However, the different rights discourses share the notion that social categories are social actors. In other words, the definition of such categories as distinct entitled entities to constitute actors. That is, the categories

TABLE 12.1

Science Practice and the Construction of Political Actors: Summary of Results, with National Development Controlled

| Theme | Measure | Method of Estimation, Model, and Time Period | Relation of Science Practice with Indicator |
|---|---|---|---|
| Women's rights | Women's status index | OLS, cross-sectional 1980–1985 | + |
| | Women's equality index | OLS, cross-sectional 1980–1985 | + |
| | Gender development index | OLS, cross-sectional 1980–1992 | + |
| Human rights | Compliance with international initiatives | OLS, cross-sectional 1970–1991 | + |
| | Human rights index | OLS, cross-sectional 1980–1985 | + |
| | Human development index | OLS, cross-sectional 1980–1992 | + |
| | Absence of repression index | OLS, cross-sectional 1980–1986 | + |
| Consumers' rights | Older rights structuring | SEM, cross-sectional 1970–1992 | + |
| | Size of organizational field | OLS, cross-sectional 1980–1992 | + |
| Gay and lesbians' rights | Gay and lesbian rights' index | SEM, panel 1980–1994 | + |
| Environmental rights | Environmental treaties ratified | SEM, panel 1980–1990 | + |

All effects are significant at the 0.05 level.

TABLE 12.2

Science Practice and Political Engagement: Summary of Results, with National Development Controlled

| Theme | Measure | Method of Estimation, Model, and Time Period | Relation of Science Practice with Indicator |
|---|---|---|---|
| Popular action | Popular mobilization factor | SEM, panel 1980–1985 | + |
| | Voting | OLS, cross-sectional 1980–1980 | + |
| Political resources | Freedom of association | OLS, cross-sectional 1980–1985 | + |
| | Access to media factor | OLS, cross-sectional 1980–1984 | + |
| Liberties | Civil liberties index | SEM, panel 1970–1990 | + |
| | Political liberties index | OLS, panel 1980–1993 | + |
| | Democracy index | OLS, cross-sectional 1980–1985 | + |

All effects are significant at the 0.05 level.

are generally constituted as having the capacity to act, the responsibility to act, and the responsibility to act for others (see Meyer and Jepperson 2000). This process of defining a single individual, a social group, or a constructed corporate entity as an actor bridges individual and collective action because members of a category can and should advocate the rights of others, and others should properly support the rights of the focal category.

This shared feature of the various rights' discourses is also manifested in their similar empirical patterns of relationship with science. The general positive effects are found despite this variety of rights discourses (Table 12.1). In the different models, we see a consistent pattern, regardless of the specific dependent variable: more intense science practice in 1970–1980 is positively and significantly associated with higher rates of the sociopolitical rights in the 1990s. Expanded national science practice in 1970–1980 is positively and significantly associated with the expansion of women's, human, consumers', gays and lesbians', and environmental rights.

This associational pattern is probably not limited to the spectrum of social categories set out in these analyses. Rather, additional social categories—such as ethnic minorities, racial groups, labor, or children—can be the focus of future studies. A wide variety of rights discourses shares a strong relationship with the process of scientization. The role of science in relation to these rights' discourses is in establishing the conceptual basis for such discourses as thinkable and practicable. Science, through its capacity as a networking mechanism to world society and through setting the conceptual base, encourages people to define themselves by these stylized categories of "rights" and thus to become actively involved in these politics of representation.

The influence of scientization processes on actorhood extends beyond the construction of political actors. It also supports agency, or the rights and capacities of actors to represent their own (and others') interests. It helps in infusing the constructed actors with both the sense and the right in the eyes of others that they have a legitimate place in the political process. Scientized societies, which are therefore more infused with this notion of actorhood, have increasingly institutionalized more participatory forms of politics, as in a more expanded liberal democracy. Overall, the global scientization process encourages popular mobilization for political action and supports a political culture of participation.

Again, empirical cross-national analyses confirm that the level of intensity of science practice in 1970 is associated with greater political engagement in the 1990s. The level of national science practice is consistently related with the empowerment of political actors in recent decades (Table 12.2). First, science practice is associated with enhanced political action—either contesting action (that is, popular mobilization for street action: demonstrations, riots, and strikes) or more institutionalized forms of political action (that is, electoral democracy). Second, expanded science is associated with the greater availability of resources for political participation—the freedom of association and access to information. Last, science practice is associated worldwide with greater liberties—civil liberties, political liberties, and democratization. Overall, science sets the basis for a worldwide political change toward greater liberal democracy.[10] Moreover, those previously communist societies that had high levels of science practice in 1970 made a rapid transition to more liberal polities during the 1990s.

In summary, the empirical evidence reviewed here confirms that scientization influences political cultures worldwide (for more detailed analyses, see Drori 1997). The scientization of society, through the introduction of science practices and their embedded modernist notions, supports a political culture of participation and mobilization.[11] Scientized societies more willingly accept standardized categories and practices, more readily employ claims for actorhood on the basis of such categorizations, and more easily mobilize for political action, thus asserting their actorhood. As Barnett and Finnemore (1999:711) write, "categorization and classification is a ubiquitous feature of bureaucratization that has potentially important implications for those who are being classified." The global process of the cultural construction of empowered actors carrying ultimate value (Strang and Meyer 1993) is hence fueled by the worldwide expansion of science practice and discourse. International discourses, which are carried by international organizations, diffuse such fixed meanings (see Keeley 1990): in the domain of science, the international science/development bureaucracy fixes meanings of actorhood worldwide (Chabbott 2002). Thus, global scientization produces a "scientific man" (and organization, and even national state)—a constructed actor, infused with the logic of scientism and drawing on scientific legitimacy for its definition (see Chapter 6 for educational implications). And "scientific man" serves as the basis for the reigning liberal model of national polities.

The effects of science on the nature of national polities worldwide are not limited to notions of actorhood. In a similar and parallel way, science-based modernist ideals of objectivity and impartiality adversely affect many traditional political habits—nepotism, for example, or informal and personalistic forms of authority. It seems likely that trends of rationalism and instrumentalism in science discourse support instrumentalism of political culture, such as "the replacement of various forms of the politics of charisma and mass enthusiasm by a cooler politics of deliberation, calculation, and public opinion disciplined by enlightenment" (Ezrahi 1988:184). Similarly, Skolnikoff (1993) shows how the globalization of information technologies alters the political organization of societies worldwide, by encouraging the forms of a more open society, a decentralized political structure, a decentralized economic structure, and more diffused (rather than concentrated) military power. In this sense, as in other dimensions that are explored below, scientization and liberal democracy complement each other.

## GLOBALIZATION PROCESSES AS "PACKAGES" OF REFORMS

### Science and Democracy as Interlinked Facets of Liberalism

Much like the classical Greek idea that the humanities are an education for democratic politics, global scientization provides a conceptual and organizational basis for contemporary politics. Since 1970, it seems that science globalization and participatory politics are interlinked and coconstitutive. Yet although expanded science is related to liberal politics in the last three decades, this does not seem to be the case historically. Nazi Germany and the communist bloc countries set examples, for other historical periods or cultural environments, where intensive scientific activity did not produce progressive politics and the empowerment of political actors. Rather, in these cases, rapid scientific growth and achievement paralleled disrespect of individual and group rights and the dominance of collectivist, or nationalist, ideologies. What, then, creates the observed connection between science and liberal progressive politics?

During the past three decades, the global world polity consolidated around liberal models, and both science and politics drew on such liberal thinking in their globalization. Liberal models define, and legitimate, nation-states, corporations, and individuals as actors. They also value freedoms in the political and economic spheres. In a general way, all these features emphasize sovereignty—of the individual, of the group, of the market. In this sense, according to liberal thinking, science and liberal democracy are regarded as complementary elements because they share the essential themes of actor-hood, freedom, and entrepreneurship. Ezrahi, for example, argues that science and liberal democracy share a common sense of authority. First, science, in its search for universal laws, rejects "claims in the name of transcendental, hier-archical, personal, or other democratically illegitimate principles of authority" and thus reinforces the principles of democratic speech and action (Ezrahi 1988:186). Second, "the authority of science and technology is consistent with liberal-democratic decentralization . . . action is not arbitrary, but guided and checked by a functional test of technical adequacy" (197). Overall, science and democracy both draw from liberalism and thus are interlinked (see Ezrahi 1990). Yet although science and democracy are interlinked through their connection with liberalism, and although science is held up as a means for achieving the utopias of liberalism (see Ben-David 1990:528–29), science is in liberal terms just one additional social sphere. The liberal model is multifaceted and is thus meaningful to, and addresses, various social spheres. The "science for development" model and calls for greater democra-tization are two examples for such global reforms. Overall, the globalization of science is the worldwide diffusion of the linkage not only between science and development, but also with democratization and with security concerns. In this sense, the globalization of science is a part of a qualitative transforma-tion of the world toward liberalism as a general world model, or a liberal *pack-age* of reforms.[12]

The rapid worldwide expansion of this liberal package reached a peak in the 1980s. Throughout the 1980s, under the leadership of Reagan and Thatcher, liberal thinking took a global stronghold, and this trend reached its pinnacle with the 1989 "victory of liberal democracy" through the col-lapse of the communist bloc. During this decade, market-oriented econo-centricity[13] is the most dominant episteme in social policy, as well as in sci-ence and in developmentalism. In this cultural atmosphere, which repeatedly employed the scripted image of the economic miracle of the Tiger economies,[14] "developing countries" turned into "emerging markets," and the terms *liberalization, privatization,* and *democratization* became the policy buzzwords worldwide. All these phrases call for similar policy measures, namely the return of Adam Smith's invisible hand to social life. Their influence permeated the field of science.[15]

*Science and Governmentality*

Scientization, as a general matter, makes modernist themes thinkable and practicable. The alliance between science and the nation-state is the Western version of the art of governance, or the principle of governmentality (Foucault

1991; see also Rose and Miller 1992). Governmentality describes a hegemonic organization and perspective—in modern thinking, the state—as transcending civil society. As an analytic concept, it joins together the discourses of state power, national identity, and instrumentality.

Governmentality, both in general and in particular through the discourse of science, is simultaneously individualizing and totalizing. It defines particularistic social entities (such as women, minorities, and racial groups) while also subjecting these entities to the totalizing effects of nation-statehood, international relations, and global processes. In other words, it concurrently permits the construction and empowerment of individualistic identities and subjects such identities to the homogenizing pressures of globalization. This is the modern-day dialectics of identity, or the close relationship between claiming individual rights and relying on universal claims of uniqueness. Globalization processes diffuse notions of identity worldwide (for example, notions of ethnic or tribal affiliation), thus homogenizing even this particularistic, or localized, notion. Roland Robertson (1994) refers to this self-contradictory process as the "universalism of particularism." Post–World War II neoliberal thought is a prime instance of such *Omnes et Singulatim*[16] form of governmentality.

This liberal mode of governmentality has two additional features: it is based on organizational rationalization, and it transforms knowledge into a form of power. First, governmentality is embodied in a set of organizational and rationalized procedures. These standardize the gathering of information, management procedures, economic models, and the actualities and perceptions of governance (see Chapter 13). Second, governmentality is characterized by transforming the definition of knowledge into an act of power. On a superficial level, scientific evidence, backed by the broad legitimacy of science, is commonly used as a justification for governmental decision making and other acts of control. On a deeper level, science itself is a mechanism of control—over nature and over social life. Scientific methodology sets the techniques of power relations[17] and the discourse of science constructs, or defines, the "thinkable" categories, whether "omnipotent" or "disenfranchised." For example, the Indians of Latin America employ the existing and legitimate category of "indigenous peoples" to establish their claims for political rights, thus employing a "thinkable" category that connotes the underprivileged. Indigenousness, like other now-thinkable and employable categories, is therefore based on global concepts and international action. When relying on science-based theories for evidence of their distinctiveness, their political voice is clearer.

Science provides the cultural scripts[18] for nation-states to act on or support action. It does so on the expansion of science itself (for example, in the universities) through the incorporation of internationally legitimate science policies (Finnemore 1996). On other social desiderata, it works through the incorporation of such global "norms" as human rights, democratization, economic policy, or mass education.

All this reflects an a-rational tendency in rational national policy and planning. Nation-states are obliged by their role as actors/agents to move toward the achievement of social goals (such as progress). In their search for successful paths to achieve these goals, they rely on the available cultural scripts (such as the "science for development" model; see Chapter 4). And they rely on perceptions of successful models, such as the story of the success of the Tiger economies. All are steps of rational planning under conditions of uncertainty, but they result in decoupled and irrational actions.

The global diffusion of policy models, such as "science for development," expresses yet another dimension of the individualizing/totalizing effects of modern governmentality, this time on a global scale. National identity and national governance are defined as local forms, but the scripts for such national actions are provided by homogenizing global forces. Science practices, for example, respond only weakly to national "needs" or interests. Rather, they respond to the prevalence of such practices in other nation-states or in response to the pressures of international organizations (Chapter 5; Castilla 1997). Nation-statehood is itself a globalized model of the modern era, carried by global discourse and international organizations.

Overall, the connections between the science and modern liberal governmentality are evident in two levels. First, science and politics both are guided through a liberal package of globalized reforms, or policies. Second, they directly reaffirm each other's social role, and their operations are intermingled.

GLOBALIZING SCIENCE AND POLITICS: CONCLUDING COMMENTS

Current research on the effects of globalization on the state revolves mostly around the question of whether "the state" will survive as a political arrangement in an environment of increasingly intensive interconnectedness among countries (for example, Ruggie 1993; Clad 1994; Smelser et al. 1994). In these discussions, the differentiating issues are the conceptualization of the

dominant world trends (homogeneity versus divergence) or varying conceptions and dimensions of sovereignty as they respond differently to global
pressures. From a world polity perspective, the concern is with what is the
changing meaning of the nation-state as it is being diffused worldwide. Thus,
for example, the concern shifts from predicting pressures on sovereignty to
understanding what tamed notion of sovereignty is being globalized.
Changed meanings of sovereignty affect nation-states worldwide (Meyer
1999). In this chapter, our concern is with the cultural institution of science,
with the meanings that it carries, and with the effects that the diffusion of
these meanings has on the nation-states.

Nation-states worldwide are affected by the globalization of science and
the diffusion of scientific practices is intensified by the expectation that the
incorporation of science into national practices will lead to greater economic
prosperity. On the basis of this discourse, nation-states allocate funds and
authority to research and development. They forcefully incorporate innova-
tive technologies in the manufacturing of goods and encourage individuals to
choose scientized professions as their vocations. Policy directives guide action
and form organizations and thus concretely shape polities worldwide.

The effect of globalization pressures on national policy regimes is recognized (for example, Johnson 1993; Schwartz 1994), but some such effects go
unnoticed and unplanned. Scientization produces such unnoticed and
unplanned political consequences by supporting actorhood. Science, adopted
as a narrow instrument, yet working as a cultural framework and a secularized
source of legitimacy and authority, affects the foundations of social life in a
much broader sense than is recognized in general, and in policy, discussions.

## Chapter 13

National Scientization, Rationalization, and Standardization

The global scientific system, as it impacts and as it is incorporated in national society, plays a broad rationalizing role. Social institutions—from corporations to education to family life—are conceived as rational responses to human needs, and their activities are framed within the logic of standardized means-ends rationality. This shows up in the greater use of standard and highly articulated units, definitions, and technologies as rules of practice. Both world-level and national standardization efforts affect corporations, governments, and not-for-profit organizations. Corporations from India to Argentina to Dubai set the accounting standards of the major accounting firms as their own (Jang 2001). Governments from Sweden to China to Kenya gather government statistics according to United Nations Education, Scientific, and Cultural Organization (UNESCO) guidelines (McNeely 1998). Publishing houses worldwide rely on the *Chicago Manual of Style* to define their specifications. Everything from apparel sizes to corporate quarterly reports to smart cards is increasingly standardized in ways that fit rational models. This trend is at the heart of much global attention—legislative efforts, industry initiatives, and social research. In this chapter, we call attention to two aspects of the trend: first, the role of science in promoting this global trend; and second, the a-rational, or highly cultural, components of the scientific pressure toward rationalized institutions.

First, the globalization of science sets the groundwork for global standardization. Science forms a conceptual basis for rationalized institutions, as well as a networking mechanism for the diffusion of the rationalistic norms of the world polity. Both the cultural ethos of science, which organized knowledge and scientific laws as universal, and the science-based web of global institutional connections that set the networking framework, promote greater rationalization and standardization of national practices. On the ethos side, if knowledge is accepted as universal and scientific laws are regarded as boundaryless, then knowledge categories are transferable from one social context to another (Strang and Meyer 1993). As patterns are identified and models are constructed, modes of operation become transferable from one context to another, too—of course, mostly from the core countries to the peripheral ones. Science creates an objective world, in which common standards are possible and a rational response to uncertainty.

Science therefore is an axis of global rationalization and standardization. Yet because the literature on globalization emphasizes economic and political forces, cross-national standardization, for example, is explained in terms of economic competition. The need to conduct business across borders in a reliable, predictable, and efficient manner is often the reason cited for the tendency of companies to adopt standard corporate operations.[1] Similarly, intergovernmental cooperation calls for compatibility among their practices, from military training to ceremonial greetings. But in fact, measures of global harmonization are promoted regardless of the current hegemonic power or the current efficiency consideration of interested parties. This calls into question explanations that stress the role of economic or political forces. In our view, over and above economic and political globalization, science—which is heavily intertwined with modernist concepts of a rational order—acts as a central impetus for countries to embrace global standards.

Second, global trends toward rationalization and standardization are cultural forces that operate beyond immediate functionality (or instrumentally manipulated consciousness). Countries adopt international accounting standards even when their prospects for business development are bleak (Jang 2001). Standardization initiatives are promoted in less developed countries, although their markets are not integrated into international production lines. On the other hand, a powerful and central United States can still cling to the nonmetric system of measures. Powerful nations often lag behind other countries in adopting international standards. Rationalization and standardization trends thus may be disconnected from both state competition and considerations of efficiency. They emerge from a modern world polity that emphasizes standards built on scientized schemes, convergence and integration around rational models, and systematic planning in a world thought to be lawful.

In this chapter, we discuss the centrality of these two features—the role of science and the primacy of cultural considerations beyond immediate power and interest—for the global process of intensifying cross-national standardization. We then turn to empirical studies of the effects of science on the spread of internationally standard forms. We show that expanded science encourages the national institutionalization of international standards ranging from the information sector to organizational practices and models.

## GLOBAL STANDARDIZATION

Standardization is a well-established global organizational field. International organizations concerned with standards development are established and expand at high rates. The earlier history includes the 1906 formation of the International Electrotechnical Commission (IEC) and the 1926 establishment of the International Federation of the National Standardizing Organizations (ISA). Now, most international standardization work is in the hands of the International Organization for Standardization (ISO), established in 1947. Operating as an umbrella organization for this field, the ISO establishes and publishes standards for a variety of fields. Its first standard, on temperature for industrial length measurement, was published in 1951, and lately, its effects have concentrated on the ISO-14000 standards on environmental management. ISO's international standardization penetrates to national and subnational levels (Mendel 2001). By 1999, sixty-two countries had at least one national standards–developing organization registered with ISO or IEC (Mendel 2001).

Although international standardization began in the electrotechnical field and for years emphasized mechanical engineering, the field has expanded to address a wide variety of issues. International standards issues range from the obvious concerns with weights and measures (for example, Bureau International des Poids et Mesures), to industry-specific standards (for example, International Bureau for Standardization of Man-Made Fibers; International Dairy Federation). They also include professional guidelines (for example, World Dental Federation; World Meteorological Organization).[2] Reflecting a similar variety, the ISO lists among its achievements the standardization of freight labeling (identifying, for example, dangerous or sensitive cargo), paper sizes (such as A4 or legal, encoded in ISO-216), and codes for country names, currencies, and languages. National standardization bodies follow: The China Standardization and Information Classifying and Coding Institute, for example, received ministerial awards in 1998 for ten standards, ranging from Classification and Codes of Occupations to Body Dimensions of Adult Chinese.[3] Standardization action, even at the global level, goes beyond the obvious industrial needs. For example, ISO's Standard for Doping Control in Sports, which controls international variation in drug testing (procedures and equipment) and in related sanctioning (penalties and appeals) in July 1999 passed its first step in becoming a fully fledged ISO standard by being declared ISO/PAS.[4]

Finally, standardization initiatives are taken by organizations whose primary goal is different than standardization. UNESCO, for example, distributes to its member countries guidelines for data reporting, thus encouraging different nation-states to gather national statistics according to these specifications, or standards (McNeely 1995). Similarly, in 1996, the International Monetary Fund (IMF) launched its "Special Data Dissemination Standard" project, focusing on the standardization of macroeconomic data among IMF member countries.[5] Such standardization initiatives join the initiatives of the international standardizing organizations in a general trend toward expanding the domain and organizational base of standardization.

This trend toward greater standardization is commonly explained as a necessity in an increasingly intertwined world. "Maintaining the nation's competitive success, industrial/economic growth and its leadership in innovation and international trade," writes Robert Mallett (1998:63), "requires a tightening of its focus on the fundamental infrastructure elements of global commerce," and standardization is just such an element. Similarly, Yates and Aniftos (1998:61) warn their audience of American civil engineers that "if new standards are not carefully monitored they could negatively affect the global competitiveness of U.S. engineering and construction firms." These fears reflect a common belief that standardization makes communication and production more efficient, thus making related businesses more profitable.

ISO policy makers have long identified financial efficiency as the key to determining the relevance of international standardization. But standardization initiatives occur regardless of their commercial viability. First, standardization is a costly venture. It requires businesses to divert funds toward administrative changes, whose profitability is—at best—projected to the future (Mendel 2001). Second, the users of proposed standards are not necessarily motivated by considerations of profit. For example, at the 1996 conference on European Standardization Strategies for Geographic Information, Jean-Claude Lammaux noted that "while the discussions in standards bodies are about global data interchange, the main economic impact of geographic information is at [the] local level. The key users of the data produced are city administrators. Such users are, however, rarely interested in being able to export their data to other countries, and rarely need to obtain data from outside their own country."[6] In general terms, standardization initiatives spring up, even in fields where interfacing across institutional environments is not a necessity.

In a similarly counterintuitive way, standardization initiatives spring up in fields where there is no powerful leader that dictates standards for all others. Much like American resistance toward the international implementation of the metric system of measurement, Britain held long to a nonmetric currency. These countries—at the core of the international system—refrained from taking the lead on obvious dimensions of cross-national standardization. As these examples reveal, standardization is better understood within the context of a global polity and culture. Rather than seeking the benefits of standardization in increased efficiency, profitability, control, and hegemony, standardization is embedded in a global culture of rational order.

Rational order, predictability, and efficiency are desirable features in current global culture. One might imagine world social arrangements where other considerations were primary. Traditional societies rely on seniority and common wisdom as determinants of social power; religious societies draw power from association with the divine.[7] Rational order, with its concentration on cause–effect relations, is a product of globalized modern, mostly Western, culture. In this global cultural environment, the role of science in encouraging the globalization of rational order institutions, such as standardization, is primary.

## SCIENTIZATION AND STANDARDIZATION

The technical methods and systemic approach of science create a basis for global standardization efforts. International standardization organizations are often essentially scientific associations. Such organizations as the Committee for European Studies on Norms for Electronics in Research (est. 1961) or the International Statistical Institute (est. 1885) are concerned with the development and application of international standards that are in turn based on the scientific methodology of their disciplines.[8] But the link between science and formal standardization does not have to be so explicit in order to show that science sets the conceptual basis for global standardization efforts.

Standardization refers to the acceptance of modes of practice that are first generally recognized as excellent and authoritative, and second are regular and routine. Standardization is not synonymous with similarity of form: it implies that there is a particular importance to the uniform structure being created across contexts or environments. There is a vision that the uniformity is beneficial for clear purposes and that it serves as a rational response toward uncertainty. Standardization is hence an exercise in the elaboration of technical and rational principles and the elimination of fatalistic or arbitrary rule making. The principles involved are justified by actual or presumed scientific knowledge and authority.

The notion of the existence of a rational and lawful order—natural and social—is at the heart of the scientific quest. Identifying patterns relies on the codification, labeling, classification, and organization of knowledge. This scientific attitude is most obvious in the categorization of social and natural realities: the scientific taxonomy of animals; the classification of plants; the labeling of objects in space as planets, suns, and galaxies; or the grouping of humans into social classes and races. In all these taxonomies, the scientific labor involves the identification and explanation of similarities or of differences across units, the labeling of the distinct categories, and the sorting of units into these categories.

"One of the quintessential aspects of modernity . . . [is] the need to compose the world as a picture," writes Arturo Escobar (1995:56). The modeling of realty is at the core of scientism. Modeling requires both abstraction and information. Information gathering, or knowledge acquisition, is necessary. Abstraction, and with it the compartmentalization of the knowledge, reconstructs such reality into generalized "sets." The process of modeling fragments reality into categories, or model components, and then reassembles them into a relational configuration. In this sense, science's dominant realist episteme encourages the objectification of reality through standardizing modeling.

The ethos of modern science is built on the assumption that the world, and the knowledge and scientific laws built on it, have a universal character. In this spirit, science calls for the worldwide application of similar categories, standards, and models. Thus, such knowledge categories as geological periods or racial groups are applicable in different places worldwide. This universalism is also applicable to modeling processes—natural and social. The laws of gravity, as well as the laws of economics, are thought to be as applicable in Africa as they are in North America. Once knowledge is accepted as universal and scientific laws are regarded as boundaryless, then knowledge categories are transferable from one social context to another. These knowledge categories are essentially standardized and hence are defined as inherently similar in spite of variations in their environments.

Different conceptions of science suggest different mechanisms by which science diffuses standardization. Realist theories often refer to science as an essentially rational and functional institution producing successes for nations that incorporate it and thus setting the basis for greater standardization. Alternatively, essentialist theories—which call attention on the inherent characteristics of science—focus on the "scientific mind," the technocratic mind-set that creates systemic solutions to problems by modeling reality, as the line between standardization and science. Institutional perspectives add emphases on science as a networking mechanism in the world polity and as carrier of modernist themes. Thus, science, in various ways, carries along the notions of rationality and inherent order and it applies them globally.

In scientized societies, social activities become more standardized—from the obvious management of national firms to governmental practices. The general trend toward greater standardization extends beyond data categories, although this is the most obvious field of change toward greater cross-national similarity. Greater standardization is also reflected in intentionally conceived similarity in such fields as education, corporate management, governmental procedures, and professional practices. Corporations worldwide therefore adopt globally accepted standards of accounting and quality control, such as ISO-9000 (Mendel 2001). Governments worldwide adopt globally accepted national reporting schemes, such as World Trade Organization definitions of gross national product and gross domestic product (McNeely 1995). Professional groups adopt similar globally accepted credentials as their membership criteria, such as MBA degrees for professional management associations or CPA certificates for professional accountancy associations. These more expansive notions of worldwide standardization are tightly related with the global trend of scientization. The prevalence of science in society results in the overall standardization of national practices and appearances. In summary, then, scientization involves the infusion of the social logic of ordered rationality into all nation-states because these notions are rooted within the scientific worldview and are embodied in globalized scientific practices.

## EMPIRICALLY CONNECTING THE GLOBAL TRENDS:
## SCIENCE PRACTICE AND NATIONAL STANDARDS

The expansion of science encourages cross-national standardization through the introduction of the concepts of rationality and order. These causal relationships between science practice and standardization efforts at the national level are demonstrable. Here, we summarize the findings from cross-national empirical analyses that causally link expanded national science activity and the standardization of various national activities during the period 1970 to 1990.

The models generally take the form of panel analyses, examining the effects of national scientific activity[9] on later standardization, with earlier standardization held constant and while a measure of national development[10] is held constant, because it is common to argue that this factor really accounts for much social rationalization. Our effort is to show that scientific expansion plays an independent role, over and above the role played by national development in the expansion of standardization efforts. Last, depending on data restrictions, we chose between panel or cross-sectional models and between structural equation modeling or ordinary least square regression as method of estimation.

The models show that greater scientization results in more intense standardization effects in three social spheres: the information sector, management procedures, and perceptions of governance. The results of the empirical models are summarized in Table 13.1.

First, the social practice most obviously affected by science practice is the management of information. Because data are a taxonomic form of representation of reality, the gathering and management of data most dramatically reflect the general processes of rationalization that are brought about by scientization. To implement the tasks of data management, national institutions of information gathering are established. These institutions (for example, National Census Bureau) work to collect, sort, and distribute information that are defined as of interest and relevant to social life. Such information gathering commonly proceeds according to international specifications (Ventresca 1996). Expanded science encourages the institutionalization of national procedures of data collection and sets such practices in the manner most prevalent in the international arena. The cross-national analyses[11] show that science practice contributes to greater efforts in gathering standardized information. These effects occur on several dimensions of the information sector: national efforts at information management,[12] the formation of an infrastructure for the information sector,[13] the link with new information

TABLE 13.1

Science Practice and Standardization of National Practices:
Summary of Results, with a Measure of National Development Controlled

| Domain Affected | Measure | Method of Estimation, Model, and Time Period | Relation of Science Practice with Indicator |
|---|---|---|---|
| Information sector | Completeness of data collection national data organizations: | SEN, panel 1980–1990 | + |
| | Years since established | Correlation | + |
| | Years since first national statistical yearbook | Correlation | + |
| | Information technology: | | |
| | Years since initial connection | SEM, cross-sectional 1980–1995 | – |
| | Expansion rate | SEM, cross-sectional 1980–1995 | + |
| | Elaborated data infrastructure | SEM, panel 1980–1990 | + |
| Corporate management | Adherence to accountancy standards | Correlation | + |
| | Managerial control, ISO-9000 certificates | Correlation | + |
| Governmental transparency | Perceived lack of corruption | SEM, panel 1980–1996 | + |

All effects are significant at the 0.05 level.

technologies,[14] and the complexity of that sector.[15] Overall, therefore, more scientized societies tend to make greater efforts to gather extensive national information, have a more expanded base for the national information sector, incorporate information technologies more quickly, and have more elaborate information-gathering procedures.

Second, science practice encourages cross-national standardization in corporate management. One such an effort to harmonize international corporate practices is that of the International Accounting Standards Committee to establish international accounting principles in order to simplify international listings in foreign stock exchanges. Similarly, ISO-9000 initiatives standardize management control and quality assurance procedures. The ISO devised this scheme as a benchmark for corporations from various economic environments to review and assess potential partners. As with the earlier findings regarding the connection between scientization and standardization, these management procedures—specifically, the use of internationally standardized procedures of financial accounting[16] and the use of internationally standardized procedures of corporate control[17]—are also related with national science practice. These analyses show that scientized societies more readily adhere to international conventions of corporate management procedures.

Third, we examine the relationship between scientization and perceptions of transparency in governance. Some countries are perceived as properly governed, whereas others are perceived as corrupt, unpredictable, or disorganized. This notion of proper governance reflects a convention, or a standard, in regards to governance. Moreover, this standard, like those of corporate and information management, is cross-national in nature. Remarkably, cross-national survey indexes of proper governance are increasingly created.[18] They turn out to be highly related with expanded science practice, even with overall national development controlled. The results indicate that science practice minimizes (at least perceived) uncontrolled governmental corruption. Science practice contributes to greater rationalization of governance procedures and to greater cross-national standardization of public conduct.

In summary, the empirical cross-national models show that science practice promotes, or is associated with, greater standardization and rationalization of national practices: information management, corporate practices, and governance ethics and appearances. Specifically, science practice in 1970 and 1980 shapes the intensity of standardization in these three social arenas during the 1990s. Overall, then, worldwide scientization contributes to the globalization of decision-making rationality.

Our evidence for the relationship between science and standardized forms of social performance accords with recent studies of social standardized practices. The various collected essays in Hopwood and Miller's (1994) book, for example, define accountancy as (1) a form of management (2) affecting the construction of entities.[19] Science and accountancy share core common features: they both are viewed as tools of management and progress, and both are embedded in modern notions of actorhood and rationality. Much like the seemingly neutral labor of science, accountancy—through the seemingly neutral labor of calculation, recording, and bookkeeping—conceives of corporations, employees, and nation-states as legitimate and standard entities (see Miller and O'Leary 1987). Also much like science, accountancy has expanded its domain to incorporate new fields under its auspices. Michael Power describes accountancy as reaching beyond the domain of money transactions. In his study *The Audit Society* (1994), Power writes, "In addition to financial audits, we now hear of environmental audits, value of money audits, management audits, quality audits, forensic audits, data audits, intellectual property audits, medical audits, and many other audits." The trend to expand the domains of accountancy reflects greater efforts for the standardization of additional social fields—relying on the professional tools and legitimacy of accountancy—and, as we demonstrate here, such standardization is highly related to global processes of scientization.

RATIONAL ORDER—SCIENCE—POLITY

Scientization supports the idea that one set of rules could apply to various nation-states, regardless of their unique histories or diverging social conditions. Natural laws are thought to be relevant worldwide. The notion of the natural world order reaches to human society as well. With this in mind, common wisdom now regards management procedures, information categories, and governance ethics as transferable from one social context to another. The logic of cross-national standardization is infused into additional fields over time. For example, the IMF's 1996 initiative "Special Data Dissemination Standard," which calls for the standardization of macroeconomic standards among its member states, resulted in the trickling down of such standardization efforts into numerous social spheres. In Slovakia, for example, the desire to adhere to these IMF standards resulted in the pooling of resources from not only the Statistical Office of Slovakia and the National

Bank of Slovakia, but also from the Ministry of Labour, Social Affairs, and Family of the Slovak Republic.[20] This ministry, although not directly related with IMF, nevertheless aligns its operations with IMF's vision. Scientization furthers such cross-national standardization of social practices, even if these practices are not directly linked with sciencelike procedures.

The trends of scientization—and therefore of rationalization and standardization—are intensifying over time. Most obviously, the institutionalization of a scientized information system, worldwide and nationally, has been most dramatic in the post–World War II period. Arturo Escobar (1995) claims that this change is due to the development of information-dependent professions and academic disciplines. In his study, Escobar stresses that the extensive efforts for information gathering and information standardization spring out of the need to expand and implement, the highly scientized model of developmentalism (1995:42). The construction of disciplines such as development economics during the late 1950s encouraged the construction and standardization of information processing. Thus, professionalized pressures supported the worldwide effort to collect data on gross national product, sector productivity, and labor-force participation. The availability of such cross-national standardized information then enabled developmentalism to elaborate more complete universalistic models, to reaffirm the validity of these models, and to call for their worldwide implementation.

Thus, the traits of modern science promote the rationalization and standardization of society, and so does the role of the sciences as organized professional groups. The highly legitimated cohesion of this professional group, bounded by a common ethos, helps makes science an influential organized force. However, scientists are not the only professional group to advance rationalization. Other modern professional groups, such as lawyers and accountants, are also central to the standardization and rationalization of society and its governance procedures. Although science practice does not directly contribute to the expansion of such rationalization-carrying occupations, it is closely associated with their growth. Together, the rationalization-promoting professions act as "moral entrepreneurs," staking professional claims and advancing related moral codes.

The relations between scientization and standardization reflect general trends toward expanded formal organization. Scientization and organizational expansion are parallel moves toward a greater rationalization of society. Rationality, as a core element in modernity, calls for the replacement of various forms of authority with the authority of deliberation, calculation, and impartial judgment (Weber 1978:212–71). Its translation into the language of organizational efficiency makes rationality serve as a justification for a wide range of modern reforms. In this sense, rationality—anchored in either scientization or bureaucratization—is being globalized with this expectation for improved efficiency. Thus, standardization and scientization are features of modern organizational systems and of governance. As we show empirically, scientization results in the attribution of rational order to a range of social spheres of governance, such as the information field or corporate management.

The greater rationalization of society has long been accepted as a central component of modern social change—from Weber's emphasis on rationalization supporting bureaucratization to Foucault's focus on its shaping of social control. Indeed, rationalization has sweeping effects over an enormous range of practices, and rationality is the essence of reasoning, or of justification. Its assumptions alter the mode of governmentality (Foucault 1991) by anchoring it in the modernist concept of rational order. In this sense, the globalization of science is associated with the general expansion of formalization—the expansion of the state, of the rationalizing professions, and of the service sector. The standardization involved—especially as it pertains to information gathering—is also a global move toward greater transparency. Transparency—or the public disclosure of information for the purpose of public review and scrutiny—is touted as the recipe du jour for various global ills: from global financial volatility emerging from the lack of financial disclosure to rampant political corruption rooted in the lack of political scrutiny to environmental pollution stemming from the lack of accountability of firms. Transparency is thus hailed as a new kind of regulation, namely "regulation by revelation." This new type of regulation constitutes a new logic of governance, or "principle of governmentality" (Foucault 1991; see Chapter 12). This new form of governmentality reorients governance toward the gathering of information, the public disclosure of such information, and the constitution of social entities that are based on such information gathering.

Although the justification for regulation by revelation, or the transparency-based mode of governmentality, is as an aid to central power, it simultaneously aids peripheral rights as well. Much like democratic governmentality (see Chapter 12), regulatory governmentality simultaneously totalizes and individualizes social actors, defining them in standard terms while also empowering them as individual units. This mode of governmentality is rooted in scientization, or in the belief in the existence of a global rational order.

# World Society and Science Globalization

Throughout this book, we have focused on the way science has emerged, expanded, and triumphed as a world institution. The authority of science is rooted in an initially Western and now worldwide culture of rationality, a culture that closely corresponds to the Weberian emphasis on the demystification of the world (Weber 2000). This culture of rationality promotes scientization and is itself promoted by the triumph of science. The whole process involves both an intensification of the authority of science (the scientization of society) and an expansion of the scope of its application (the socialization of science). Alternative bases of authority are undercut by science, and alternative accounts of reality are adjusted to conform to scientific accounts. Scientific authority and science-based accounts expand to make sense of nature, society, and the social actors in society and to guide more and more aspects of individual and collective action. Scientific accounts constitute models of progress within which science is expected and scripted to play a pivotal role.

Science globalization is thus best explained as an outcome of a world society that fostered diffuse beliefs in progress, a progress often imagined to involve the development of the capacities of individuals, organizations, nation-states, and increasingly the world itself. Progress is thus often equated with development. In some analyses, development in turn is cast in technical and economic terms: development as human capital formation, as organizational goal setting and effective resource management, as national technological and economic growth, and as world production capacity building. But this is too narrow; modern conceptions of both development and science take a much broader form, encompassing dimensions of life and society far beyond production and exchange.

In fact, there is much worldwide consensus as to the meaning of progress and development. And despite persistent underdevelopment in many different domains, there continues to be confidence in development as a goal that can be attained through rational means couched in universalistic terms. However complex, development is presumed to have a lawlike character that can be discerned and promoted through rational scientific analysis. This is true whether one thinks in terms of the development of learners, firms, and other organizations, or of economies and polities. Furthermore, positive links are presumed between individual and organizational development on the one hand, and national and world development on the other.

Social progress has always been defined to include a broader emphasis than technological development and wealth accumulation. This more diffuse emphasis is historically evident in efforts to discover the universal laws of nature as well as in attempts to scientifically upgrade the human condition. Early scientific and educational developments were touted as evidence of general social and political progress: for example, the formation of learned societies (Wuthnow 1980), the establishment of national population censuses (Ventresca 1996), and the creation of schools and scientific training for the masses (Kamens and Benavot 1992). From the late eighteenth century on, this broader emphasis involved a concern for justice and equality, often linked to democracy and to an expanding set of citizenship rights. After World War II, citizenship rights increasingly became global human rights, and these rights (and their abuses) were also increasingly dramatized in the form of human rights proclamations, organizations, conferences, journals, and education (McNeely 1998; Soysal 1994). The long-term rise in importance of health and environmental issues as indicators of quality of life further reveals the expanded scope of established concepts of progress. Scientific authority is prominent in the display of health, environmental, and other social problems as well as in the identification of policies and programs needed to properly address these problems. As with the emphasis on economic development, the broader modern focus on social and political progress has entailed a universalistic discourse. All entities, it is thought, can attain more progress and can do so by adhering to principles that apply everywhere. Such a focus on a broad conception of progress underlies the generation of the rights of individuals in organizations, societies, and the wider world. It gives rise to calls for democratic regimes and to an elaboration of what constitutes democracy, including a strong emphasis on equality at the national and world levels (Chapters 11 to 13). And notions of progress as including justice are evident in national and international organizations that emphasize the need for development with equity. Throughout the world, scientific authority is also utilized to discern and promote progress as justice by debunking ideologies and policies of an undemocratic or elitist character. As is the case with respect to narrow notions of progress as economic growth, the broader perspectives also presume positive links among individual, organizational, societal, and world levels.

With respect to the achievement of all aspects of progress, a crucial feature of world scientization is its capacity to create for persons, organizations, and nation-states a strong sense of legitimate agency that validates, in these entities, the pursuit of development goals through rational, purposive action. Belief in progress goes hand in hand with belief in individual and collective human agency. Such beliefs underlie the enormous modern interest in individual and social planning, training, and learning. Strong concepts of the empowerment of human actors create strong needs for appropriately socialized individuals, organizations, and nation-states—that is, for entities with the "right" goals and with the "correct" strategies for attaining them.

Not surprisingly, world scientization also creates the putatively disinterested experts who roam the nation and the world, providing advice on how individuals, organizations, and nation-states may achieve development goals. This or that piece of concrete policy advice may be questioned, but the overall value of scientific authority and expertise remains intact in a world in which "consulting" is a dramatically successful industry. For example, even if a given strategy for economic development—or the solution for an AIDS epidemic— is questioned, the outcome is likely to be the adoption of a new and expanded scientific solution or strategy, not the discarding of scientific authority.

The extraordinary modern faith in rationally attainable progress for all and in "science for progress" can be mind-boggling, but it seems routine to moderns and to those who aspire to contemporary modern identity. Survival may have seemed to be enough of a goal in earlier world historical eras; the children of this extraordinary faith must act as if "man will not merely endure; he will prevail."[1] And they must proceed as if the category "man" included all men, women, and children throughout the world. So the bold universalistic sentiment captured in the phrase "the rivers of the Ganges flow in Walden Pond"[2] is now likely to be considered a sound ecological and sensible international relations principle. How individual, organizational, and societal progress is to happen is quite standardized: for example, through education and through science in both education and the wider society. Much confidence in purposeful learning and in goal-oriented activity is required to be a true believer—or at least, to operate smoothly under the sacred canopy of science for progress. This confidence is buttressed by local, national, and international actors that read similar scripts and affirm the centrality of science in the quest for progress (Chabbott 2002).

The worldwide triumph of this extraordinary faith is evident when we consider how little worldwide support there is for less universalistic and less rationalized alternative cultures. For example, accounts that emphasize the inherent superiority of some cultures or peoples, some organizational leaders, or one's gender are suspect. The principles of effective teamwork and entrepreneurial creativity can be scientifically analyzed and learned in school; so can democratic and civic values. Men can learn to care and women to compete: both caring and competing can be dissected scientifically. The "learning society" is lionized, rather than the celebration of civilization, race, and even charisma. Much of what is to be learned is seen, in scientized fashion, as divorced from local circumstance and presumed to be of value and learnable throughout the world. So there is much evidence of growing world homogeneity in formal school and university curricula (Meyer et al. 1992b; Frank et al. 2000c). But beyond individual education, there may be growing consensus in organizations and in societies on what is to be learned and why and how this is to be learned, so beyond textbook isomorphism lies the standardization of national and organizational development plans. The authority of science underlies these plans and shapes confidence in planning, quite apart from any compelling empirical evidence supporting this worldview. It is mostly a matter of faith.

World society is thus characterized by a culture of formal rationality, which generates models of science as a broad instrument for progress. In this book, we have sought to clarify the content of these models, their empirical manifestations at the world and nation-state levels, and their impacts on various aspects of social progress. Here, we highlight key concepts of the world society perspective, contrasting this orientation with the more instrumentalist and microsociological ideas that are emphasized in some theories and almost all the policy literature. Next, we reiterate our key findings, paying special attention to world trends, to national adaptations of world emphases on science, and to national-level effects of scientization. Last, we identify further research directions addressing issues that are raised in our work.

WORLD SOCIETY ASSUMPTIONS AND CONCEPTS

From a world society perspective, the triumph of science for progress models is first and foremost an institutional triumph. This means that the centrality of science is not a simple function of its utility in a variety of contexts. This is not to deny that science and scientific applications are useful. But much of what science is about falls well outside the scope of the utilitarian. For example, the knowledge requirements of progress as social justice are difficult to make sense of from the point of view of costs and benefits. To be sure, even in this domain, some justifications for the extension of the principle of equality to women or for the formation of environmental ministries are rationalized accounts of the value of equality in society or the importance of sustainable development as sound ecological policy. But once a principle or policy has gained the scientific high ground, its affirmation and acceptance is more driven by the authority of science than by an assessment of the results of its application.

The case for women's rights has thus best been made by invoking rationalized principles of personhood and citizenship and by noting that the adoption of these principles shows a societal commitment to progress and justice. Whether this or that aspect of women's upgraded status positively affects the economy or the polity becomes a secondary matter. Women's right to higher education is not undermined by evidence of a lack of positive economic effects of increased female participation in higher education (Benavot 1989). Likewise, the case for environmentalism is driven by ecological and medical scientific principles, and again, by what their adoption symbolizes for the adopting individual, organization, and nation-state. The earlier and pragmatic opposition spearheaded by India and China, which was based on both claims of sovereignty and national industrialization plans, has weakened. At the level of principle and policy, both women's rights and environmental protection have become tenets of progress, with standardized cross-national ratings on the status of women and on the state of the environment. Thus, a broad range of implementation failures can be seized on as evidence of bad faith or incompetence.

An emphasis on the authority of science distinguishes our perspective from the market logic that underlies a range of utilitarian arguments. The value of science is dependent on a wider culture and does not merely mirror its benefits for individuals, organizations, or societies. Our perspective also differs from another set of realist explanations that perceive scientific expansion as rooted in the power of the particular interest groups that support science. Here the emphasis is not on whether science really works, but rather whether those in power—industrialists, military leaders, and in some versions even scientists themselves—support the scientific enterprise, perhaps because it aids in strategies for social control and governmentality (Foucault 1991; see also Chapters 12 and 13). Such realist ideas are helpful in explaining why and how some aspects of science are favored and spread. But too many other aspects of the triumph of science lie outside their scope. These include the recurring scientific interests in cosmological issues far removed from mundane matters, the symbolic value attached to having a national science policy, the degree to which all sorts of social issues are subjected to scientific analysis, and the relative ease with which science is activated to solve problems, even when these problems are created by science. When science fails, contemporary moderns do not turn to other gods. Instead, with deep Augustinian faith, they embrace science anew to solve the problems: environmental science to deal with the pollution of industrial science, or rationalized peace education programs to cope with international tensions fueled by military science.

The detailed field studies of constructivist social scientists (for example, Knorr-Cetina and Mulkay 1983; Latour and Woolgar 1986) challenge the standard realist accounts by describing how scientific work is produced and managed. They show how different these production and management processes are from both popular understandings of how science is done as well as from the formally constituted methods and logics favored by the scientists themselves. These studies are important in their own right, and some of their underlying assumptions about the role of culture in facilitating the making and presenting of science parallel our own work. The institutionalist perspective seeks to make explicit the degree to which the wider culture legitimates laboratory and other scientific organizational cultures and thus their practices and stylized accounts of practices. What makes the realist research reports of the scientists so plausible is not the thespian qualities of scientists and technicians, and certainly not their raw power, but the taken-for-granted status of the scripts that guide them. Not surprisingly, even the microconstructivists themselves adhere to scientific rules of the game—as we do here—in their deconstruction of science production, and in this fashion, they maintain their credibility in scientific terms. This irony simply illustrates the extent to which the social construction of science in the modern system flows from a wider culture downward and inward, rather than arising from successful (or fraudulent) specific practice. What makes studies of laboratory life more compelling than studies of library life are the legitimated standing of science as an important activity and scientists as important participants in the broader society

Our perspective focuses on the scripts of science and on their carriers across the world. Our perspective is thus both constructivist and macrosociological. Isomorphism and decoupling are two key concepts within this perspective. Isomorphism refers to the fact that similar principles, policies, and practices arise in different countries and organizations and among different peoples. This suggests to us that these principles, policies, and practices are global and universalized scripts that are then locally enacted. Decoupling refers to the fact that there are massive gaps between principles and policies and between policies and practices, or that policies and practices that are supposed to go hand in hand often do not appear in tandem. Hard-core social scientific realists might expect actors to change their policies to accommodate their practices or to upgrade their practices to bring them into line with their often loftier principles and policies. Much of the persistent loose coupling has been depicted as "organized hypocrisy" (Brunsson 1989; Krasner 1999). But a willful or a malicious posture need not be in place for decoupling to operate. Limited resources and lack of skills may account for decoupling, even when the best of intentions can be assumed. And decoupling may be brought about by actors cutting and pasting from a range of acceptable but not identical constitutions, plans, and other blueprints for promoting progress. The absence of a coercive world state and the presence of a pervasive world culture facilitate the enactments of science and scientized social accounts that lead to both isomorphism and loose coupling.

This enactment leads to both the consolidation of the authority of science (scientization of society) and the expansion of the scope of its application (the socialization of science). It is a worldwide process in three distinctive respects. First and most obviously, nation-states and other entities increasingly adapt the science for progress models and the corresponding scientized scripts. Second, a standardized and universalized picture of the world is increasingly reified in these scripts. Human capital formation processes are expected to benefit the world economy in addition to individual earners and national economies. Democratic values and orientations are expected to show positive individual and societal effects and also to contribute to a better world order. Education for all would enhance progress and justice not only at the individual and societal levels, but also at the world level itself. The standards invoked to protect air, water, and a range of habitats are world standards. Human rights issues are discussed in terms of world standards that make up, or are promulgated as, an international regime. Third, in all these domains, the carriers of science for progress models increasingly operate on a worldwide scale, as international organizations and putatively disinterested experts, affirming and promoting the preferred discourse and plans for action in world arenas.

The global institutional triumph of science does not produce or presuppose a world without conflict. On the contrary, many generative tensions are created in a world culture of rationality that fosters both development and justice, and does so simultaneously for individuals, organizations, and nation-states (Meyer et al. 1997a). In fact, the pervasiveness of world culture, together with the absence of a world state, fuel these generative tensions. Highly legitimated economic growth planners are pitted against equally empowered environmentalists, and the productive capacities of firms are weighed against the rights of workers to safe working conditions. These tensions no longer operate solely within the domain of a nation-state. The plight of child laborers in Central American or Asian countries is documented and televised for consumption in Western Europe and the United States. The lowered costs of labor, and more generally of doing business, continue to attract capital to some less developed countries, as theories of economic globalization emphasize. But these lowered costs cannot be simply cited to justify child labor exploitation or more generally poor working conditions. Transnational social movements are triggered to uphold world standards of human rights, and these movements put a brake on unbridled marketization. Scientific expertise is invoked by all contesting parties, thus strengthening the common scientized frames within which much contestation takes place. In such confrontations, all the parties bring not only lawyers and political leaders, but also squadrons of scientists and reams of scientific documentation. Questions are quickly turned into scientific ones—economists testify as to whether a market is really free or not; medical researchers assess whether specific working conditions do or do not produce disease; and several disciplines weigh in on the question of rising ocean levels and the liabilities involved.

WORLD SOCIETY: RESEARCH FINDINGS

Throughout this book, we describe massive expansions in world-level scientific organization and discourse organized around expanding claims for scientific relevance in every social domain (Chapters 3 and 4). Since the nineteenth century, international science and technology organizations have proliferated through the world. More and more of these organizations embrace development goals. The triumph of scientific rationales has at times led to the popular misconception that abstract scientific breakthroughs preceded technological developments in the past and were motivated to meet technological needs. On the contrary, earlier scientific developments were rooted in the rationalizing religious and political cultures of the Western world, rather than its technical or economic system. Throughout the current period, science policy discourse becomes more universalistic in tone and more international in its organizational settings. More aspects of technical and economic development and more dimensions of social and political progress are specified in the discourse and serve as the basis for the formation of different science for progress organizations (Chapter 4).

At the national level, we find homogeneous cross-national trends, and some cross-national comparisons, that make little sense if only the endogenous characteristics of societies are considered and much sense if countries are seen as deeply embedded in a world sociocultural context. Countries with weak economies or inadequate infrastructures nevertheless move to embrace scientific authority, creating the appropriate ministries and adopting the preferred "science for development" discourse (Chapter 5 and elsewhere). The most obvious case is the diffusion of ministries of science, technology, and the environment. These ministries and corresponding national science policies emerge, even in countries with limited scientific and technical resources. Their emergence is facilitated by the advice of international organizations enacting their status as disinterested experts. Finnemore (1993), for example, shows how United Nations Education, Scientific, and Cultural Organization (UNESCO) aided Lebanon in drafting its national science policy.

Nation-states of every sort build science—and expansive versions of science at that—into their educational curricula and stress the relevance of science for every citizen in all parts of society (Chapter 6). The international system clearly plays an important role here. All sorts of national virtues and shortcomings are attributed to achievements or failures in science and mathematics. The relatively poor performance of the United States in these international tests has lead to publications with titles such as *A Nation at Risk* and *Facing the Consequences*. The achievement winners are regarded as having world-class standards with respect to curriculum or teacher training. There is much universalism in the whole enterprise: countries can and should learn from one another. Furthermore, much of ongoing curricular reform and change involves a focus on the individual as highly empowered: it is child-centric, with a strong emphasis on active learning (Chapter 6). In a scientized world, children are expected to learn that they can actively learn and even produce science and that all of this can take place in many settings, not just the traditional classroom. The change indicates the triumph of science within a world culture, which assigns agency to persons in general, very much including children.

Countries with religious traditions often apparently at odds with scientific accounts also move in scientized directions. They open the doors of higher and scientific education to women, as in the case of Saudi Arabia (Chapter 8). International organizations clearly play a role in this process. They highlight the relative absence of women in science and engineering in higher education (European Technology Assessment Network on Women and Science 2000) and suggest policy remedies. Recent nation-level increases in female enrollments in these fields of study appear related to these scientized discussions and the degree to which the status of women has become an international agenda item (Berkovitch 1999).

Moreover, the findings suggest that the external forces that matter are not economic globalization or military competition. The changes are little attuned to the power of the market or to the gaining of military advantage; instead, they are more aligned to the authority of science as a world institution: linkages to world society and organizations and professions are more important factors (Chapters 5 to 8). Coercion is not the overriding dynamic. Nor is it the simple imitation of the concrete science establishments of a few dominant powers. For instance, the private nonprofit character of the National Academy of Sciences in the United States (est. 1863) is not much copied around the world. The isomorphism involved occurs at a deeper and more professionalized level.

As a result of the loose intranational linkages—and the strong international ones—of the whole modern science package, we observe that components of this package are rather weakly correlated across countries—particularly in the developing world (Chapter 7). Nation-states are enacting scripts from a larger environment, rather than fitting together tightly coupled domestic systems. This sort of observation of massive loose coupling at the national level is, in institutional theories, the result that should be expected when local models are enacted from a wider environment rather than built up locally.

The patterns of relationships that we find between the variables of interest raise many questions about the viability of a realist orientation in explaining scientific development. The national characteristics that determine patterns of scientific emphasis reflect difference in political and cultural orientations more than immediate functional requirements or interest patterns (Chapter 9). The Western countries emphasize the broadest aspects of science, including those focused on individual welfare (medicine, social science). The communist countries, following a deeply set ideological commitment to material production, emphasized the physical sciences and engineering. Some countries deeply committed to hard-line economic growth follow similar patterns. But overall, similarities, paralleling the system as a whole, seem more striking than do differences among countries.

Beyond identifying world and national trends that reveal the triumph of science, our research directly examines some important causal issues. We analyze the influence of scientific developments on different measures of socioeconomic progress. Much of the literature concerns itself with issues of economic growth: consistent with much conventional modernization theory, we find that scientific development qua human capital formation as regards scientists and engineers promotes economic growth (Chapter 10). But the same analysis also shows that more elite forms of scientific development impair economic growth. Moreover, countries that manage the narrower forms of scientific development that lead to economic growth are subsequently under much pressure to encourage broader emphases in their educational system and to consider social and political goals initially ignored. The ("Asian tiger") economic winners of the 1980s are in varying ways altering the educational and occupational systems that appeared to have reaped many economic gains and experimenting with more transparent and democratic ways of managing organizations and society.

The peculiarly limited economic effects of the modern broad package of expanded scientific activity reflects the breadth of the associated conceptions of social progress. So expanded scientific activity in a country clearly affects things such as environmental policies (many of which obviously constrain short-term economic growth; Chapter 11).

In the same vein, we further find that the whole broad science package positively influences democratization and the achievement of a higher level of administrative rationality (Chapters 12 and 13). Scientization involves more than a narrow technical focus with economic growth as its target. Our analyses show that broader sociopolitical dimensions of progress are affected by indicators of scientific development.

FURTHER RESEARCH DIRECTIONS

We conclude by identifying three broad areas of further research: first, the impact of global scientization on world society itself; second, the role of organizations and of experts in promoting scientization, especially in domains previously buffered from science; and third, the influence of science globalization in the construction of agency at the national level. Some studies along the lines we discuss are already under way, examining the expansion of world culture, science globalization, and the effects of science on progress. Other research questions are more exploratory but seem worthy of further pursuit.

Research on all these issues can build on the growth of quantitative cross-national studies in sociology. Earlier substantive and methodological skepticism has clearly faded. Many studies have facilitated the development of relevant methodological tools and data analysis strategies (see Chapter 2); these can be extended to the study of change and stability in the state of the world itself. And global society must be conceptualized as including dramatic political, cultural, and organizational components, and thus as much more than a world economy. Even what is called the "world economy" is in fact a complex organizational and ideological system, with implications far beyond traditional economic exchange.

*Scientization and the World*

Consider the multiple dimensions of world society that are now commonly discussed. The world economy, of course, is no longer just the object of scholarly discourse but also a popular topic, repeatedly covered in the mass media throughout the world. But so too are the world's environment, population, and health conditions. At the organizational level, there is clearly a growing sense of world-class standards, best practices, and benchmarking: conditions everywhere in the world are evaluated, and unsatisfactory conditions are seen as social problems demanding national and international action.

At the individual level, world human rights regimes emerge. Education, for example, is set forth as a human right (Chabbott and Ramirez 2000; Chabbott 2002) in a number of crucial international depictions, and its importance is frequently discussed in terms of human capital and national economic growth, but also in terms of human rights and basic justice. This discussion has a universalistic tone and is expected to apply everywhere. Standards about the organization of school curricula and teacher development, and their implications for academic achievement, also acquire worldwide currency. More and more countries participate in international assessments to assess the status of their student achievement (and by implication, human capital) in comparison to other countries. National educational reports are replete with references to such world standards (Chabbott and Ramirez 2000).

These processes are facilitated by the extent to which international data gathering is standardized. This standardization is built on a growing consensus on the relevant dimensions of the world about which countries and international bodies should collect data. These data not only allow cross-national comparisons and international "races" on a growing number of dimensions, but also facilitate a rationalized sense of "the state of the world" and comparisons of this state across time. The latter development goes hand in hand with an increased world consciousness, giving rise to optimistic notions such as world order, world citizens, international public servants, and global stewardship, but also to pessimistic ones such as world plagues, global warming, and a world of "haves" and "have-nots."

Further studies are needed to examine the institutionalization of the state of the world as an increasingly dramatized cultural and political frame. These studies can proceed with content analyses of the mass media, educational curricular coverage of the world, academic scholarship with world foci, and official world documents generated by international organizations. Relevant studies could track the cultural growth of varying dimensions of the state of the world across these different sources. Varying state of the world reports are already in circulation, but instead of contributing to the reification of the state of the world, the proposed studies would clarify the change and stability in the content of these reports. It is of some importance to study the processes involved because it is quite clear that world agendas greatly affect national and subnational ones in many domains.

It is also possible to directly assess the degree to which scientization influences what constitutes the state of the world in popular, scholarly, and official discourse. Examples of lines of work that move in this proposed direction include the rise in concern with the status of women in international agendas (Berkovitch 1999), the emergence and expansion of regimes of international statistics (McNeely 1995), and population censuses and polices (Ventresca 1996; Barrett and Frank 1999), environmental policies (Frank et al. 1999, 2000a), and the world rise of development discourse and professionals (Chabbott 2002). But more research is needed that directly focuses on what counts as a relevant world dimension, world value, or world danger. It is important to clarify the timing of the emergence of a world property, its status as a good or a threat, and the stability and change of both properties and their cultural merit. To illustrate, consider the frequency with which population is now seen as a potential danger, although historically the dearth of population was the more common fear (Barrett and Frank 1999). Similarly, energy consumption was routinely treated as an indicator of industrialization a few decades ago, but energy conservation is now more likely to be seen as a measure of sound social policy; and in any event, the whole energy enterprise is now commonly discussed as a world matter. It is also important to trace the rise and fall of what constitutes a suitable indicator of a world property (see Block and Burns 1986 on the rise of domestic product as an indicator of national productivity in the United States).

*Role of Organizations and Experts*

Much of our work has assumed that nation-state and organizational isomorphism has resulted from the enactment of world models of individual and collective progress. We have argued that the more nation-states and organizations are attuned to the broader world, the more they enact the appropriate models. We have also argued that these models were articulated and disseminated via carriers that took the form of international organizations and experts (see also Meyer et al. 1997a). Finnemore's work (1993, 1996) on UNESCO as a carrier of science policy norms illustrates this general process.

Two different kinds of further studies would be directly relevant on these questions. First, we need research on how world standards in a given domain are set. Studies could examine the role of professional associations, media, and conferences in the formation of agendas and dissemination of principles and policies. One example involves the rise of the ideas about the importance of benchmarking, a notion that is spreading from business to education to administrative structures of government. An earlier worldwide wave of efforts emphasized Japanese business practices. More recently, the Asian tigers were the focus of much professional discourse and proposed models for imitation. There is much scientized confidence that one can identify best practices in different domains and adapt these best practices in different organization and in different national states. The flow is often, but by no means always, from the West to the rest. At the individual level, studies could examine the worldwide proliferation of self-improvement books, built on the implication that the universalistic standards that operate at organizational and societal levels also operate at individual levels.

A second direction is to examine emergent and contested domains of world standardization. For example, one can focus on the category of indigenous peoples and on the rise of language rights and other related emphases on respecting indigenous peoples. This recent development challenges the right of states to promote a national language through schooling to create national identification and common citizenship. The historical pattern of state-sponsored nationalism via the creation and diffusion of a national culture is well documented (Anderson 1991). The exclusion of groups from access to a national language was a mechanism of social control that further stigmatized these groups. Progressive thought then favored the construction of a literate citizenry in which all would enjoy the advantages of the national language. In the current milieu, though, mandatory common language policies and practices are criticized as undercutting valued multiculturalism and violating human rights (Skutnabb-Kangas 2000). So although organizations and experts speak with one voice on the value of schooling for all, the medium of instruction has become a contested issue. Studies of contestation along such lines would help clarify both institutionalization and deinstitutionalization processes.

## *Science Globalization and the Construction of Agency*

Agency and empowerment are increasingly becoming fashionable concepts, not only in the social sciences, but also in popular parlance. All sorts of actors need to be empowered, it turns out, and their empowerment would not only benefit them but ultimately their organizations, communities, and societies. Much of the talk is optimistic and does not view relationships between different actors as zero-sum games. To be sure, emergent tensions and conflicts are recognized, but in the proper and progressive scientific spirit, these are discussed as problems that can be solved, not as irreconcilable differences that will surely lead to violence and bloodshed. In this frame, empowered actors are envisioned as rational actors with valid purposes to be pursued in reasonable ways. There is much optimism that rules facilitating the empowerment of all sorts of parties simultaneously could be established—rules that could operate within nation-states and other entities as well as between nation-states. Such rules would, of course, rely on the expertise of scientists but also on the consent of a growing number of actors who need to have appropriately accepted the scientized frames involved.

The social science literature often approaches these issues from implicit normative perspectives, arguing the pros and cons of expanded scientific authority. Much needed, however, is research that directly examines policies and practices that presuppose the value of empowered actors and contribute to their construction. These studies could, for instance, examine the rise of professional and human resource development policies and practices in all sorts of organizations across the world (for example, Luo 2000). These developments presuppose active people engaged in active learning. They are inconsistent with older Taylorist worker-proof (and teacher-proof and child-proof) organizing strategies and arrangements. These developments frequently go hand in hand with a focus on learning to learn, emphasizing general decision-making and management skills less attuned to specific occupational requirements. Research is needed to clarify which aspects of the model of empowered agency acquire worldwide currency and which ones vary more dramatically across types of countries and organizations. Similar studies of the roles of experts, consultants, advisors, and gurus in science for progress scripts can help distinguish between common and variable qualities and how these expertly qualities change and stabilize over time (see, for example, Ikenberry 1992; and specific chapters in Boli and Thomas 1999 for historical or field-specific examples). Educational credentials and abstract expertise have clearly become more desirable qualities in a rapidly expanding world of consultants, just as expanded capacities for goal-setting and for planning in a means/ends fashion are expected of modern empowered individuals, organizations, and nation-states. Taken as a whole, these studies could deal with the construction of both the scientific expert and the uses of scientific expertise to generate the empowered actor.

These research directions are informed by the institutional perspective set forth in this book and involve lines of inquiry that take advantage of the methodological strategies employed in much of our work. Some of the research directions noted here involve continuities with present and previous work, as in the call for further studies of organizations and experts as carriers of world culture and scientific authority (see Boli and Thomas 1997). But different and more challenging work is needed to directly analyze the ongoing rationalization of the world itself. And the proposed studies of the construction of agency constitute a direct theoretical challenge to social scientists who invoke a more or less culture-free agency as a starting point. A comparative cultural analysis of agency is much needed, even in a world increasingly devoted to its unqualified celebration.

# Reference Matter

# Notes

## Chapter 2

The authors contributed equally to this chapter.

1. In this chapter, we emphasize institutionalism as a theory of globalization and nation-state behavior. Institutional research on organizations is not addressed. See Schneiberg and Clemens (forthcoming) for a more comprehensive review of institutional research methodology.

2. Conceptually, the world is the level of analysis. However, this does not require that all institutional research must explicitly employ transnational variables. National-level or even individual-level information may be aggregated and used as an indicator of world-level properties. For example, researchers often use individual-level densities as an indicator of population-level legitimacy and competition. Likewise, institutional researchers may at times use individual-level or nation-level data to measure the structure and content of the world polity (Jepperson 1994).

3. See Boli and Thomas (1999) for a thorough treatment of international nongovernmental organizations (INGOs) as world culture.

4. It is thought that INGOs act more as a global civil society, supporting grassroots mobilization and generating lobbying pressure on nations, whereas international governmental organizations (IGOs) deal primarily or exclusively with the state and directly influence the formation of policies. However, more work needs to be done on the differential effects of INGOs versus IGOs as mechanisms through which the world polity affects nations.

5. This research approach borrows heavily from the field of organizational ecology (Hannan and Freeman 1989). However, the theoretical approach and statistical modeling strategies of institutionalism are rather different. In particular, institutional researchers have little interest in showing the typical density dependence processes emphasized by organization ecologists. International organizations are quite unlike capitalist firms and often do not "compete" in any normal sense of the word. For institutionalists, the growth of organizational populations is a representation of world polity structuration, to be explained by

substantive historical changes in the world system rather than by competitive processes.

6. In this example, $Y_t$ is always a nonnegative integer and thus could also be modeled by means of event-count models, described below, particularly if the number of environmental associations tends to be small. For the time being, let us assume that $Y_t$ is a continuous measure.

7. For models in which a lag of the dependent variable is not included as an explanatory variable, the Durbin-Watson $d$ statistic is usually used to identify significant serial correlation.

8. Other approaches, such as Hildreth-Lu, employ various search routines to determine an appropriate estimate of $\rho$ (Pindyck and Rubinfeld 1991: 141–43).

9. Often, the exact date is not available, and so the month or year of founding is used instead. This is reasonable, as long as the time aggregation ("rounding") is small compared with the overall time period under study. However, such rounding (e.g., to the year) may produce excessive numbers of ties (i.e., events at the exact same time), which is problematic for event-history models. In practice, this can be avoided through additional assumptions—for example, that the founding events were randomly distributed throughout a year, rather than rounding them all to the end of the year. However, if multiple events fall within many time intervals, the use of count models such as Poisson regression is preferable.

10. Wu (1990) outlines a graphical diagnostic technique for selecting a functional form of $q(t)$.

11. See Ragin (1987) for a discussion of such approaches.

12. But for reasons we discuss below, they are often measured by a single indicator that reflects both processes, such as the level of national participation in INGOs.

13. Again, the *Yearbook of International Organizations* (UIA 1949–2000) is an invaluable source, reporting data on national membership in INGOs, IGOs, and many treaties. Also, Bowman and Harris (1984, 1993) provide systematic information on when nations sign international treaties.

14. If the dependent variable is not measured as a continuous variable but instead as a count, then ordered categories, or binary categories, approaches such as Poisson/negative binomial regression, ordered logit, or logistic regression can be substituted, respectively, for OLS.

15. Additional explanatory and control variables could be added to equation 9. This modification can be concisely written by making both $b$ and $X_i$ vectors.

16. Dffit and DFBETA analysis, now available in most statistics packages, can be useful in identifying so-called influential cases. See Stokes (1997) for a detailed explanation.

17. Unlike pooled time series, however, event-history analysis is fairly sensitive to "left censoring," in which measurement begins after the event-generating process has begun (see Carroll and Hannan 2000:149–50). It is problematic for an analysis of, say, the adoption of a policy to begin at a point in time in which some nations had already adopted the policy. Data should ideally be collected far back enough in time to begin the analysis before the policy was first adopted anywhere.

### Chapter 3

A version of this chapter appeared as Evan Schofer's chapter, "The Rationalization of Science and the Scientization of Society: International Science Organizations, 1870–1995," in *Constructing World Culture: International Nongovernmental Organizations Since 1875*, edited by John Boli and George Thomas, 249–66 (Stanford, Calif.: Stanford University Press, 1999).

1. The notable exception is the International Council of Scientific Unions, a central professional association that forged tight links with many international associations, particularly UNESCO.

2. Examples include labor unions (Boli and Thomas 1999) and patent organizations (Hironaka 2002), which are highly organized in many nations but remain relatively weak in the international sphere.

3. Such processes do not affect professional science INGOs for the simple reason that very few professional science international governmental organizations exist.

4. The *Yearbook of International Organizations* categories used were Science, Fundamental Science, Research, and Research Standards. Organizations only tangentially related to science (on the basis of their descriptions) were excluded.

5. Inactive organizations were not included in these analyses. The analysis of socially oriented science INGOs is not likely to be affected because they have had an extremely low mortality rate. The analysis of professional science INGOs should be interpreted with more caution, however, because the absence of inactive organizations may bias the sample. Preliminary examination indicates, though, that inactive science INGOs do not appear to be particularly different in character or founding pattern from their active counterparts. Thus, indications are that bias is minimal.

6. A more comprehensive measure reflecting scientific organization in all core nations would be preferable. However, historical evidence suggests that professionalization occurred at nearly the same time in core nations (McClellan 1985). Thus, the use of data for one nation as a proxy for others seems appropriate.

7. Results with this aggregated variable are virtually unchanged compared with results in which these dummy variables were introduced individually.

*Chapter 4*

This chapter is adapted from Gili Drori's dissertation (Drori 1997).

1. This approach to national progress argues that development, like all social change, is caused by the change in the normative orientation of the members of society. Focusing on the role of values, motivations, and psychological forces to bring societal changes about, it draws extensively from Weber's canonized work *The Protestant Ethic and the Spirit of Capitalism* (published 1904). More recent applications of this notion are the characterization of the "modern man" (e.g., Inkeles and Smith 1974) and the delineation of the components of the "mental virus" that "transmits" modernity (e.g., McClelland 1961, 1969).

2. See, e.g., the framing of science goals in the 1999 World Conference on Science (held in Budapest, Hungary), its Declaration, and its Framework for Action.

3. For characterization of the discursive regime of developmentalism and an analysis of its constitution and its effects, see Escobar (1983, 1995) and Ferguson (1990).

4. Although here we emphasize the subordination of the science discourse to developmentalism, these discursive regimes, in fact, are mutually supportive. The faith in science and technology is one of the main factors shaping developmentalism and its policy plans (Escobar 1995:32, 35–36). See also Escobar (1983).

5. This is true whether social development is seen as a unidirectional progression (along the lines of Toennies and Durkheim) or whether it is seen as a continuous process (along the lines of Parsons' pattern variables).

6. For example, in the number of reports about "science for development" and the number of pages devoted to "science for development" matters in summary volumes.

7. For example, the first Development Decade brought the 1963 Conference on the Application of Science and Technology for the Benefit of Less Developed Countries (held in Geneva), and the 1965–1970 preparatory work of the Committee on the Application of Science and Technology to Development shaped the series of resolutions for the second Development Decade, adopted by the UN General Assembly on October 24, 1970.

8. Kenneth King (1989), in his review of international sponsorship of science education, concurs that the beginning of this policy "fashion" dates to the 1960s and that its intensification occurs during the 1980s. But he sees 1970s as a period of contraction, mainly as a result of shifts in aid policies (not necessarily antiscience in themselves).

9. Such as the Rehovot, Israel, 1960 conference, which was aimed at newly formed states; see Gruber (1961).

10. Ranging from proceedings of expert conferences to case-specific studies on the effects of scientific projects on development or the effects of science in certain countries and regions (e.g., Clarke 1985; Ajeyalemi 1990; Eisemon and Davis 1991).

11. Such as the UN Conference on Trade and Development (UNCTAD), the UN Industrial Development Organization (UNIDO), and UNESCO.

12. Such organizations include, for example, the Intergovernmental Committee on Science and Technology for Development (est. 1971) and the UN Center for Science and Technology for Development (est. 1980). This organizational basis for "science for development" was altered several times to accommodate both parallel structural changes in UN organizations and concerns for efficiency. Today, all "science for development" matters are concentrated in the hands of the UN Commission on Science and Technology for Development (under the auspices of the Economic and Social Council [ECOSOC]), which is a fifty-three-member body set up on a governmental basis.

13. Like the administration of "science for development," the funding channels of "science for development" also went through organizational changes. For example, the Interim Fund on Science and Technology for Development (est. 1980) was reorganized in 1982 into the UN Financing System for Science and Technology for Development, and reorganized yet again in 1986 into the UN Fund for Science and Technology for Development. For a review of the efforts of UN agencies to promote "science for development," as summarized by the UN, see the following UN texts: 1968:214–20, 1979:82–84, 1986:208–13.

14. Founded in 1977 and composed of organizations and individuals from twenty-three countries.

15. Founded in 1981, membership includes individuals, organizations, and research groups from thirty-one countries.

16. The Islamic Foundation for Science, Technology, and Development, founded in 1979 at the 10th Islamic Conference of Foreign Ministers, is a subsidiary organ of the Organization of Islamic Conference; members are governmental representatives from fifty-one countries.

17. The Research and Development Forum for Science-Led Development in Africa was founded in 1992 at the instigation of the African Academy of Sciences. The classification of these organizations is judged by their goals, as appearing in the *Yearbook of International Organizations* (UIA 1995). Although the description of organizational aims is short and cryptic, it highlights the main concerns of the organization and thus serves as a useful basis for categorical definition.

18. Such as this text from a World Bank (1988:62) document, *Education in Sub-Saharan Africa*: "Governments that are interested in laying the groundwork for a more technical oriented economy . . . should place heavy emphasis on

general mathematics and science . . . These subjects are relatively inexpensive to teach and are likely to promote economic growth more efficiently than can in-school vocational education."

19. From UN General Assembly Resolution 2318 (XXII) of December 15, 1967, declaring that "The General Assembly [is] convinced that science and technology can make an outstanding contribution to economic and social progress" to the World Conference on Science 1999 Declaration on Science and the Use of Scientific Knowledge stating "that scientific research and its application may yield significant returns towards economic growth and sustainable human development" (item 10).

20. Numerous governmental reports and plans from around the world—such as the United Kingdom's 1993 Government White Paper "Realising Our Potential" or Finland's 1993 policy guidelines "Finland, A Country of Knowledge and Skills"—convey similar "science for development" themes. Such world polity themes are hence directly translated into national texts. In Israel, for example, governments repeatedly proclaim a commitment to science, education, and research and development as necessary infrastructure for building a thriving national economy. Israel's 1996 government, headed by Benjamin Netanyahu, declares that "The Government will initiate publicly and privately funded projects for investment in . . . education and research and development, with the aim of creating in Israel the environment necessary for the Israeli economy to join these of the developed countries of the world" ("Basic Principles for Government Policy 1996," sec. Vd.1).

21. The Technical University in Helsinki, Finland, for example, describes "science for development" themes as its goals: "In particular, the University will support the competitiveness of national industries. . . . It is responsible through scientific research for producing new technological knowledge, for taking care of the transfer of scientific knowledge and technology to the use of society, and for promoting the creation of new enterprises" (cited in Häyrinen-Alestalo 1998:12).

22. The following excerpts from the definition of aims by various science international nongovernmental organizations demonstrate the discursive similarity.

"Strengthen the earth sciences and their effective application in the progress of developing countries" (International Commission on the Lithosphere, founded 1980, headquartered in Ottawa, Canada; membership, individuals from sixty-six countries).

"Promote establishment and use of chemical information systems to solve specific problems, particularly those related to national development" (International Chemical Information Network, founded 1988, headquartered in Paris, France; membership, institutions, commercial businesses, and individuals).

"Promote and foster the growth of scientific community in Africa, and stimulate and nurture the spirit of scientific discovery and technical innovation as to serve socio-economic development and regional integration and world peace and security" (African Academy of Sciences, est. 1985, headquartered in Nairobi, Kenya; membership, individuals from twenty-three African nations and five foreign fellows).

23. In Part IV, we claim and empirically demonstrate that scientific expressions in fact support the expansion of various rights (women's, environmental, consumers', gay and lesbian, and human in general) and democracy (various forms of political participation; see Chapter 11). We argue that the observed connection between science, rights, and democracy is one of the consequences of science globalization. Here, however, we claim that in policy circles, the positive effect of science is less recognized; on the contrary, policy talk reflects a cautionary attitude regarding the effect of scientific advances on the realization of rights. There is a clear distinction between policy talk and actual activity.

24. UN General Assembly Resolution 3384 (XXX), November 10, 1975; emphasis added. Available at <http://www.unhchr.ch/html/menu3/b/70.htm>. Accessed April 30, 2002).

25. Pugwash Aims, 1957; emphasis added. Available at: <http://www.pugwash.org/about/conference.htm>. Accessed April 30, 2002.

26. Namely, the Declaration on the Use of Technological Progress in the Interests of Peace for the Benefit of Mankind (formalized as UN General Assembly Resolution 3384 [XXX], 1975) and the two human rights covenants—International Covenant on Civil and Political Rights (signed December 1966) and International Covenant on Economic, Social, and Cultural Rights (also signed December 1966).

27. This initiative called for the study of this right by a UN subcommission and for an amendment to the 1975 Declaration to include this newly fashioned human right.

28. The UN delayed the establishment of the UN Center for Human Rights, located in Geneva, until 1990 and created the post of High Commissioner for Human Rights only in 1994.

29. Western countries consistently abstained from voting on UN resolutions in support of the 1975 Declaration on the Use of Scientific and Technological Progress in the Interests of Peace and the Benefit of Mankind because this Declaration, as well as most of the action toward greater UN human rights work, was initiated by socialist bloc countries.

30. For a review of UN efforts in the matter of "science and human rights," see Ogata (1990); see also "Who's top?", *The Economist*, March 29, 1997, 15–16, 21–23.

31. Nongovernmental organizations also refer to science as a profession

that requires special attention by addressing the matter of academic freedom and its violations in various countries. Such an initiative was taken, for example, by Human Rights Watch, with its establishment of the Academic Freedom Initiative. This cause is also taken up by science international nongovernmental organizations. For example, the American Association for the Advancement of Science's Science and Human Rights Program aims to protect the human rights of scientists, to advance scientific methods and skills for documenting and preventing human rights abuses, and to develop scientific methodologies for monitoring and implementing human rights. Clearly, international nongovernmental organizations address the specific matter of scientific and technological effects on rights in their reports. For example, international nongovernmental organizations, such as Amnesty International and Human Rights Watch, routinely mention the use of scientific methods in torture in such places as Israel, South Africa, Turkey, and Latin American countries.

32. For example, numerous science international organizations aim at peace, and, as the Arab Atomic Energy Agency states, will work for a "peaceful application of atomic energy." To such organizations, concerns regarding human rights seem marginal to their core operations and take the form of ritualized window dressing.

33. In comparison to the lack of an organizational field around science-*human* rights themes (with only two international organizations), science-*animal* rights themes are explicitly the core goal of five international organizations.

34. The notion of science globalization as ritualistic participation in the discourse of national identity is explored further in Chapter 12.

35. By *instrumentalism*, we mean applying a purely technical approach and assuming a cause–effect relationship. We do not mean *instrumentation*, which refers to the focus of scientists on the device aspect, or apparatus, of their activity (see Knorr-Cetina 1981; Woolgar 1988:88).

## Part III

1. And other relevant subunits, such as international organizations, revolutionary movements, and regional bodies.

2. Such as a professional association, a policy agency, or a learned society.

3. For example, expansion in tertiary-level science enrollments (Ramirez and Lee 1995), women's share in these disciplines (Chapter 8), and science and math instruction time in primary and secondary schools (Kamens and Benavot 1991; Benavot 1992a:156).

4. As the number of scientists, engineers, and technicians rise; see Strath (1998).

5. As measured by article publication, citation counts of scientific articles, published book titles in the sciences, and registered technical patents.

6. As measured, for example, by the rates of national participation in scientific conferences (Schubert et al. 1983) and national memberships in international scientific organizations (Castilla 1997).

7. An increasing number of nation-states have a ministry of science affairs (Chapter 5), a national body for science policy (Finnemore 1993, 1996), and a university as the institution of higher learning (Riddle 1989), and national science organizations increase both in number and specialization.

## Chapter 5

An earlier version of this chapter appeared as "The Worldwide Founding of Ministries of Science and Technology, 1950–1990," *Sociological Perspectives* 43, no. 2 (2000): 247–70.

1. Symbolic interactionism, which stresses the constitutive importance of the "generalized other" in social interaction and the significance of the wider community's model of the internal structure of the "self," provides part of the theoretical basis for the distinction between modern "actors" and "others" (Strang and Meyer 1993).

2. Because OECD and UNESCO have dramatically different membership bodies, UNESCO's activities are much more globalized and more instrumental in transferring the notions of science and national development originating in the core states to the third world. Whereas OECD recommended that nation-states should set up their science policy structures outside the official-line authority of the central administration (OECD 1963), UNESCO, by contrast, actively advised its members to integrate these new science policy bodies directly into the highest levels of government (Finnemore 1993, 1996).

3. An interesting example is the case of the United States. Although the United States has a large and stable scientific infrastructure, its liberal polity style (stressing private over public action) has made it reluctant to create a ministerial unit at the level of the federal government to control and coordinate various scientific activities. Several earlier science bodies such as the National Research Council (est. 1917) were created privately outside the state. Other bodies (e.g., the office of Scientific Research and Development, est. 1941) had an explicitly narrow orientation toward science and technology related to the military. In recent times, the United States has somewhat changed its orientation toward broader governmental responsibility over a wider range of scientific resources and activities, yet it still does not have a ministerial unit (or "department") that directly controls science. For current international purposes (e.g., OECD and UNESCO reports), American officials consistently identify the National Science Foundation, founded in 1950, as the primary science policy-making agency for the U.S. government (Finnemore 1991, 1996).

4. A more broadly qualified functional consideration could be taken into account for science ministry foundings in developing nation-states after 1970. Establishing a science ministry in African countries, for example, could be beneficial to recruiting aid and funding from various international groups as well as to encouraging and developing science and technology. Having a rational organizational structure with proper modern ministerial units symbolically presents the country as a sound and reliable governmental actor, which is advantageous for attracting investment and aid from international society.

5. Including all current expenditure for purchases of goods and services by all levels of government.

6. The two indicators for the functional argument are collinear so that the absence of the GDPPC effect in the earlier period does not negate the argument of hypothesis 2.1. Although they are collinear even after the conventional treatments (e.g., logging) are conducted, I present the models including both functional indicators because I want to examine the direct effect of scientific development as well as that of general economic development. When the models include only the GDPPC indicator without the tertiary-level science student variable, the results consistently show the positive and significant effect of GDPPC for the earlier period and the negative and significant effect for the later period with no major changes in the other coefficients.

7. In population ecology models, it is argued that organizational density is a measure of prevalence or legitimacy, which reflects the extent to which an organizational form is taken for granted (Scott 1995). However, the density effect in the present analysis most likely does not work because it is a more indirect measure of legitimacy and normative pressures than the others.

## Chapter 6

1. Recall, for example, the horrified and bipartisan public reaction to reports that Nancy Reagan (and at least indirectly Ronald) regularly consulted an astrologer during their time in the White House.

2. See Chapter 8 for further discussion of issues of access to science.

3. Perhaps it is true that only university-level research can completely eschew relevance. Subject areas in lower levels of educational systems have, for a long time, needed to claim relevance in order to garner legitimacy.

## Chapter 7

1. Beekum and Ginn (1993:1296) discuss three possible "junctions" where decoupling may occur: (1) between the normative and the operational level of organizations (i.e., between intentions and actions); (2) between the technical core and the peripheral operations (i.e., between subsystems); and (3) between

actors. From an institutional perspective, the third category is collapsed into systemic units.

2. See discussion in Chapter 11.

3. For example, the difference between the Canadian International Development Agency (CIDA) and the U.S. Agency for International Development (USAID) in their aid programs to developing nations.

4. Such as the establishment of CIRIT (Comissió Interdepartamental de Recerca i Innovació Tecnològica, or the Interdepartment Commission for Research and Technological Innovation) in 1980, and the 1986 approval by the Spanish government of the Catalan "Science Act," which centralized research funds and funding and which established CICYT (Comisión Interministerial de Ciencia y Tecnología, or the Interdepartment Commission for Science and Technology).

5. The division of science into these dimensions is consistent with the dimensions documented in Chapters 9 and 10.

6. The second and third dimensions (science research activity and scientific labor force) are often referred to as basic and applied science and are distinguished on this basis.

7. We also employed a more sophisticated multivariate methodology—namely, analysis of covariance structures—to ascertain the validity of the model as a whole. Yet although this methodology is more technically sensitive to the indicator–concept relationship, it is technical and cumbersome. Because it produced substantively similar results, both in terms of relating indicators to each dimension and in terms of the relationships between the dimensions, or latent variables, the simpler method of indexing is used for the purposes of this chapter.

8. A model that uses a constant set of developing countries, namely twenty-nine cases, provided substantively identical results. The Pearson correlation score between policy and research equals 0.51, $r^2$ between policy and labor force equals 0.56, and $r^2$ between research and labor force equals 0.43, whereas all such scores are significant at a level higher than 0.02.

9. Again, a model that uses a constant number of cases, namely twenty-six developing countries, provides substantively identical results. The only non-significant correlation is between science labor force and research work ($r^2 = -0.20$), whereas the Pearson correlation scores between policy and labor force and between policy and research work are significant ($r^2 = -0.36$ and 0.77, respectively). Also, as in the general model, the only negative relationship is between policy and labor force.

10. This finding does not negate the pervasiveness of loose coupling in Western science institutions, as most of us working in such environments can attest to. Loose coupling is a common organizational ailment everywhere.

Nevertheless, this finding focuses on loose coupling as a general national trait and points out that in Western countries, the tendency toward it in general is lower than in other countries.

11. In its list of goals, the present II Research Plan lists not only the development of knowledge-based technology for the use of local industry, but also culture and society, health and quality of living, environmental and natural resources, agriculture and food, industrial technologies, and, finally, building technologies and public works (Bellavista and Renobell 1998:4). This wide scope of goals for science reflects a hodgepodge attitude toward policy objectives.

12. "Despite the targets, purposes and instruments, priorities can't be applied within the terms of the official discourse. There is a traditional dynamics to distribute public finance to research, everybody treated the same way, (coffee for all irrespective of preference)," according to Bellavista and Renobell (1998:4).

13. Local science policy agencies reorient their operations toward such processes as the monitoring of activities rather then toward the creation of an actual link between science and industry. Hence, for example, CIRIT is putting extra efforts into monitoring the publication of journal articles and their citations as an indication of scientific output, rather then zeroing on the contribution, if any, of Catalan science to corporations. Moreover, two (!) special agencies for monitoring such trends in scientific performance, namely CONACIT and CAS, are currently being established to assist CIRIT in these efforts.

*Chapter 8*

1. Both men and women could be capable of producing a feminized science that focuses less on the traditional uses of science (e.g., science for the creation of war machines, science as interventionist) and more on issues of concern to women (Nichols et al. 1998).

2. For detailed analyses of the Third International Math and Science Study (TIMSS) results for science achievement, see Mullis et al. (1998), Martin et al. (1997), and Beaton et al. (1996).

3. The International Association for the Evaluation of Educational Achievement sponsored the Second International Science Study (SISS) and the Second International Mathematics Study, similar to TIMSS. For more details on the First and Second International Association for the Evaluation of Educational Achievement Science Studies, see IEA (1988); Keeves (1992); Keeves and Kotte (1992); Postlethwaite and Wiley (1992); and Rosier and Keeves (1991).

4. This does not assume that older and returning students are ineligible to pursue tertiary-level education or S&E studies, although in many parts of the world, such nontraditional students are more likely to enroll in practice-based

institutes of higher learning, which do not grant degrees.

5. Nonetheless, the greater participation of women in natural science higher education needs clarification. Although the number of degrees awarded to women increased by only 1 percent between 1975 and 1991, there was a concomitant decline in the number of degrees awarded to men (NSB 1998; Johnson 1994). Johnson (1994) reports that this trend began in 1977 so that by 1991, the numbers of degrees awarded to men in the natural sciences had declined by 3 percent annually.

6. In a world that is increasingly scientized, the world increasingly revolves around money. This may suggest the possibility of a change in the regime of (economic) developmentalism from science to business as the source of expertise or knowledge.

7. The NSF (1995) reported that women employed in the S&E workforce are relatively young. Among S&E occupations other than computer science and math, well over half of the women at every degree level had received their degrees after 1984.

8. Few comparative studies exist that examine women in scientific careers or academia. For a comparison of American, Israeli, and Greek women in scientific careers (broadly defined), see Etzion et al. (1993).

9. We employed panel regression analyses with a cross-national and longitudinal design whereby the dependent variable measured at time $t2$ (1992) is regressed against the independent variables at $t1$ (1972) in addition to the lagged dependent variable. Panel analyses have been used widely in order to examine the causes of societal change over time (Hannan 1979) as well as educational change (Rubinson and Ralph 1986).

10. Cross-national analyses were conducted as well for 1972 and 1992. The results were consistent over time and suggest that women's participation in higher education as well as male participation in S&E positively and significantly influenced women in S&E education for each time period, net of economic development and state power.

11. These data come from appendix B in UNCSTD (1995). For a more detailed analysis of twenty-four multilateral organizations associated with the United Nations and their gender, science, and technology activities, see UNIFEM (1994, 1999).

12. These include *Winning with Women in Trades, Technology, Science and Engineering* (Canada 1993) and *The Rising Tide: A Report on Women in Science, Engineering and Technology* (UK 1993).

## Chapter 9

1. Or of main universities, in nations where there is no national system (see Frank and Gabler 2000).

2. The Institute for Scientific information (ISI) compiles data on publications in scientific journals and on their citations, known as SCI (*Science Citation Index*) and SSCI (*Social Science Citation Index*). In our subset of these data, the information is tabulated by country of affiliation of the authors and by scientific discipline, per year. These data, specifying the number of articles published in a particular country in a particular discipline in a particular year, permits us to refer to, for example, worldwide trends in disciplinary productivity or to disciplinary emphasis per country (defined later in this chapter).

3. Calling for countries to develop their relative advantage in science by carving a niche of scientific specialization and emphasis.

4. ISI/SCI offers a disciplinary breakdown for up to one hundred fourteen different disciplines, listing such specific scientific fields as physiological psychology and neuropsychology as subfields of psychology and offering information on annual publication rates in each. In most developing countries, the number of publications in these refined scientific disciplines is minuscule.

5. The structure of these data somewhat corrects for bias in data gathering. In other words, although SCI information may be biased toward publication in Western scientific journals, the percentage is calculated within each group of countries.

6. We tried various combinations of disciplinary publications to delineate the indicators of disciplinary emphases—for example, adding chemistry to the hard sciences. The use of either combination of disciplines in these ordinary least squares models produced substantively similar results.

7. This is, in part, due to the zero-sum nature of our measures of disciplinary concentration (Table 9.1), which reflect a *relative* emphasis on specific disciplines.

8. Because these characteristics—namely, democracy and membership in the socialist and newly industrialized country groups—are overlapping (or, in statistical terms, there is high collinearity), we could not introduce all these factors as independent variables into a single regression model.

9. Triple Helix policy refers to the pattern of joint efforts and cooperation between three nodes of scientific activity—academia, government, and industry. It is the connection among the three components that firmly applies scientific knowledge toward technological innovation and industrial productivity (see Etzkowitz and Leydesdorff 1997).

10. At the risk of being literal about "talk," the language used to discuss climate change in Senegal hints at the source of this policy regime. Apparently, although Senegal is a French-speaking nation and is tied to the French organizational network for climate change and control, the specific terms used in professional and policy circles in Senegal are in English (Engels 1999).

11. In the early 1980s, the Russian scientific community was by far the

greatest in the world, surpassing the American scientific community by 10 to 30 percent (depending on the definition of degrees and fields; Dezhina and Graham 1999:1303).

12. The traditional cooking stove, which is still popular in many Kenyan households today.

*Chapter 10*

A version of this chapter was published as Evan Schofer, Francisco Ramirez, and John Meyer, "The Effects of Science on National Economic Development, 1970–1990," *American Sociological Review* 65 (2000): 877–98.

1. We use a colloquial, or common sense, definition of scientific activity. We focus on the conventionally defined scientific fields, with their basic and applied research and related components of training and labor-force structures.

2. Even within the conventional tradition, criticisms are put forward and corrective policies proposed. Incentives to produce new knowledge may be too limited (Arrow 1962). Linkages from science to technical development may be too weak. Also, much contemporary world science may be useless or misdirected for developing countries (Shenhav and Kamens 1991). Moreover, investment in science may produce increased dependence (Cooper 1973). Correctives are proposed for each of these concerns. For the most part, these lines of thought predict reduced benefits of scientific expansion on economic development but do not propose negative effects.

3. See Drori (1998, 2000) for a review of the inconsistent and inconclusive empirical comparative research on the effects of science education on economic development.

4. As discussed in Chapter 9, data from the *Science Citation Index* (Institute for Scientific Information 1982) show that developing nations, on average, behave quite similarly to developed nations in terms of patterns (although not magnitude) of scientific output. In fact, the only major differences are that developed nations produce proportionally more social science and medical research, whereas the developing nations put more emphasis on chemistry and physics. This suggests that developing nations are not creating dramatically different science infrastructures compared with the developed world. Only in a few regions (Asia, communist Eastern Europe) do patterns of science output stand out—with more emphasis on "technically relevant" sciences (math, physics, chemistry, engineering). But even then, the differences are not great.

5. The categories "socially" and "technically" relevant science are not distinct in all cases. For example, certain types of medical research could simultaneously produce new valuable technical innovations (e.g., new pharmaceuticals) and also legitimate constraints on economic activity. Nevertheless, it is reasonable to

argue that certain fields (e.g., environmental sciences) tend to be more socially relevant than others (e.g., mathematics).

6. Economists have discussed processes of technology licensing and transfer at great length (e.g., Fransman 1986; Evenson and Ranis 1990).

7. One mechanism is the copying of intellectual property. Another mechanism, suggested by Evenson and Ranis (1990), is that market inefficiencies make it difficult to license technology at fair value.

8. See Frank et al. (2000b) for a discussion of the role of local scientists as "receivers" of worldwide social concerns.

9. Another dimension of science infrastructure mentioned in the literature is the strength of science professional associations. We do not discuss them here because there are few arguments linking such infrastructures to economic growth—particularly if one controls for the level of scientific research and the size of the scientific labor force. Indeed, in separate analyses (not presented here), we find that measures of scientific professional activity have no effect on measures of economic growth.

10. One criticism of the SCI data is that they overemphasize journals in the English language, leading to underestimation of science activity in non-English-speaking nations. We acknowledge this possibility. However, to our knowledge, no one has used other measures to show that such bias exists. As a precaution, we checked to see if the inclusion of an "English-speaking nation" dummy variable in our analyses affected the results. It did not have a significant effect, nor did it alter the sign or significance of any other variables in the analyses.

11. Publications with multiple authors in different nations are attributed fractionally. That is, for an article with four authors from four different nations, each nation is counted as having produced 0.25 publications.

12. Ideally, we would use data from 1970, but the earliest data available in computerized form are from 1973. This is unlikely to be a substantial source of bias because national scientific output is highly autocorrelated over time. The 1973 data should be a close proxy for the same variable in 1970.

13. All science data are derived from the *Science Citation Index* (as opposed to the *Social Science Citation Index*). Consequently, this measure of psychology research reflects physiological psychology and neuropsychology, rather than the entire academic field of psychology.

14. In smaller-sample analyses, these variables have similar effects on economic development as the publication and citation measures.

15. A better measure would be the number of people with science or engineering degrees per capita. However, such measures are not available. Number of people currently in school or number of degrees granted in a given year are the best data available.

16. For the purposes of this chapter, we focus on national economic development—the expansion of productive economic activity within a given national boundary. We focus on national, rather than world or regional, economic development because the nation-state is the primary focus of both science and economic development policy, projects, and resources, and the presumed beneficiary of such efforts. We do not address the benefits of science that accrue at the sub- or supranational level. Nor do we measure the benefits of science that are extremely long term or noneconomic in nature. These are important issues—indeed, we speculate that many benefits of science are not captured in measures of short-term economic growth. Nevertheless, those effects are beyond the scope of the current analyses.

17. Our sample varies from 80 to 112 cases, depending on which variables are included in the analysis. Certain variables have more missing data than others, requiring more cases to be removed from the analysis.

18. The choice of twenty years is somewhat arbitrary. We have explored growth over other time spans. Results were generally consistent. As one would expect, shorter time spans yield less stable effects of independent variables because there is less time for the effects of independent variables to cumulate. Also, GDP growth fluctuates substantially from year to year, making GDP growth measures over a short time span relatively more "noisy."

19. Although Levine and Renelt (1992) do not find consistent effects of political factors or trade, these variables are commonly used in analyses of GDP growth. We include them to be consistent with earlier studies and to show that our results are robust and stable even when these control variables are included.

20. In our panel models, previous level of development is the lagged dependent variable and is measured with the same three indicators: log GDP per capita, log energy consumption per capita, and percentage of labor force outside of agriculture. In panel models, the expected effect is positive.

21. We also measured investment as a three-year average from 1969 to 1971 and as a ten-year average from 1965 to 1975. All of these measures produced similar results.

22. The UNESCO statistical yearbooks are the primary data source. Missing values are filled in from the World Bank (1992) data set.

23. Development theorists, often studying Asia, suggest that one should focus on exports, rather than total trade or imports. However, Levine and Renelt (1992) show that the effects of total trade, exports, and imports are nearly identical. Thus, there is no reason to suppose a particular effect of exports.

24. Less developed nations were defined as having per capita GDP of $2,000 or less in 1970, which is nearly half our sample. We also used cutoffs that reflected the lowest quartile and lowest three-fourths of nations in our sample. Results were not affected.

25. In addition to the analyses presented here, we have explored a variety of other control variables including: civil unrest, coups, assassinations, foreign investment, foreign debt, regime type, population growth rate, and primary school enrollment. Findings regarding scientific research and scientific labor force are similar to those in the models presented.

26. That is, GDP in 1970 was regressed on each science indicator, and the residual was saved for use in the factor analysis.

27. Variables are *not* residualized in later regression analyses, presented below.

28. Related to this, we wondered if our findings applied to the very poorest of nations, many of which were dropped out of our analysis as a result of missing data. To address this, we made efforts to expand our case base by substituting missing variables with data from nearby years. Also, because our science indexes are made up of several measures, single missing measures could be crudely estimated on the basis of the others. This allowed us to increase our sample to 103 cases, compared with 87 in model 4. Results were nearly identical to those presented here.

29. To identify outliers and influential cases, we examined partial plots, studentized residuals, leverage, Cook's $D$ and DFBETAs (see Bollen and Jackman 1985 for a review).

30. As noted above, the measure for "technical" science activity consists of two indicators: a nation's total journal article publications in the fields of engineering, chemistry, and physics, and the total number of citations in those same fields (per capita, logged). "Socially relevant" science was similarly measured with publications and citations in clinical medical science, psychology, and biology. Results are not affected by the exact fields chosen to represent "technical" and "socially relevant" science. "Technical" sciences can also be measured with publications in fields such as mathematics and computer science, and "socially relevant" sciences can be measured by means of biomedical science, ecological science, and so on.

31. This finding also suggests that processes of brain drain are not so extreme as to erase the benefits of expanded tertiary-level science education.

*Chapter 11*

1. For instance, between 1952 and 1964, Britain passed legislation on industrial emissions and land use such as the Navigable Waters Act (1953), the Protection of Birds Act (1954), the Rural Water Supplies and Sewage Act (1955), the Litter Act (1958), the Radioactive Substances Act (1960), the Estuaries and Tidal Waters Act (1960), the Rivers (Prevention of Pollution) Act (1961), the Deer Act (1963), and the Water Resources Act (1963) (Vogel 1986). These acts were perceived as part of the state's responsibility for sensible

management of the land, rather than comprehensive protection for the environment as we now understand it.

2. By 1992, 78 countries had signed the Convention on International Trade in Endangered Animals, 121 had signed the Montreal Convention, and an amazing 166 countries had signed the Biodiversity Convention, practically every country in the world.

3. An event-history analysis examines the rate at which events occur. The dependent variable is the instantaneous likelihood that an event such as a national park founding or a treaty signing will occur, given the set of countries at risk of having such an event (Tuma and Hannan 1984).

4. Data comes from the International Union for the Conservation of Nature (1990).

5. Data come from the *Statesman's Yearbook* (Steinberg 1900–1995). Note that the U.S. Environmental Protection Agency is not included in this count because it is not a cabinet-level department.

6. Data from Bowman and Harris (1995).

7. Data from Bowman and Harris (1995).

8. Data from Zils (1998).

*Chapter 12*

An expanded version of this text, by Gili Drori, received the International Sociological Association's Bielefeld Prize for the Internationalization of Sociology 1999; see *Zeitschrift für Soziologie* 28, no. 4 (1999): 318. The chapter and the awarded text are adapted from Gili Drori's dissertation (Drori 1997).

1. Such as the attitude in the social sciences as regards actors, whether human or organizational entities, as analytic focal points.

2. These three indicators of scientific activity were combined into a factor or an index, depending on the type of method used (see Tables 12.1 and 12.2). Also, these components of national science activity were used repeatedly in all models. Occasionally, a fourth indicator—a dummy variable for the existence of a national science policy body (Finnemore 1991)—was added to the factor or index of national science activity.

3. National development is a factor score, combining gross national product per capita, secondary education ration, and a dummy variable for noncore countries.

4. The analyses are too detailed to be completely reported here; they are described in full in Drori (1997). These analyses follow the tradition of cross-national and empirical studies (see Chapter 1); statistical methods are used to evaluate the relationships between science practice and various political out-

come variables. Each such concept is indicated by a variety of measures, often merged as factors but also examined as single indicators in parallel models.

5. Indicated by indexes of women's status, women equality, and gender development.

6. Frank and McEneaney (1999).

7. Indicated by the history of the local consumer-related organizational field and the size of this field by 1992.

8. Indicated by Humana's human rights index, the human development index, violations of human rights as cited by Amnesty International, and Sivard's index of repression.

9. Indicated by a country's commitment to international environmental treaties.

10. Although the conceptualization of democracy varies between liberal notions (centering on accountability) and critical notions (centering on participation or on social equity), the indicators reflect merely the enactment of practices that are commonly acknowledged as democratic. This, for us, is both an empirical and a theoretical choice: our interest is in the worldwide adoption of legitimated practices and not the investigation of the validity of such practices.

11. This claim as to the empirical, post-1970 link between science and democracy should not be confused with the discussion in the social studies of science literature, on the democratization of science. Although that discussion centers on the accessibility to, and accountability of, science toward its audiences, our point is that global processes of scientization and democratization are bound together by a common world culture and international policy regime.

12. For a discussion of epistemic linkages among global policies, see Haas (1992).

13. What came to be known as the "Washington consensus," which is described by Wade as "reflecting the demise of Keynesianism and the ascendancy of supply-side economics . . . [and being] based on the twin ideas of the state as the provider of the regulatory framework for private-sector exchanges . . . and of the world economy as open to movements of goods, services, and capital, if not labour" (1996:5).

14. During the 1980s and early 1990s, economists, development professionals, and students of science were describing the economic surge of the East Asian newly industrialized countries as dependent on these countries' technological advances, emphasis on science education, and scientifically trained labor force. See King (1989:108) for a discussion of the common belief in the validity of this explanation.

15. Science policies were recently amended to reflect this latest trend in developmentalism: State-centered and state-sponsored science initiatives were rewritten to reflect more market-oriented ventures. They are manifested in

market-oriented changes to the university system and research (Gibbons et al. 1994; Etzkowitz and Leydesdroff 1997; Clark 1998; Nowotny et al. 2000). Such amendments highlight both the agency of actors within society and the science-based recipes they are to use for achieving economic success. They reflect the tight discursive link between science and notions of national progress.

16. Meaning "each and all." Foucault used this as the title for a series of lectures he gave during the late 1960s on matters of governmentality.

17. As Foucault summarizes, "The production of effective instruments for the formation and accumulation of knowledge—methods of observation, techniques of registration, procedures for investigation and research, apparatus of control . . . all this means power" (1980:102).

18. What Meyer and Jepperson (2000) call "cultural technology," or scientized policy models, such as the "science for development" model.

*Chapter 13*

This chapter is adapted from Gili Drori's dissertation (Drori 1997).

1. ISO documents conceive of nonharmonized standards as "technical barriers to trade," or obstacles to rationalization of the international trading process. Available at: <http://www.iso.ch/iso/en/aboutiso/introduction/whyneeded. html>. Accessed April 30, 2002.

2. Information retrieved from directories of the Union of International Associations (UIA).

3. Information retrieved from World Standards Services Network (WSSN). Available at: <http://www.wssn.net/wssn/gen_inf.htm/>. Accessed April 30, 2002.

4. Publicly Available Specification, which is the first step to becoming a new ISO-sponsored and -monitored field; see <http://www.iso.ch/iso/en/stds-development/whowhenhow/proc/deliverables/iso_pas.html>. Accessed April 30, 2002.

5. For general information on IMF initiative and its criteria, see <http://dsbb.imf.org/>. Accessed April 30, 2002.

6. Available at: <http://www.diffuse.org/oii/en/eurogi.html>. Accessed April 30, 2002.

7. One could also imagine an alternative vision of global standardization, drawing on nationalistic criteria in its justification for such initiatives, as well as for the logic of the specifics of standard settings. For example, a recent dramatization of the "invention" of the unit of measure the meter, which was based on the diaries of Jean-Baptiste Delambre and Pierre Mechain, describes the meter as a product of the political spirits of the French Revolution and French nationalism. The play, *Les Monts du Metre* by Jean-Claude Bastos, tells of the seven

years that the two French astronomers traveled across France in order to measure the first meter; their attempt to decimalize distance was fueled by calculations that the meter is a decimal fraction of the size of France, thus linking nationalistic rationales with distance standardization.

8. Information retrieved from directories of the UIA.

9. Like the analyses in Chapter 12, national science activity is indicated by a combination (factor or index) of the following cross-national measures: citations of scientific article publications, science book publications, and membership in International Council of Scientific Unions organizations. Occasionally, a fourth indicator—the existence of a national science policy body (Finnemore 1991)—is added.

10. Again, like the analyses offered in Chapter 12, national development is indicated by a combination (factor or index) of the following cross-national measures: gross national product per capita (or alternatively, energy consumption per capita), secondary education ratio, and a dummy variable for noncore countries.

11. The results of the empirical analyses and more data specifications are presented in full in Drori (1997).

12. Indicated by such cross-national variables as the capability to gather data, as indicated by the number of missing data in UN statistical publications (Drori 1997). This scale describes the number of missing data cells per country in the total number of data tables in UN annual statistical yearbooks.

13. Indicated by the dates of the first population census and the publication of first national statistical yearbook (Ventresca 1996).

14. Indicated by the duration until first connection with the Internet and the rate of national Internet growth. Data compiled from <gopher://nic.merit.edu:7043/00/nsfnet/statistics/nets.by.country/>. Accessed April 30, 2002.

15. Indicated by deviations from standard reporting specifications, namely the number of comments added to national reporting in the total number of data tables in UN annual statistical yearbooks (Drori 1997).

16. Indicated by a scaled variable for adherence to international standards of accounting, as devised in summary texts of PriceWaterhouse (Fitzgerald et al. 1979).

17. Indicated by the number of ISO-9000 certifications given to corporations, by nation (Mendel 2001).

18. Indicated by a scale of perceived government corruption, as collected, analyzed, and complied by Transparency International. Available at: <http://www.transparency.org/>. Accessed April 30, 2002.

19. For a review of similar recent studies of accountancy, see Porter (1996).

20. See IMF documents regarding the Slovak Republic. Available at: <http://www.imf.org/external/country/SVK/>. Accessed April 30, 2002.

*Conclusion*

1. William Faulkner, 1950 acceptance speech for the 1949 Nobel Prize in Literature, Stockholm. Available at: <http://ww.mcsr.olemiss.edu/egibp/faulkner/lib_essays.html>. Accessed April 30, 2002.

2. Adapted from Henry David Thoreau's *Walden* (chap. 9). Available at: <http://www.americanliterature.com/wp/wp09.html>. Accessed April 30, 2002.

# Bibliography

Abbott, Andrew. 1988. *The system of professions*. Chicago: University of Chicago Press.

Abramovitz, Moses. 1956. Resource and output trends in the U.S. since 1870. *American Economic Review* (papers and proceedings, May 1956).

———. 1991. *Thinking about growth and other essays on economic growth and welfare*. Cambridge: Cambridge University Press.

Adams, Walter. 1968. *The brain drain*. New York: Macmillan.

African Primary Science Project (Kenya). 1966. *Kenya primary science*. Nairobi, Kenya: African Primary Science Project.

African Primary Science Project (Ghana). 1970. *African primary science*. Accra, Ghana: Ghana Publishing Company.

Aikin, Charles, and Louis W. Koenig. 1949. Introduction to Hoover Commission: A symposium. *American Political Science Review* 43:933–40.

Ajeyalemi, Duro. 1990. Science and technology education in Africa: A comparative analysis and future prospects. In *Science and technology in Africa: Focus on seven sub-Saharan countries*, edited by Duro Ajeyalemi, 114–27. Lagos, Nigeria: University of Lagos Press.

Alderson, Arthur S. 1999. Explaining deindustrialization: Globalization, failure, or success. *American Sociological Review* 64:701–21.

Allwood, Carl Martin. 1998. The creation and nature(s) of indigenized psychologies from the perspective of the anthropology of knowledge. *Knowledge and Society* 11:153–72.

Alper, Joe. 1993. The pipeline is leaking women all the way along. *Science* 260:409–11.

Alter, Peter. 1987. *The reluctant patron: Science and the state in Britain 1850–1920*. New York: St. Martin's Press.

Alvares, Claude. 1992. *Science, development, and violence: The revolt against modernity*. New Delhi: Oxford University Press.

American Association for the Advancement of Science (AAAS). 1989. *Science for all Americans*. Washington, D.C.: American Association for the Advancement of Science.

————. 1999. *Project 2061: Science benchmarks.* Available at:
<http://project2061.aaas.org/tools/benchol>. Accessed May 2, 2002.

Amsden, Alice H. 1989. *Asia's next giant: South Korea and late industrialization.* New York: Oxford University Press.

Anderson, Benedict. 1991. *Imagined communities: Reflections on the Origin and Spread of Nationalism.* New York: New Left Books.

Anderson, C. A., and M. J. Bowman. 1976. Education and economic modernization in historical perspective. In *Schooling and society,* edited by L. Stone, 3–19. Baltimore, Md.: Johns Hopkins University Press.

Andrews, Richard N. L. 1999. *Managing the environment, managing ourselves: A history of American environmental policy.* New Haven, Conn.: Yale University Press.

Approtech. 1993. Mainstreaming women in science and technology. *Report of the International Symposium of Women in Science and Technology Development and Transfer,* Thailand, July 1992. Asian Alliance of Appropriate Technology Practitioners (Approtech-Asia) and Women in Science and Engineering Forum of Thailand. Bangkok: WISE-T.

Archibold, Randal C. 1999. Poor scores by U.S. students lead to 10-state math efforts. *New York Times,* 6 May.

Arnold, David. 1996. *The problem of nature: Environment, culture and European expansion.* Cambridge: Blackwell.

Aronowitz, Stanley. 1988. *Science as power: Discourse and ideology in modern society.* Minneapolis: University of Minnesota Press.

Arrow, Kenneth. 1962. Economic welfare and the allocation of resources for invention. In *The rate and direction of inventive activity: Economic and social factors,* edited by R. R. Nelson, 609–626. Princeton, N.J.: Princeton University Press.

Arvanitis, Rigas, and Yvon Chatelin. 1988. National scientific strategies in tropical soil sciences. *Social Studies of Science* 18:113–46.

Ashby, Eric, and Mary Anderson. 1981. *The politics of clean air.* Oxford: Clarendon Press.

Auger, Pierre. 1961. *Current trends in scientific research.* Paris: UNESCO.

Australia. 1994. Women in science, engineering, and technology: A discussion paper. Prepared by the Women in Science, Engineering and Technology Advisory Group, Canberra, September 1994.

Baker, David. 1993. Compared to Japan, the U.S. is a low achiever . . . really: New evidence and commentary on Westbury. Response to Baker by Westbury and rejoinder by Baker. *Educational Researcher* 19:18–26.

Banks, Arthur S. 1950–1990. *Political handbook of the world.* New York: CSA Publications.

———. 1976. *Cross-national time-series data, 1815–1973*. Ann Arbor, Mich.: Inter-University Consortium for Political and Social Research.

Barinaga, Marcia. 1994. Surprises across the cultural divide. *Science* 263:1468–472.

Barnes, Barry. 1985. *About science*. Oxford: Basil Blackwell.

Barnes, Barry, David Bloor, and John Henry. 1996. *Scientific knowledge: A sociological analysis*. London: Athlone.

Barrett, Deborah A. 1995. Reproducing persons as a global concern: The making of an institution. Ph.D. diss., Stanford University.

Barrett, Deborah, and David J. Frank. 1999. Population control for national development: From world discourse to national policies. In *Constructing world culture: International nongovernmental organizations since 1875*, edited by John Boli and George Thomas, 198–221. Stanford, Calif.: Stanford University Press.

Barnett, Michael N., and Martha Finnemore. 1999. The politics, power, and pathologies of international organizations. *International Organization* 53:699–732.

Barro, Robert J. 1991. Economic growth in a cross section of countries. *Quarterly Journal of Economics* 106:407–43.

Barro, Robert J., and Xavier Sala-i-Martin. 1995. *Economic growth*. New York: McGraw-Hill.

Barron, David N. 1992. The analysis of count data: Overdispersion and autocorrelation. *Sociological Methodology* 22:179–220.

Basalla, George. 1967. The spread of Western science. *Science* 156:611–22.

Bates, Ralph S. 1965. *Scientific societies in the United States*. Cambridge, Mass.: MIT Press.

Baumann, Erich. 1967. *Lehrbuch der Physik*. Munich: Ehrenwirth Verlag.

Beaton, Albert E., Eugenio J. Gonzalez, Dana L. Kelly, Michael O. Martin, Ina V. S. Mullis, and Teresa A. Smith. 1996. *Science achievement in the middle school years: IEA's third international mathematics and science study (TIMSS)*. Chestnut Hill, Mass.: TIMSS International Study Center.

Becker, Gary. 1964. *Human capital*. New York: Columbia University Press.

Beekum, I., and G. O. Ginn. 1993. Business strategy and interorganizational linkages within the acute care hospital industry: An expansion of Miles and Show typology. *Human Relations* 46:1291–318.

Bellavista, Joan, and Victor Renobell. 1998. Science, technology, and innovation in Catalonia: Actors, knowledge and paradigms. Paper presented at the European Association for Studies of Science and Techology meeting, Lisbon, Portugal, October 1998.

Ben-David, Joseph. 1990. *Scientific growth*. Berkeley: University of California Press.

Benavot, Aaron. 1985. Education and economic development in the modern world. Ph.D. diss., Stanford University.

———. 1989. Education, gender, and economic development: A cross-national study. *Sociology of Education* 62:14–32.

———. 1992a. Curricular content, educational expansion, and economic growth. *Comparative Education Review* 36:150–174.

———. 1992b. Educational expansion and economic growth in the modern world, 1913–1985. In *The political construction of education*, edited by B. Fuller and R. Rubinson, 117–34. New York: Praeger.

Benavot, Aaron, Yun-Kyung Cha, David Kamens, John W. Meyer, and Suk-Ying Wong. 1991. Knowledge for the masses: World models and national curricula, 1920–1986. *American Sociological Review* 56:85–100.

Benedick, Richard Elliot. 1991. *Ozone diplomacy*. Cambridge, Mass.: Harvard University Press.

Bentler, Peter M. 1989. *EQS: Structural equations program version 3.0*. Los Angeles: BMDP Statistical Software.

Berger, Peter. 1968. *The sacred canopy: Elements of a sociological theory of religion*. New York: Doubleday.

Berger, Peter, and Thomas Luckmann. 1967. *The social construction of reality*. New York: Doubleday.

Berkovitch, Nitza. 1999. *From motherhood to citizenship: Women's rights and international organizations*. Baltimore, Md.: Johns Hopkins University Press.

Bijker, Wiebe E., Thomas P. Hughes, and Trevor J. Pinch, eds. 1989. *The social construction of technological systems*. Cambridge, Mass.: MIT Press.

Bix, Amy Sue. 1997. Diseases chasing money and power: Breast cancer and AIDS activism challenging authority. *Journal of Policy History* 9(1):5–32.

Blau, Francine D., and Marianne A. Ferber. 1992. *The economics of women, men, and work*. 2nd ed. Englewood Cliffs, N.J.: Prentice-Hall.

Block, Fred, and Gene Burns. 1986. Productivity as a social problem: The uses and misuses of social indicators. *American Sociological Review* 51:767–80.

Blute, M. 1972. The growth of science and economic development. *American Sociological Review* 37:455–64.

Boli, John. 1987. World polity sources of expanding state authority and organization, 1870–1970. In *Institutional structure*, by G. M. Thomas, J. W. Meyer, F. O. Ramirez, and J. Boli, 71–91. Newbury Park, Calif.: Sage.

———. 1993. Sovereignty from a world polity perspective. Paper presented at the American Sociological Association Annual Meeting, Miami, Fla., August 1993.

Boli, John, and George M. Thomas. 1997. World culture in the world polity: A century of international non-governmental organization. *American Sociological Review* 62:171–90.

———. 1999. *Constructing world culture: International nongovernmental organizations since 1875.* Stanford, Calif.: Stanford University Press.

Bollen, Kenneth A. 1989. *Structural equations with latent variables.* New York: John Wiley and Sons.

Bollen, Kenneth A., and Robert W. Jackman. 1985. Regression diagnostics: An expository treatment of outliers and influential cases. *Sociological Methods and Research* 13:510–42.

Bollen, Kenneth A., and J. Scott Long. 1993. *Testing structural equation models.* Newbury Park, Calif.: Sage.

Bornschier, Volker, and Christopher Chase-Dunn. 1985. *Transnational corporations and underdevelopment.* New York: Praeger.

BOSTID. 1994. Barriers faced by developing country women entering professions in science and technology. Report from the workshop sponsored by the Board on Science and Technology for International Development (BOSTID), National Research Council, Washington, D.C., March 1994.

Boudon, Raymond. 1974. *Education, opportunity, and social inequality: Changing prospects in western society.* New York: John Wiley and Sons.

Bowles, Samuel, and Herbert Gintis. 1976. *Schooling in capitalist America.* New York: Basic Books.

Bowman, M. J., and David John Harris. 1984. *Multilateral treaties: Index and current status.* London: Butterworths.

———. 1995. *Multilateral treaties: Index and current status: Cumulative supplement.* London: Butterworths.

Boyle, Elizabeth Heger. Forthcoming. *The measure of mothers' love: Female genital cutting in global context.* Baltimore, Md.: Johns Hopkins University Press.

Boyle, Elizabeth H., Barbara McMorris, and Mayra Gómez. 2002. Local conformity to international norms: The case of female genital cutting. *International Sociology* 17(1):5–33.

Bracey, Gerald W. 1996. International comparisons and the condition of American education. *Educational Researcher* 25:5–11.

———. 1998. Are U.S. students behind? *American Prospect* 3:64–72.

Bradley, Karen. 2000. The incorporation of women into higher education: Paradoxical outcomes? *Sociology of Education* 73(1):1–18.

Bradley, Karen, and Francisco O. Ramirez. 1996. World polity and gender parity: Women's share of higher education, 1965–1985. *Research in Sociology of Education and Socialization* 11:63–91.

Braun, T., and A. Schubert. 1988. Scientometric versus socio-economic indicators: Scatter plots for 61 countries, 1978–1980. *Scientometrics* 13:3–9.

Brimblecombe, Peter. 1987. *The big smoke: A history of pollution in London since medieval times.* London: Methuen.

Broadbent, Jeffrey. 1998. *Environmental politics in Japan.* New York: Cambridge University Press.

Bromley, D. Allan. 1995. Science advice in a global context. In *Worldwide science and technology advice to the highest levels of governments,* edited by W. Golden, 6–9. New York: Pergamon Press.

Bruchhaus, E. 1985. Improved cooking stoves: Miracle weapon in the fight against the desert? *Development and Cooperation* 1:24–26.

Brunsson, Nils. 1989. *The organization of hypocrisy.* Chichester, U.K.: John Wiley and Sons.

Brunsson, Nils, and Kerstin Sahlin-Andersson. 2000. Constructing organizations. *Organization Studies* 21:721–46.

Bryk, Anthony, and Stephen W. Raudenbush. 1992. *Hierarchical linear models for social and behavioral research: Applications and data analysis methods.* Newbury Park, Calif.: Sage.

Burke, Claire, and Derek de Solla Price. 1981. The distribution of citations from nation to nation on a field by field basis: A computer calculation of the parameters. *Scientometrics* 3:363–77.

Buttel, Frederick H. 2000. World society, the nation-state, and environmental protection. *American Sociological Review* 65:117–21.

Caillods, Francoise, Gabriele Gottelmann-Duret, and Keith Lewin. 1996. *Science education and development: Planning and policy issues at the secondary level.* Paris: UNESCO.

Caldwell, Lynton K. 1982. *Science and the national environmental policy act.* Alabama: University of Alabama Press.

Callicott, J. Baird. 1998. The wilderness idea revisited. In *The great new wilderness debate,* edited by J. Baird Callicott and Michael P. Nelson, 337–66. Athens, Ga.: University of Georgia Press.

Callon, Michel, John Law, and Arie Rip. 1986. *Mapping the dynamics of science and technology: Sociology of science in the real world.* Houndmills, U.K.: Macmillan.

Cameron, A. Colin, and Pravin K. Trivedi. 1998. *Regression analysis of count data.* New York: Cambridge University Press.

Canada. 1993. Winning with women in trades, technology, science, and engineering. Report of the National Advisory Board on Science and Technology (NABST) by the Human Resources Committee. Presented to the prime minister, Ottawa, January 1993.

Carnoy, Martin, and Henry H. Levin. 1985. *Schooling and work in the democratic state.* Stanford, Calif.: Stanford University Press.

Carroll, Glenn R., and Michael T. Hannan. 2000. *The demography of corporations and industries.* Princeton, N.J.: Princeton University Press.

Castilla, Emilio. 1997. Institutional aspects of national science activity: The factors influencing nation-state entry into the International Council of Scientific Unions, 1919–1990. Paper presented at the Society for Social Studies of Science Annual Meeting, Tucson, Ariz., October 1997.

Cerny, Philip. 1997. International finance and the erosion of capitalist diversity. In *Political economy of modern capitalism,* edited by Collin Crouch and Wolfgang Streeck, 173–81. Thousand Oaks, Calif.: Sage.

Chabbott, Colette. 1999. Development INGOs. In *Constructing world culture: International nongovernmental organizations since 1875,* edited by J. Boli and G. Thomas, 222–48. Stanford, Calif.: Stanford University Press.

———. 2002. *Constructing education for development: International organizations and education for all.* London: Taylor & Francis.

Chabbott, Colette, and Francisco O. Ramirez. 2000. Development and education. In *Handbook of sociology of education,* edited by M. T. Hallinan, 163–87. New York: Kluwer.

Chase-Dunn, Christopher. 1989. *Global formation: Structures of the world economy.* Cambridge: Basil Blackwell.

Choi, Hyung-Sup. 1995. Policy mechanisms and development strategy for science and technology: The approach of the Republic of Korea (South Korea). In *Worldwide science and technology advice to the highest levels of governments,* edited by W. Golden, 296–301. New York: Pergamon Press.

Clad, J. 1994. Disappearing states: Collapsing governance in the Third World. In *The international system after the collapse of the East/West order,* edited by Armand Cleese, Yoshikaza Sakamoto, and Richard Cooper, 667–74. New York: Library Binding/Kluwer.

Clark, Burton R. 1998. *Creating entrepreneurial university: Organizational pathways of transformation.* New York: Elsevier.

Clarke, Robin. 1985. *Science and technology in world development.* New York: Oxford University Press/UNESCO.

Cohen, Patricia. 1982. *A calculating people: The spread of numeracy in early America.* Chicago: University of Chicago Press.

Cole, Robert. 1989. *Strategies for learning: Small-group activities in American, Japanese, and Swedish industry.* Berkeley: University of California Press.

Cole, Stephen. 1992. *Making science.* Cambridge, Mass.: Harvard University Press.

Cole, Jonathan R., and Stephen Cole. 1973. *Social stratification in science.* Chicago: University of Chicago Press.

Cole, Stephen, and Thomas J. Phelan. 1999. The scientific productivity of nations. *Minerva* 37(1):1–23.

Coleman, James. 1974. *Power and the structure of society.* New York: Norton.

Collins, Randall. 1979. *The credential society: A historical sociology of education and stratification.* New York: Academic Press.

———. 1986. *Weberian sociological theory.* Cambridge: Cambridge University Press.

Cooper, Charles. 1973. *Science, technology, and development: The political economy of technical advance in underdeveloped countries.* London: Frank Cass.

Corrin, Chris. 1992. Women's studies in central and eastern Europe. In *Working out: New directions for women's studies,* edited by Hilary Hinds, Ann Phoenix, and Jackie Stacey, 124–33. London: Falmer Press.

Cozzens, Susan. 1997. The discovery of growth. In *Science in the twentieth century,* edited by J. Krige and D. Pestre, 127–42. Amsterdam: Harwood Academic.

Craig, John. 1981. The expansion of education. *Review of Research on Education* 9:151–213.

Crane, Diana. 1972. Invisible colleges: Diffusion of knowledge in scientific communities. Chicago: University of Chicago Press.

Crawford, Elisabeth. 1992. *Nationalism and internationalism in science, 1880–1939: Four studies of the Nobel population.* Cambridge: Cambridge University Press.

Crawford, Elisabeth, Terry Shinn, and Sverker Sorlin. 1993. *Denationalizing science: The contexts of international scientific practice.* Dordrecht, the Netherlands: Kluwer.

Curriculum Centre, Ministry of Education and Culture (Israel). 1969a. *The animal and his environment.* Jerusalem, Israel: Curriculum Centre, Ministry of Education and Culture.

———. 1969b. *The structure of matter.* Jerusalem, Israel: Curriculum Centre, Ministry of Education and Culture.

Davies, Scott, and Neil Guppy. 1997. Globalization and educational reforms in Anglo-American democracies. *Comparative Educational Review* 41:435–59.

De Oliveira Faria, Manuel. 1973. *Botanica.* Braga, Portugal: Livraria Cruz.

Denison, Edward. 1962. *The sources of economic growth in the United States and the alternatives before us.* New York: Committee for Economic Development.

———. 1967. *Why growth rates differ? Post war experiences in nine western countries.* Washington, D.C.: Brookings Institute.

Deyo, Fredric C. 1989. *Beneath the miracle: Labor subordination in the new Asian industrialism*. Berkeley: University of California Press.

Dezhina, Irina, and Loren Graham. 1999. Science and higher education in Russia. *Science* 286:1303–4.

Dillon William R., and Mathew Goldstein. 1984. *Multivariate analysis, methods and application*. New York: John Wiley and Sons.

DiMaggio, Paul, and Walter Powell. 1983. The iron cage revisited: Institutional isomorphism and collective rationality in organizational fields. *American Sociological Review* 48:147–60.

———. 1991. Introduction to *The newinstitutionalism in organizational analysis*, edited by W. W. Powell and P. J. DiMaggio, 1–39. Chicago: University of Chicago Press.

Dobbin, Frank. Forthcoming. The institutional construction of economic ideas: On the history of railroad finance in the United States and France. In *How institutions change*, edited by W. W. Powell and D. Jones. Chicago: University of Chicago Press.

Dominick, Raymond H. 1992. *The environmental movement in Germany*. Bloomington: Indiana University Press.

Drori, Gili. 1993. The relationship between science, technology, and the economy of lesser developed countries. *Social Studies of Science* 23:201–15.

———. 1996. Science as an infrastructure for the Internet: A cross-national analysis of the effects of science practice on global computer connectedness. Paper presented at the annual meeting of the American Sociological Association, New York, August 1996.

———. 1997. National science agenda as a ritual of modern nation-statehood: The consequences of national "Science for National Development" projects. Ph.D. diss., Stanford University.

———. 1998. A critical appraisal of science education for economic development. In *Socio-cultural perspectives on science education: An international dialogue*, edited by W. W. Cobern, 49–74. Dordrecht, the Netherlands: Kluwer.

———. 2000. Science education and economic development: Trends, relationships, and research agenda. *Studies in Science Education* 35(2):27–58.

Dunlap, Thomas R. 1985. "The Coyote Itself": Ecologists and the value of predators, 1900–1972. In *Environmental history: Critical issues in comparative perspective*, edited by Kendall E. Bailes, 594–618. Lanham, Md.: University of America Press.

Earle, John, Don Henning, Ray Holmes, and George Roberts. 1993. *The world around us*. Kenwyn, South Africa: Juta.

Eaton, Jonathan, and Samuel Kortum. 1993. *International technology diffusion*. Boston: Boston University, Memo.

Edquist, Charles. 1992. *Technological and organizational innovations, produc-tivity and employment*. Working paper. Geneva: World Employment Pro-gramme Research, International Labour Organization.

Eisemon, Thomas Owen, and Charles H. Davis. 1991. University research and the development of scientific capacity in sub-Saharan Africa and Asia. In *International higher education: An encyclopedia*, edited by Philip G. Altbach, 1:275–95. New York: Garland.

———. 1992. Universities and scientific research capacity. *Journal of Asian and African Studies* 27:68–93.

Eistenstadt, Samuel N. 1966. *Modernization: Protest and change*. Englewood Cliffs, N.J.: Prentice-Hall.

Ellul, Jacques. 1964. *The technological society*. New York: Knopf.

Elzinga, Aant, and Catharina Lanstrom, eds. 1996. *Internationalism and sci-ence*. London: Taylor Graham.

Engels, Anita. 1999. Globaler Umweltdiskurs und lokale Umweltkrisen-Gesellschaft und anthropogener Klimawandel im Senegal [Global envi-ronmental discourse and local environmental crises—Society and anthro-pogenic climate change in Senegal]. Ph.D. diss., University of Bielefeld, Germany.

———. 2000. Is the IPCC the ultimate model of the worldwide diffusion of science? Paper presented at Society for Social Studies of Science/European Association for Studies of Science and Techology Conference, Vienna, Austria, October 2000.

Escobar, Arturo. 1983. *Power, knowledge, and discourse as domination: The for-mation of development discourse, 1945–1955*. Paris: UNESCO.

———. 1995. *Encountering development: The making and unmaking of the Third World*. Princeton, N.J.: Princeton University Press.

Etzion, Dalia, Litsa Nicolaou-Smokoviti, and Lotte Bailyn. 1993. A cross-cultural comparison of American, Israeli and Greek women pursuing tech-nical and scientific careers. *International Review of Sociology* 1–2:76–94.

Etzkowitz, Henry, and Loet Leydesdorff. 1997. *Universities and the global knowledge economy: A triple helix of university–industry–government relations*. London: Pinter.

European Technology Assessment Network (ETAN) on Women and Science. 2000. *Science policies in the European Union: Promoting excellence through mainstreaming gender equality*. Brussells: European Commission.

Evenson, Robert E., and Gustav Ranis. 1990. *Science and technology: Lessons for development policy*. Boulder, Colo.: Westview Press.

Ezrahi, Yaron. 1988. Changing political functions of science in the modern liberal-democratic state. In *Cultural traditions and worlds of knowledge:*

*Explorations in the sociology of knowledge*, by S. N. Eisenstadt and Ilana Friedrich Silber, 181–202. Greenwich, Conn.: JAI Press.

———. 1990. *The descent of Icarus: Science and the transformation of contemporary democracy.* Cambridge, Mass.: Harvard University Press.

Faulkner, Joe, and Jacqueline Senaker. 1995. *Knowledge frontiers: Public sector research and industrial innovation in biotechnology, engineering ceramics, and parallel computing.* Oxford: Clarendon Press.

Ferguson, James. 1990. *The anti-politics machine: "Development," depoliticization, and bureaucratic power in Lesotho.* Cambridge: Cambridge University Press.

Fesquet, Alberto E. J. 1947. *Curso de ciencias naturales 5o. grado.* Buenos Aires: Editorial Kapelusz Y Ci'a.

Fiala, Robert, and Audrey Gordon-Lanford. 1987. Educational ideology and the world educational revolution, 1950–1970. *Comparative Education Review* 31:315–32.

Finnemore, Martha. 1991. Science, the state, and international society. Ph.D. diss., Stanford University.

———. 1993. International organization as teachers of norms: The United Nations Educational, Scientific, and Cultural Organization and science policy. *International Organization* 47:567–97.

———. 1996. *National interests in international society.* Ithaca, N.Y.: Cornell University Press.

Firebaugh, Glenn. 1992. Growth effects of foreign and domestic investment. *American Journal of Sociology* 98:105–30.

———. 1996. Does foreign capital harm poor nations? New estimates based on Dixon and Boswell's measures of capital penetration. *American Journal of Sociology* 106:563–75.

Fitzgerald, R. D., Alan D. Stickler, and Thomas R. Watts. 1979. *International survey of accounting principles and reporting practices.* Scarborough, Ontario: PriceWaterhouse International.

Foreman, Dave. 1998. Wilderness: From scenery to nature. In *The great new wilderness debate*, edited by J. Baird Callicott and Michael P. Nelson, 568–84. Athens, Ga.: University of Georgia Press.

Forje, John. 1988. In search of a strategy for national science and technology policy in Africa. In *Science, technology, and development*, edited by Atul Wad. Boulder, Colo.: Westview Press.

Forthney, Judith A. 1970. International migration of professionals. *Population Studies* 24.

Foucault, Michel. 1970. *The order of things.* New York: Random House.

———. 1971. *Discipline and punish: The birth of the prison.* New York: Vintage Books.

————. 1972. *The archeology of knowledge*. New York: Random House.

————. 1980. *Power/Knowledge*. Brighton, U.K.: Harvester Press.

————. 1991. Governmentality. In *The Foucault effect: Studies in governmentality*, edited by Graham Burchell, Colin Gordon, and Peter Miller, 87–104. Chicago: University of Chicago Press.

Frame, J. Davidson, Francis Narin, and Mark P. Carpenter. 1977. The distribution of world science. *Social Studies of Science* 7:501–16.

Frank, David J., John W. Meyer, and David Miyahara. 1995. The individualist polity and the prevalence of professionalized psychology: A cross-national study. *American Sociological Review* 60:360–77.

Frank, David J., Ann Hironaka, John Meyer, Evan Schofer, and Nancy Tuma. 1999. The rationalization and organization of nature in the world culture. In *Constructing world culture*, edited by J. Boli and G. Thomas, 81–99. Stanford, Calif.: Stanford University Press.

Frank, David J., and Elizabeth McEneaney. 1999. The individualization of society and the liberalization of state policies on same-sex sexual relations, 1984–1995. *Social Forces* 77:911–44.

Frank, David J., and Jason Gabler. 2000. Social sciences in the university: Change and variation over the twentieth century. Paper presented at the annual meeting of the American Sociological Association, Washington, D.C., August 2000.

Frank, David J., Ann Hironaka, and Evan Schofer. 2000a. The nation-state and the environment, 1900–1995. *American Sociological Review* 65:96–116.

————. 2000b. Environmental protection as a global institution. *American Sociological Review* 65:122–27.

Frank, David J., Suk Ying Wong, John W. Meyer, and Francisco O. Ramirez. 2000c. What counts as history: A cross-national and longitudinal study of university curricula. *Comparative Education Review* 44:29–53.

Fransman, Martin. 1986. *Technology and economic development*. Boulder, Colo.: Westview Press.

Fuller, Bruce, and Richard Rubinson. 1992. *The political construction of mass education: School expansion, the state, and economic change*. New York: Praeger.

Gaillard, Jacques. 1991. *Scientists in the third world*. Lexington, Ky.: University Press of Kentucky.

Gaillard, Jacques, and Roland Waast. 1992. The uphill emergence of scientific communities in Africa. *Journal of Asian and African Studies* 27:41–67.

Galbraith, John K. 1973. *Economics and the public purpose*. Boston: Houghton Mifflin.

Galison, Peter, and Bruce Herly. 1998. *Big science: The growth of large-scale research*. Stanford, Calif.: Stanford University Press.

Gallas, Karen. 1995. *Talking their way into science*. New York: Teachers College Press.

Garcia, Ana Antonia. 1988. *Ciencias naturales y matematica*. Santo Domingo, Dominican Republic: Secretara de Estado de Educacion, Bellas Artes y Cultos.

Garnier, Maurice, and Jerald Hage. 1990. Education and economic growth in Germany. *Research in Sociology of Education and Socialization* 9:25–53.

Gastil, Raymond D. 1978. *Freedom in the world*. New York: Freedom House.

Geertz, Clifford. 1973. *The interpretation of cultures: Selected essays*. New York: Basic Books.

Gibbons, Michael, Camille Limoge, and Helga Nowotny. 1994. *The new production of knowledge: The dynamics of science and research in contemporary societies*. Thousand Oaks, Calif.: Sage.

Gibitz, Anton, and Karl Kern. 1945. *Naturgeschichte*. Innsbruck, Austria: Verlagsanstalt Tyrolia.

Gieryn, Thomas. 1983. Boundary work and the demarcation of science from nonscience: Strains and interests in professional ideologies of scientists. *American Sociological Review* 48:781–95.

Gilbert, N. G., and Steve Woolgar. 1974. The quantitative study of science: An examination of the literature. *Science Studies* 4:279–94.

Gilford, Dorothy M. 1993. *A collaborative agenda for improving international comparative studies in education*. Washington, D.C.: National Academy Press.

Glacken, Clarence. 1985. Culture and environment in Western civilization during the nineteenth century. In *Environmental history: Critical issues in comparative perspectives*, edited by Kendall E. Bailes, 46–57. Lanham, Md.: University of America Press.

Goffman, Erving. 1959. *The presentation of self in everyday life*. New York: Anchor.

Greene, William H. 1993. *Econometric analysis*. Englewood Cliffs, N.J.: Prentice-Hall.

Greve, Henrich R., David Strang, and Nancy Brandon Tuma. 1995. Specification and estimation of heterogeneous diffusion models. *Sociological Methodology* 25:377–420.

Grinder, R. Dale. 1980. The battle for clean air: The smoke problem in post–Civil War America. In *Pollution and reform in American cities, 1870–1930*, edited by Martin V. Melosi, 83–104. Austin: University of Texas Press.

Grossman, Gene M., and Elhanan Helpman. 1994. Endogenous innovation in the theory of growth. *Journal of Economic Perspectives* 8(1):1–23.

Gruber, Ruth. 1961. *Science and the new state: The proceedings of the international conference on science in the advancement of the new states at Rehovot, Israel 1960.* New York: Basic Books.

Grumbine, R. Edward. 1998. Using biodiversity as a justification for nature protection in the U.S. In *The great new wilderness debate*, edited by J. Baird Callicott and Michael P. Nelson, 595–616. Athens, Ga.: University of Georgia Press.

Guha, Ramachandra. 1998. Radical American environmentalism and wilderness preservation: A Third World critique. In *The great new wilderness debate*, edited by J. Baird Callicott and Michael P. Nelson, 231–45. Athens, Ga.: University of Georgia Press.

Gurr, Ted R. 1988. *Polity II: Political structures and regime change 1800–1986.* Ann Arbor, Mich.: ICPSR.

Haas, Peter. 1992. Introduction: Epistemic communities and international policy coordination. *International Organization* 46:1–35.

Habermas, Jürgen. 1987. *The philosophical discourse of modernity.* Cambridge, Mass.: MIT Press.

———. 1993. *Justification and application.* Cambridge, Mass.: MIT Press.

———. 1996. *The Habermas reader.* Cambridge: Polity Press.

Haddad, Wadi D., Martin Carnoy, Rosemary Rinaldi, and Omporn Regel. 1990. *Education and development.* Discussion paper 95. Washington, D.C.: World Bank.

Hage, Jerald, Maurice Garnier, and Bruce Fuller. 1988. The active state, investment in human capital, and economic growth. *American Sociological Review* 53:824–37.

Hakala, Johanna. 1998. Internationalisation of science: Views of the scientific elite in Finland. *Science Studies* 11:52–74.

Hall, A. Rupert. 1983. *The revolution in science, 1500–1750.* 3rd ed. London: Longman.

Hall, Peter A., ed. 1989. *The political power of economic ideas.* Princeton, N.J.: Princeton University Press.

Hankins, Thomas. 1985. *Science and the enlightenment.* Cambridge: Cambridge University Press.

Hannan, Michael T. 1979. Issues in panel analysis of national development: A methodological overview. In *National development and the world system*, by J. W. Meyer and M. T. Hannan, 17–33. Chicago: University of Chicago Press.

Hannan, Michael T., and John Freeman. 1989. *Organizational ecology.* Cambridge, Mass.: Harvard University Press.

Hanson, Sandra L. Maryellen Schaub, and David P. Baker. 1996. Gender stratification in the science pipeline: A comparative analysis of seven countries. *Gender and Society* 10:271–90.

Hanushek, Eric A., and John E. Jackson. 1977. *Statistical methods for social scientists.* San Diego, Calif.: Academic Press.

Hanushek, Eric A., and Dennis Kimko. 2000. Schooling, labor force quality, and the growth of nations. *American Economic Review* 90:1184–1208.

Haraway, Donna J. 1976. *Crystals, fabrics, and fields.* New Haven, Conn.: Yale University Press.

———. 1989. *Primate visions.* New York: Routledge.

———. 1996. *Modest-witness@Second-Millenium: FemaleMan-Meets-Onco-Mouse: Feminism and technoscience.* New York: Routledge.

Harbison, F., and C. Myers. 1964. *Education, manpower, and economic growth.* New York: McGraw-Hill.

Harmon, David. 1998. Cultural diversity, human subsistence, and the national park ideal. In *The great new wilderness debate,* edited by J. Baird Callicott and Michael P. Nelson, 217–30. Athens, Ga.: University of Georgia Press.

Harris, Joseph P. 1946. Wartime currents and peacetime trends. *American Political Science Review* 40:1137–54.

Häyrinen-Alestalo, Marja. 1998. Is the university able to respond to the demands of technology policy? Paper presented at the meeting of the European Association for Studies of Science and Technology, Lisbon, Portugal, October 1998.

Hayworth, Rex. 1975. *Integrated science For Hong Kong.* Hong Kong: Heineman Educational Books (Asia).

Heckman, James. 1979. Sample selection as a specification error. *Econometrica* 47:153–61.

Heil, David, Maureen Allen, Timothy Cooney, and Angie Matamoros. 1994. *Discover the wonder.* Glenview, Ill.: Scott Foresman.

Herbert, Michael, Hans-Dieter Bunk, Bernhard Klotz, and Annegret Knauf. 1991. *Mein Entdeckerbuch 4.* Stuttgart, Germany: Ernst Klett Schulbuchverlag.

Hicks, Alexander M. 1994. Introduction to pooling. In *The comparative political economy of the welfare state,* edited by Thomas Janoski and Alexander M. Hicks, 169–88. New York: Cambridge University Press.

Hironaka, Ann. 1998. The institutionalization and organization of a global environmental solution: The case of environmental impact assessment legislation. Paper presented at the American Sociological Association Annual Meeting, San Francisco, Calif., August 1998.

————. 2002. Changing meanings, changing institutions: An institutional analysis of patent legislation. *Sociological Inquiry* 72(1):108–30.

Hoerl, Arthur E., and Robert W. Kennard. 1970. Ridge regression: Biased estimation for nonorthogonal problems. *Technometrics* 12:55–67.

Holmes, Brian, and Martin McLean. 1989. *The curriculum: A comparative perspective.* London: Unwin Hyman.

Home, R. W., and Sally Gregory Kohlstedt. 1991. *International science and national scientific identity: Australia between Britain and America.* Dordrecht, the Netherlands: Kluwer.

Honig, Benson, and Francisco Ramirez. 1996. Technicians, technical education, and global economic development: A cross-national examination. Unpublished paper.

Hopwood, Anthony G., and Peter Miller, eds. 1994. *Accounting as social and institutional practice.* Cambridge: Cambridge University Press, 299–316.

International Bank Reconstruction and Development (IBRD). 1971. *World tables.* Washington, D.C.: IBRD.

Ikenberry, John G. 1992. A world economy restored: Expert consensus and the Anglo-American postwar settlement. *International Organization* 46:289–321.

Inhaber, Herbert. 1977. Scientists and economic growth. *Social Studies of Science* 7:517–24.

Inkeles, Alex. 1977. The international evaluation of education achievement: A review. *Proceedings of the National Academy of Education* 4:139–200.

Inkeles, Alex, and David Smith. 1974. *Becoming modern: Individual change in six developing countries.* Cambridge, Mass.: Harvard University Press.

Institute for Scientific Information. 1973. *Science citation index.* Philadelphia: Institute for Scientific Information.

————. 1982. *Science citation index.* Philadelphia: Institute for Scientific Information.

International Association for the Evaluation of Educational Achievement (IEA). 1988. *Science achievement in seventeen countries: A preliminary report.* Oxford: Pergamon Press.

International Council of Scientific Unions (ICSU). 1970. *Yearbook.* Paris: ICSU Secretariat.

————. 1987. *Advice to Organizers of International Scientific Meetings.* Paris: ICSU Secretariat.

International Union of Air Pollution Prevention Associations (IUAPPA). 1988. *Clean air around the world.* Brighton, U.K.: International Union of Air Pollution Prevention Associations.

————. 1991. *Clean air around the world.* 2nd ed. Brighton, U.K.: International Union of Air Pollution Prevention Associations.

International Union for the Conservation of Nature (IUCN). 1990. *1990 United Nations list of national parks and protected areas.* Gland, Switzerland: IUCN.

Jackson, Robert H., and Carl G. Rosberg. 1982. Why Africa's weak states persist: Empirical and juridical statehood. *World Politics* 35:1–24.

Jacobs, Jerry A. 1996. Gender inequality and higher education. *Annual Review of Sociology* 22:153–85.

Jaggers, Keith, and Ted Robert Gurr. 1995. *Polity III: Regime change and political authority, 1800–1994.* Ann Arbor, Mich.: ICPSR.

Jamison, Ellen. 1993. *Scientists and engineers in advanced industrial countries.* Washington, D.C.: International Office, U.S. Bureau of the Census.

Jang, Yong Suk. 2000. The worldwide founding of ministries of science and technology, 1950–1990. *Sociological Perspectives* 43(2):247–70.

———. 2001. The expansion of modern accounting as a global and institutional practice. Ph.D. diss., Stanford University.

Janicke, Martin, and Helmut Weidner. 1997. *National environmental policies: A comparative study of capacity-building.* New York: Springer.

Jardine, Evelyn. 1923. *Practical science for girls as applied to domestic subjects.* London: Methuen.

Jasanoff, Sheila. 1990. *The fifth branch: Science advisers as policymakers.* Cambridge, Mass.: Harvard University Press.

Jepperson, Ronald L. 1991. Institutions, institutional effects, and institutionalism. In *The new institutionalism in organizational analysis,* edited by W. W. Powell and P. J. DiMaggio, 143–63. Chicago: University of Chicago Press.

———. 1994. Individualism in social science: Forms and limits of a methodology. *Contemporary Sociology* 23:898–900.

Jepperson, Ronald L., and John W. Meyer. 1991. The public order and the construction of formal organizations. In *The new institutionalism in organizational analysis,* edited by W. W. Powell and P. J. DiMaggio, 204–31. Chicago: University of Chicago Press.

Johnson, R. J. 1993. The rise and decline of the corporate-welfare state: A comparative analysis in global context. In *Political geography of the twentieth century,* edited by P. J. Taylor, 116–69. New York: Bellhaven.

Johnson, Jean M. 1994. International comparisons of women in higher education in science and engineering. Paper presented at the Annual Conference of the Comparative and International Education Society, San Diego, Calif., March 21–24.

Jones, Eric L. 1981. *The European miracle.* Cambridge: Cambridge University Press.

Jones, Graham. 1971. *The role of science and technology in developing countries.* London: Oxford University Press.

Joselin, F. E. 1948. *Science at work: Machines.* London: Oxford University Press.

Josephson, Paul R. 1996. *Totalitarian science and technology.* Atlantic Highlands, N.J.: Humanities Press.

Kahneman, Daniel, P. Slovic, and Amos Tversky. 1982. *Judgment under uncertainty.* Cambridge: Cambridge University Press.

Kalland, Arne, and Gerard Persoon. 1998. An anthropomorphic perspective on environmental movements. In *Environmental movements in Asia,* edited by A. Kalland and G. Persoon, 1–43. Surrey, U.K.: Curzon Press.

Kamens, David H., and Aaron Benavot. 1991. Elite knowledge for the masses: The origins and spread of mathematics and science education in national curricula. *American Journal of Education* 99:137–80.

———. 1992. A comparative and historical analysis of mathematics and science curricula, 1800–1986. In *School knowledge for the masses,* by J. Meyer, D. Kamens, and A. Benavot, 101–23. London: Falmer.

Kamens, David, John W. Meyer, and Aaron Benavot. 1996. Worldwide patterns of academic secondary education curricula. *Comparative Education Review* 40:116–38.

Kamien, Morton I., and Nancy L. Schwartz. 1982. *Market structure and innovation.* New York: Cambridge University Press.

Keeley, James. 1990. Toward a Foucaldian analysis of international regimes. *International Organization* 44:83–105.

Keeves, John P., ed. 1992. *The IEA Study in Science III: Changes in science education and achievement, 1970–1984.* Oxford: Pergamon Press.

Keeves, John P., and Dieter Kotte. 1992. Disparities between the sexes in science education: 1970–84. In *The IEA study in science III: Changes in Science Education and Achievement, 1970–1984,* edited by John P. Keeves, 141–64. Oxford: Pergamon Press.

Kelly, Mary, and Mary Ryan-Enright. 1990. *Discover our world: Environmental studies book 3.* Dublin: Folens.

Keohane, Robert O., and Helen V. Milner. 1996. *Internationalization and domestic politics.* Cambridge: Cambridge University Press.

Kerr, Clark, J. T. Dunloop, F. Harbison, and C. A. Myers. 1960. *Industrialism and industrial man.* New York: Oxford University Press.

Kim, Young Soo. 1996. The expansion of ministries in modern nation-states. Ph.D. diss., Stanford University.

King, Kenneth. 1989. Donor aid to science and technology education: A state of the art review. *Studies in Science Education* 17:99–122.

Kline, Stephen, and Nathan Rosenberg. 1986. An overview of innovation.

In *The Positive sum strategy: Harnessing technology for economic growth*, edited by R. Landau and Nathan Rosenberg, 275–306. Washington, D.C.: National Academy Press.

Knoll, Karl, and Joachim Knoll. 1965. *Naturlehre*. Munich: Verlag Martin Lurz.

Knorr-Cetina, Karin. 1981. *The manufacture of knowledge: An essay on the constructivist and contextual nature of science*. Oxford: Pergamon Press.

Knorr-Cetina, Karin, and Michael Mulkay, eds. 1983. *Science observed: Perspectives on the social studies of science*. London: Sage.

Kormondy, Edward J., ed. 1989. *International handbook of pollution control*. Westport, Conn.: Greenwood Press.

Krasner, Stephen D., ed. 1983. *International regimes*. Ithaca, N.Y.: Cornell University Press.

Krasner, Stephen D. 1999. *Sovereignty: Organized hypocrisy*. Princeton, N.J.: Princeton University Press.

Krippendorff, Ekkehart. 1975. Towards a class analysis of the international system. *ACTO POLITICA* 10:3–13.

Kristapsons, Janis, Ina Dagyte, and Helle Martinson. 1998. The Baltic way of science system transformation: Methods and results. Paper presented at the European Association for Studies of Science and Technology meeting, Lisbon, Portugal, October 1998.

Kuhn, Thomas S. 1977. *The essential tension*. Chicago: University of Chicago Press.

Lange, Lydia. 1985. Effects of disciplines and countries on citation habits: An analysis of empirical papers in behavioral sciences. *Scientometrics* 8:205–15.

Latour, Bruno. 1987. *Science in action: How to follow scientists and engineers through society*. Cambridge, Mass.: Harvard University Press.

Latour, Bruno, and Steve Woolgar. 1986. *Laboratory life: The construction of social facts*. 2nd ed. Princeton, N.J.: Princeton University Press.

Lauren, Paul. 1998. *The evolution of international human rights*. Philadelphia: University of Pennsylvania Press.

Lee, Molly. 1990. Structural determinants and economic consequences of science education: A cross-national study, 1950–1986. Ph.D. diss., Stanford University.

Levine, Ross, and David Renelt. 1992. A sensitivity analysis of cross-country growth regressions. *American Economic Review* 82:942–63.

Levine, Shar, and Leslie Johnstone. 1995. *Everyday science: Fun and easy projects for making practical things*. New York: John Wiley and Sons.

Lichtenberg, Frank R. 1992. Investment and international productivity differences. Working paper 41. Cambridge, Mass.: National Bureau of Economic Research.

Linnman, Gunnel, Gasta Rodhe, Birgid Wennerberg, and Nils Linnman. 1981. *Biologiboken*. Nacka, Sweden: Esselte Studium.

Locke, Robert R. 1984. *The end of practical man: Entrepreneurship and higher education in Germany, France, and Great Britain—1880–1940*. Greenwich, Conn.: JAI Press.

Lockheed, Marlaine, and Adriaan Verspoor. 1990. *Improving primary education in developing countries*. Washington, D.C.: World Bank.

Long, J. Scott. 1997. *Regression models for categorical and limited dependent variables*. Thousand Oaks, Calif.: Sage.

Lubrano, Linda L. 1993. The hidden structure of Soviet science. *Science, technology and human values* 18:147–75.

Luo, Xiaowei. 2000. The rise of the social development model: Institutional construction of international technology organizations, 1856–1993. *International Studies Quarterly* 44:147–75.

Lynch, M. 1985. *Art and artifact in laboratory science: A study of shop work and shop talk in research laboratory*. London: Routledge and Kegan Paul.

Maddison, Angus. 1995. *Explaining the economic performance of nations: Essays in time and space*. Aldershot, U.K.: Edward Elgar Publishing.

MacKenzie, Donald A. 1981. *Statistics in Britain 1865–1930: The social construction of scientific knowledge*. Edinburgh: Edinburgh University Press.

Mallett, Robert. 1998. Why standards matter? *Issues in science and technology* 15(2):63.

March, James. 1988. *Decisions and organizations*. Oxford: Blackwell.

March, James, and Johan Olsen. 1976. *Ambiguity and choice in organizations*. Bergen, Norway: Universitetsforlaget.

Markham, Adam. 1994. *A brief history of pollution*. New York: St. Martin's Press.

Marshall, Robert. 1998. The problem of the wilderness. In *The great new wilderness debate*, edited by J. Baird Callicott and Michael P. Nelson, 85–96. Athens, Ga.: University of Georgia Press.

Martin, Brian. 1991. *Scientific knowledge in controversy: The social dynamics of the fluoridation debate*. Albany, N.Y.: State University of New York Press.

Martin, Michael O., Ina V. S. Mullis, Albert E. Beaton, Eugenio J. Gonzalez, Teresa A. Smith, and Dana L. Kelly. 1997. *Science achievement in the primary school years: IEA's third international mathematics and science study (TIMSS)*. Chestnut Hill, Mass.: TIMSS International Study Center.

Marquardti, D. W., and R. D. Snee. 1975. Ridge regression in practice. *American Statistician* 29(1):3–19.

Marx, Leo. 2000. *The machine in the garden.* New York: Oxford University Press.

Mateus, Francisco. 1960. *O homem e as coisas.* Lisbon, Portugal: Directorate of Primary Instruction.

Matthews, R. C. O. 1973. The contribution of science and technology to economic development. In *Science and technology in economic growth,* edited by B. R. Williams, 1–30. London: Macmillan.

Mazuri, Ali A. 1975. The African university as a multinational corporation: Problems of penetration and dependency. *Harvard Educational Review* 45:191–210.

McClellan, James E. 1985. *Science reorganized: Scientific societies in the eighteenth century.* New York: Columbia University Press.

McClelland, David C. 1961. *The achieving society.* New York: Free Press.

———. 1969. *Motivating economic achievement.* New York: Free Press.

McEneaney, Elizabeth H. 1998. The transformation of primary school science and mathematics: A cross-national analysis, 1900–1995. Ph.D. diss., Stanford University.

McEneaney, Elizabeth, and John Meyer. 2000. The content of the curriculum: An institutionalist perspective. In *Handbook of sociology of education,* edited by M. Hallinan, 189–211. New York: Plenum.

McKnight, Curtis, F. J. Crosswhite, J. A. Dossey, E. Kifer, J. O. Swafford, K. J. Travers, and T. J. Cooney. 1987. *The underachieving curriculum: Assessing U.S. school mathematics from an international perspective.* Champaign, Ill.: Stipes Publishing.

McKnight, Stephen, A., ed. 1992. *Science, pseudo-science, and utopianism in early modern thought.* Columbia, Mo.: University of Missouri Press.

McNeely, Connie. 1995. *Constructing the nation-state: International organizations and prescriptive action.* Westport, Conn.: Greenwood Press.

———, ed. 1998. *Public rights, public rules: Constituting citizens in the world polity and national policy.* New York: Garland.

McNeely, Jeffrey A., Kenton R. Miller, Walter V. Reid, Russell A. Mittermeier, and Timothy B. Werner. 1990. *Conserving the world's biological diversity.* Gland, Switzerland: World Conservation Union (IUCN).

Mead, Herbert. 1934. *Mind, self, and society.* Chicago: University of Chicago Press.

Mendel, Peter. 2001. Global models of organization: International management standards, reforms, and movements. Ph.D. diss., Stanford University.

Merton, Robert K. 1938/1970. *Science, technology and society in seventeenth century England.* New York: Ferting Howard.

————. 1942/1973. *The sociology of science: Theoretical and empirical investigations.* Chicago: University of Chicago Press.

Meusel, Heinz, and Gerhard Meyendorf. 1960. *Chemie.* Berlin: Volk und Wissen Volkseigner Verlag.

Meyendorf, Gerhard, Gunter Wegner, and Lothar Fritsch. 1991. *Chemie 7.* Berlin: Volk und Wissen.

Meyer, John. 1977. The effects of education as an institution. *American Journal of Sociology* 63:55–77.

————. 1983. Institutionalization and the rationality of formal organizational structure. In *Organizational environments: Ritual and rationality,* by J. W. Meyer and W. R. Scott, 261–82. Thousand Oaks, Calif.: Sage.

————. 1986. Myths of socialization and personality. In *Reconstructing individualism,* edited by T. Heller, M. Sosna, and D. Wellbery, 212–25. Stanford, Calif.: Stanford University Press.

————. 1987. World polity and the authority of the nation-state. In *Institutional structure: Constituting state, society, and the individual,* by George Thomas, John W. Meyer, Francisco O. Ramirez, and John Boli, 41–70. Newbury Park, Calif.: Sage.

————. 1994a. The Evolution of stratification systems. In *Social Stratification,* edited by David Grusky, 730–37. Boulder, Colo.: Westview Press.

————. 1994b. Rationalized environments. In *Institutional environments and organizations,* by W. R. Scott and J. W. Meyer, 28–54. Thousand Oaks, Calif.: Sage.

————. 1999. The changing cultural content of the nation-state: A world society perspective. In *State and culture: New approaches to the state after the cultural turn,* edited by George Steinmetz, 123–43. Ithaca, N.Y.: Cornell University Press.

Meyer, John, and Brian Rowan. 1977. Institutionalized organizations: Formal structure as myth and ceremony. *American Journal of Sociology* 83:340–63.

Meyer, John, Francisco Ramirez, Richard Rubinson, and John Boli-Bennett. 1977. The world educational revolution, 1950–1970. *Sociology of Education* 50:242–58.

Meyer, John, and Brian Rowan. 1978. The structure of educational organizations. In *Environments and organizations,* by Marshal Meyer et al., 78–109. San Francisco: Jossey-Bass.

Meyer, John W., and Michael T. Hannan. 1979. *National development and the world system.* Chicago: University of Chicago Press.

Meyer, John W., Michael Hannan, Richard Rubinson, and George Thomas.

1979. National economic development, 1950–1970: Social and political Factors. In *National development and the world system,* edited by J. Meyer and M. Hannan, 85–116. Chicago: University of Chicago Press.

Meyer, John W., John Boli, and George M. Thomas. 1987. Ontology and rationalization in the western cultural account. In *Institutional structure: Constituting state, society, and the individual,* edited by G. M. Thomas, J. W. Meyer, F. O. Ramirez, and J. Boli, 12–40. Newbury Park, Calif.: Sage.

Meyer, John W., W. Richard Scott, David Strang, and Andrew Creighton. 1988. Bureaucratization without centralization: Changes in the organizational system of U.S. public education, 1940–1980. In *Institutional patterns and organizations: Culture and environment,* edited by Lynne Zucker, 139–68. Cambridge: Ballinger.

Meyer, John W., David Kamens, Aaron Benavot, and Young Kyoung Cha, and Suk-Ying Wong. 1992a. *School knowledge for the masses: World models and national primary curricular categories in the 20th century.* London: Falmer Press.

Meyer, John W., Francisco O. Ramirez, and Yasemin Soysal. 1992b. World expansion of mass education, 1870–1980. *Sociology of Education* 65:128–49.

Meyer, John W., Joane Nagel, and Conrad Wesley Snyder Jr. 1993. The expansion of mass education in Botswana: Local and world society perspectives. *Comparative Education Review* 37:454–75.

Meyer, John W., John Boli, George Thomas, and Francisco O. Ramirez. 1997a. World society and the nation-state. *American Journal of Sociology* 103:144–81.

Meyer, John W., David Frank, Ann Hironaka, Evan Schofer, and Nancy B. Tuma. 1997b. The rise of an environmental sector in world society. *International Organization* 51:623–51.

Meyer, John, and Ronald Jepperson. 2000. The "actors" of modern society: The cultural construction of social agency. *Sociological Theory* 18:100–120.

Meyer, Marshall, John Freeman, Michael Hannan, John Meyer, William Ouchi, Jeffrey Pfeffer, and W. Richard Scott. 1978. *Environments and Organizations.* San Francisco: Jossey-Bass.

Miller, Michael K., and Ken R. Smith. 1980. Biased estimation in policy research: An illustrative example of ridge regression in a health system model. *Rural Sociology* 45:483–50.

Miller, Peter, and Timothy O'Leary. 1987. Accounting and the construction of the governable person. *Accounting, Organizations, and Society* 12:235–65.

Ministry of Education (China). 1989. *Zi ran*. Beijing: Ministry of
    Education.
Ministry of Education (Fiji). 1978. *Physical science: Causing and controlling
    change*. Suva, Fiji: Ministry of Education.
Ministry of Education (Japan). 1995. *New science grade 4*. Tokyo: Ministry
    of Education.
Ministry of Education (South Korea). 1991. [*Science*]. Seoul, South Korea:
    Ministry of Education.
Ministry of Education (Soviet Union). 1970. *Fizika*. Moscow: Ministry of
    Education.
Mittleman, James H. 1997. The dynamics of globalization. In *Globalization:
    Critical reflections*, edited by James H. Mittleman, 1–20. Boulder, Colo.:
    Lynne Rienner.
Moon, Hyeyoung, and Evan Schofer. 1998. The globalization of geological
    science. Paper presented at the American Sociological Association meet-
    ing, San Francisco, August 1998.
Moore, Kelly. 1996. Organizing integrity: American science and the
    creation of public interest organizations. *American Journal of Sociology*
    101:1592–627.
Morell, Virginia. 1993a. Called trimates: Three bold women shaped their
    field. *Science* 260:420–25.
———. 1993b. Seeing nature through the lens of gender. *Science* 260:428–
    29.
Morita-Lou, Hiroko. 1985. *Science and technology indicators for development*.
    Boulder, Colo.: Westview Press.
Mulkay, Michael J. 1983. *Science observed*. London: Sage.
———. 1991. *Sociology of science*. Bloomington: Indiana University Press.
Mullis, Ina V. S., Michael O. Martin, Albert E. Beaton, Eugenio J. Gonza-
    lez, Dana L. Kelly, and Teresa A. Smith. 1998. *Mathematics and science
    achievement in the final year of secondary school: IEA's third international
    mathematics and science study (TIMSS)*. Chestnut Hill, Mass.: Center for
    the Study of Testing, Evaluation, and Educational Policy, Boston
    College.
Nagel, Joane, and Conrad Wesley Snyder, Jr. 1989. International funding of
    education development: External agendas and internal adaptation, the
    case of Liberia. *Comparative Education Review* 33:3–20.
Nandy, Ashis. 1988. *Science, hegemony, and violence: A requiem for modernity*.
    New Delhi: Oxford University Press.
Nash, Roderick Frazier. 1989. *The rights of nature: A history of environmental
    ethics*. Madison, Wis.: University of Wisconsin Press.
———. 1998. The international perspective. In *The great new wilderness*

*debate*, edited by J. Baird Callicott and Michael P. Nelson, 207–16. Athens, Ga.: University of Georgia Press.

National Council of Educational Research and Training (India). 1969. *Physics for middle schools 1*. New Delhi, India: National Council of Educational Research and Training.

National Science Board (NSB). 1998. *Science and engineering indicators—1998*. Arlington, Va.: National Science Foundation (NSB 98-1).

National Science Foundation (NSF). 1995. SESTAT surveys of science and engineering college graduates. Available at: <http://srsstats.sbe.nsf.gov/>. Accessed May 2, 2002.

National Science Foundation. 1998. *Women, minorities, and persons with disabilities in science and engineering*. Arlington, Va.: National Science Foundation.

Nayar, B. K. 1976. *Science and development: Essays in various aspects of science and development*. Bombay: Orient Longman.

Nelson, Richard R. 1992. U.S. technological leadership: Where did it come from and where did it go? In *Entrepreneurship, technological innovation, and economic growth*, edited by F. M. Scherer and Mark Perlman. Ann Arbor, Mich.: University of Michigan Press.

Neter, John, William Wasserman, and Michael H. Kutner. 1990. *Applied linear statistical models*. Boston: Richard D. Irwin.

Newey, Whitney K., and Kenneth D. West. 1987. A simple, positive semi-definite, heteroskedasticity and autocorrelation consistent covariance matrix. *Econometrica* 55:703–8.

Nichols, M. Louise. 1934. *Science for boys and girls*. Philadelphia: J.B. Lippincott.

Nichols, Sharon E., Penny J. Gilmer, Anthony D. Thompson, and Nancy Davis. 1998. Women in science: Expanding the vision. In *International handbook of science education*, edited by B. J. Fraser and K. G. Tobin, 967–78. Dordrecht, the Netherlands: Kluwer.

Nicholson, John. 1989. *Macmillan primary science, standard 5*. Gaborone, Botswana: Macmillan Botswana.

Novy, Stanislav. 1970. *Pokusn pracovn listy k Pruprave k fyzikalnimu vyucovani, ve. 4 rocniku ZDS*. Prague: Stàtnì Pedagogickè Nakladatelstvì.

Novy, Stanislav, Josef Hofman, and Danuše Kvasnikov. 1979. *Pìrodovda*. Prague: Stàtnì Pedagogickè Nakladatelstvì.

Nowotny, Helga, Peter Scott, and Michael Gibbons. 2000. *Re-thinking science: Knowledge and the public in an age of uncertainty*. Oxford: Polity Press.

Nuffield Foundation. 1966. *Nuffield biology text 1: Introducing living things*. London: Longman.

Nuss, Shirley A. 1980. The position of women in socialist and capitalist countries: A comparative study. *International Journal of Sociology of the Family* 10:1–13.

Odhiambo, Thomas R., and T. T. Isoun. 1988. *Science for development in Africa*. Nairobi: ICIPE Science Press.

Organization for Economic Cooperation and Development (OECD). 1963. *Science and the policies of governments*. Paris: OECD.

———. 1993. *OECD economic outlook 53*. Paris: OECD.

———. 1971. *Science growth and society: A new perspective (report of the Secretary General's ad hoc group on new concepts in science policy)*. Paris: OECD Publication.

Ogata, Sadako. 1990. Introduction: United Nations approaches to human rights and scientific and technological development. In *Human rights and scientific and technological development*, edited by G. G. Weeramantry, 1–30. Tokyo: UN University Press.

Olzak, Susan. 1989. Analysis of events in the study of collective action. *Annual Review of Sociology* 15:119–41.

Opie, John. 1985. Environmental history: Pitfalls and opportunities. In *Environmental history: Critical issues in comparative perspective*, edited by K. E. Bailes, 22–35. Lanham, Md.: University of America Press.

Ornstein, Martha. 1928. *The role of scientific societies in seventeenth century Europe*. Chicago: University of Chicago Press.

Pardo Miller, L. P., M. A. Trujillo, C. F. Castillo, and M. Gaviria de Gomez. 1993. *Viva la ciencia! Ciencias naturales y salud*. Bogota, Colombia: Editorial Normal.

Pemberton, W. E. 1979. *Bureaucratic politics: Executive reorganization during the Truman administration*. Columbia, Mo.: University of Missouri Press.

Perrow, Charles. 1991. A society of organizations. *Theory and Society* 20:725–62.

Piccoli, Edoardo. 1928. *Scienze, albo e letture per le classi 5o elementari e rurali*. Florence, Italy: R. Bemporad.

Pindyck, Robert S., and Daniel L. Rubinfeld. 1991. *Econometric models and economic forecasts*. 3rd ed. New York: McGraw-Hill.

Poggi, Gianfrano. 1978. *The development of the modern state: A sociological introduction*. Stanford, Calif.: Stanford University Press.

Poor vs. rich: A new global conflict. 1975. *Time Magazine*, December 22, 20–27.

Porter, Theodore M. 1996. Accounting made visible. *Social Studies of Science* 26:712–15.

Postlethwaite, T. Neville, and David E. Wiley. 1992. *The IEA study of science II: Science achievement in twenty-three countries*. Oxford: Pergamon Press.

Power, Michael. 1994. The audit society. In *Accounting as social and institutional practice*, ed. Anthony G. Hopwood and Peter Miller, 299–316. Cambridge: Cambridge University Press.

Prasad, Eswara K. V. 1979. Education and unemployment of professional manpower in India. *Economic and Political Weekly* 14(20).

Price, Derek J. de Solla. 1963. *Little science, big science*. New York: Columbia University Press.

———. 1986. *Little science, big science—And beyond*. New York: Columbia University Press.

Psacharopolous, George. 1973. *Returns to education: An international comparison*. Amsterdam: Elsevier and Jossey-Bass.

———. 1984. The contribution of education to economic growth: International comparisons. In *International productivity comparisons and the causes of the slowdown*, edited by J. Kendrick, 335–55. Cambridge: Ballinger.

Ragin, Charles C. 1987. *The comparative method: Moving beyond qualitiative and quantitative strategies*. Berkeley: University of California Press.

Ramasubban, Radhika. 1996. Women in science: The case of India. In *World Science Report 1996*, 334–35. Paris: UNESCO.

Ramirez, Francisco O., and Richard Rubinson. 1979. Creating members: The political incorporation and expansion of public education. In *National development and the world system*, edited by J. W. Meyer and M. T. Hannan, 72–82. Chicago: University of Chicago Press.

Ramirez, Francisco O., and John Boli. 1982. Global patterns of educational institutionalization. In *Comparative education*, edited by P. Altbach, R. Arnove, and G. Kelly, 15–36. New York: Macmillan.

Ramirez, Francisco O., and John Boli. 1987. Global patterns of educational institutionalization. *Institutional structure: Constituting state, society and the individual*, by G. M. Thomas, J. W. Meyer, F. O. Ramirez, and J. Boli, 150–72. Newbury Park, Calif.: Sage.

Ramriez, Francisco O., and John Boli. 1987. The political construction of mass schooling: European origins and worldwide institutionalization. *Sociology of Education* 60:2–17.

Ramirez, Francisco O., and Marc Ventresca. 1992. Institutionalizing mass schooling: Ideological and organizational isomorphism in the modern world. In *The political construction of education: School expansion, the state, and economic change*, edited by Bruce Fuller and Richard Rubinson, 47–60. New York: Praeger.

Ramirez, Francisco O., and Gili S. Drori. 1992. The globalization of science: An institutionalist perspective. Paper presented at the American Sociological Association Annual Meeting, Pittsburgh, Penn., August 1992.

Ramirez, Francisco O., and Molly Lee. 1995. Education, science, and development. In *Social change and educational development: Mainland China, Taiwan and Hong Kong*, edited by Gerard A. Postiglione and Lee Wing On, 15–39. Hong Kong: University of Hong Kong Press.

Ramirez, Francisco O., Yasemin Soysal, and Suzanne E. Shanahan. 1997. The changing logic of political citizenship: Cross-national acquisition of women's suffrage rights, 1890–1990. *American Sociological Review* 62:735–45.

Ramirez, Francisco O., Xiaowei Luo, Evan Schofer, and John W. Meyer. 1998. Science and math achievement and economic growth. Paper presented at the American Educational Research Association Annual Meeting, San Diego, Calif., April 1998.

Ramirez, Francisco O., and Christine Min Wotipka. 2001. Slowly but surely? The global expansion of women's participation in science and engineering fields of study, 1972–92. *Sociology of Education* 74:231–51.

Ranis, Gustav. 1977. *Science, technology, and development: A retrospective view.* New Haven, Conn.: Economic Growth Center, Yale University.

Rauner, Mary. 1998. The worldwide globalization of civics education topics, 1955–1995. Ph.D. diss., Stanford University.

Reinders, E. 1920. *Leerboek der Plantkunde*. Groeningen, the Netherlands: J.B. Wolters.

Riddle, Phyllis. 1989. University and state: Political competition and the rise of universities, 1200–1985. Ph.D. diss., Stanford University.

Rittberger, Volker. 1982. International policymaking for development through the United Nations: Science and technology before and after UNCSTD. In *Science and technology in the changing international order: The UN conference on science and technology for development*, edited by Volker Rittberger, 213–56. Boulder, Colo.: Westview Press.

Robertson, Roland. 1990. Mapping the global condition: Globalization as a central concept. In *Global culture: Nationalism, globalization and modernity*, edited by Mike Featherstone, 15–30. London: Sage.

———. 1992. *Globalization: Social theory and global culture*. London: Sage.

———. 1994. Globalization and glocalization. *Journal of International Communication* 1:33–52.

Rohwer, Gotz. 1994. *TDA working papers*. Unpublished working papers.

Rose, Hilary. 1995. Good-bye truth, hello trust: Prospects for feminist science and technology studies at the millennium? In *Science and the construction of women*, edited by M. Maynard, 15–36. London: UCL Press.

Rose, Nicholas, and Peter Miller. 1992. Political power beyond the state. *British Journal of Sociology* 43:173–205.

Rosenau, James. 1970. *The adaptation of national societies: A theory of political system behavior and transformation.* New York: McCaleb-Seiler.

Rosenberg, Nathan. 1994. *Exploring the black box: Technology, economics and history.* Cambridge: Cambridge University Press.

Rosier, Malcolm J., and John P. Keeves. 1991. *The IEA study of science I: Science education and curricula in twenty-three countries.* Oxford: Pergamon Press.

Rossum, Wouter, and Esther K. Hicks. 1997. Social and economic scientists in sub-Saharan Africa and the flow of scientific information: A network study. Paper presented at the Sunbelt 17th International Social Network Conference, San Diego, Calif., November 1997.

Roth, Wolff-Michael. 1997. From everyday science to science education: How science and technology studies inspired curriculum design and classroom research. *Science and Education* 6:373–96.

Rowling, F. 1928. *A primer of simple science for African schools, book 2.* London: Sheldon Press.

Rubinson, Richard, and John Ralph. 1984. Technical change and the expansion of schooling in the United States, 1890–1970. *Sociology of Education* 57:144–52.

———. 1986. Methodological issues in the study of educational change. In *Handbook of theory and research for the sociology of education,* edited by John G. Richardson, 275–304. New York: Greenwood Press.

Rubinson, Richard, and Bruce Fuller. 1992. Specifying the effects of education on national economic growth. In *The political construction of education,* edited by Bruce Fuller and Richard Rubinson, 101–15. New York: Praeger.

Rubinson, Richard, and Irene Brown. 1994. Education and the economy. In *The handbook of economic sociology,* edited by N. Smelser and R. Swedberg, 583–99. Princeton, N.J.: Princeton University Press.

Ruggie, John Gerard. 1993. Territory and beyond. *International Organization* 47(1):149–74.

Runte, Alfred. 1987. *National parks: The American experience.* 2nd rev. ed. Lincoln, Nebr.: University of Nebraska Press.

Russell, Colin A. 1983. *Science and social change 1700–1900.* New York: Macmillan.

Sachs, Michael, ed. 1990. *World guide to scientific associations and learned societies.* 5th ed. Munich: Saur.

Sagasti, Francisco. 1973. Underdevelopment, science and technology: The point of view of the underdeveloped countries. *Science Studies* 3:47–59.

Saka, E. C., C. B. Nyavor, and E. S. Brenya. 1988. *Science course for primary schools, book 4.* Accra, Ghana: Pearl Publications.

Sanielevici, Emil, and Alexandru Dabija. 1969. *Botanik.* Bucharest, Romania: Editura Didactici Pedagogica

Sarewitz, Daniel. 1996. *Frontiers of illusion: Science, technology, and the politics of progress.* Philadelphia: Temple University Press.

Scheid-Cook, T. L. 1990. Ritual conformity and organizational control: Loose coupling or professionalization? *Journal of Applied Behavioral Science* 26:183–99.

Schneiberg, Marc, and Elisabeth S. Clemens. Forthcoming. The typical tools for the job: Research strategies in institutional analysis. In *How institutions change*, edited by W. W. Powell and D. Jones. Chicago: University of Chicago Press.

Schofer, Evan. 1999a. The rationalization of science and the scientization of society: International science organizations, 1870–1990. In *Constructing world culture: International nongovernmental organizations since 1875*, edited by J. Boli and G. Thomas, 249–66. Stanford, Calif.: Stanford University Press.

———. 1999b. The expansion of science as social authority and institutional structure in the world system, 1700–1990. Ph.D. diss., Stanford University.

School Council Publications (Scotland). 1982. *Materials.* Edinburgh: School Council Publications.

Schott, Thomas. 1988. International influence in science: Beyond center and periphery. *Social Science Research* 17:219–38.

———. 1991. The world scientific community: Globality and globalisation. *Minerva* 29:440–62.

———. 1992a. Soviet science in the scientific world system: Was it autarchic, self-reliant, distinctive, isolated, peripheral, central? *Knowledge: Creation, Diffusion, Utilization* 13:410–39.

———. 1992b. Scientific research in Sweden: Orientation towards the American center and embeddedness in Nordic and European environments. *Science Studies* 5:13–37.

———. 1993. World science: Globalization of institutions and participation. *Science, Technology, and Human Values* 18:196–208.

———. 1994. Institutional convergence and deepening inequality among nations in science and technology. Paper presented at the American Sociological Association Annual Meeting, San Francisco, Calif., August 1994.

Schubert, A., S. Zsindely, and T. Braun. 1983. Scientometric analysis of attendance in international scientific meetings. *Scientometrics* 5(3):177–88.

Schultz, Theodore W. 1961. Investment in human capital. *American Economic Review* 51:1–16.

———. 1963. *The economic value of education*. New York: Columbia University Press.

Schwartz, Herman. 1994. Small states in big trouble: State reorganization in Australia, Denmark, New Zealand, and Sweden in the 1980s. *World Politics* 46(4):526–55.

Science Education Center (Philippines). 1979a. *Money in rabbits*. Quezon City: University of the Philippines.

———. 1979b. *From rice washing to delicious nata de arroz*. Quezon City: University of the Philippines.

Scott, W. Richard. 1987. *Organizations: Rational, natural, and open systems*. Englewood Cliffs, N.J.: Prentice-Hall.

———. 1995. *Institutions and organizations*. Thousand Oaks, Calif.: Sage.

Secretary of Public Education (Mexico). 1990. *Ciencias naturales, cuarto grado*. Mexico City: Secretary of Public Education.

Selznick, Philip. 1949. *TVA and the grassroots*. Berkeley: University of California Press.

Shapin, Steven, and Simon Shaffer. 1985. *Leviathan and the air-pump: Hobbes, Boyle, and the experimental life*. Princeton, N.J.: Princeton University Press.

Shenhav, Yehouda, and David Kamens. 1991. The "costs" of institutional isomorphism in non-western countries. *Social Studies of Science* 21:427–545.

Singer, J. David, and Melvin Small. 1990. *National material capabilities data 1816–1985*. Ann Arbor, Mich.: Inter-University Consortium for Political and Social Research.

Skolnikoff, Eugene. 1993. *The elusive transformation: Science, technology and the evolution of international politics*. Princeton, N.J.: Princeton University Press.

Skutnabb-Kangas, Tove. 2000. Linguistic human rights and teachers of English. In *The sociopolitics of English language teaching*, edited by J. Kelly Hall and W. G. Eggington, 22–44. Clevedon, U.K.: Multilingual Matters Ltd.

Slotten, Hugh R. 1994. *Patronage, practice, and the culture of American science: Alexander Dallas Bache and the U.S. coast survey*. Cambridge: Cambridge University Press.

Smelser, Neil J. 1963. Mechanics of change and adjustment to change. In *Industrialization and society*, edited by B. F. Hoselitz and W. E. Moore, 32–54. Paris: UNESCO/Mouton.

Smelser, Neil, B. Badie, and P. Birnbaum. 1994. The sociology of the state revisited. In *Sociology*, edited by N. Smelser, 65–74. New York: Blackwell.

Snyder, David, and Edward Kick. 1979. Structural position in the world-system and economic growth 1955–1970. *American Journal of Sociology* 84:1096–126.

Solomon, Joan, and Glen Aikenhead. 1994. *STS education: International perspectives on reform*. New York: Teachers College Press.

Sorlin, Sverker. 1992. Introduction: The international contexts of Swedish science: A network approach to the internationalism of science. *Science Studies* 5:5–12.

Soysal, Yasemin. 1994. *The limits of citizenship*. Chicago: University of Chicago Press,.

Spaey, Jaques. 1971. *Science for development: An essay on the origin and organization of national science policies*. Paris: UNESCO.

Spence, Michael. 1973. Job market signaling. *Quarterly Journal of Economics* 87:355–75.

Starr, Paul, and Ellen Immergut. 1987. Health care and the boundaries of the political. In *Changing boundaries of the political*, edited by Charles Maier, 221–54. Cambridge: Cambridge University Press.

Stepan, Nancy. 1978. The interplay between socio-economic factors and medical science: Yellow fever research, Cuba and the United States. *Social Studies of Science* 8:397–423.

Steinberg, Sigfrid Henry. 1950–1995. *Statesman's yearbook*. London: Macmillan St. Maren's Press.

Stichweh, Rudolf. 1997. Science in the system of world society. Paper presented at the Society for Social Studies of Science Annual Meeting, Tucson, Ariz., October 1997.

Stokes, Houston. 1997. *Specifying and diagnostically testing econometric models*. 2nd ed. Westport, Conn: Quorum Books.

Strang, David. 1991. Adding social structure to diffusion models: An event history framework. *Sociological Methodology and Research* 19:324–52.

Strang, David, and John W. Meyer. 1993. Institutional conditions for diffusion. *Theory and Society* 22:487–511.

Strath, Annelie. 1998. Scientization and economic development: A cross-national comparative analysis. Ph.D. diss., Stanford University.

Summers, Robert, and Alan Heston. 1991. The Penn world table (mark 5): An expanded set of international comparisons, 1950–1988. *Quarterly Journal of Economics* 106:1–41.

Sweeny, Jill, David Relph, and Larry DeLacey. 1991. *Science 3, books 1 and 2*. Auckland, New Zealand: New House Publishers.

Tarr, Joel A. 1996. *The search for the ultimate sink: Urban pollution in historical perspective.* Akron, Ohio: University of Akron Press.

Taylor, Charles L., and Joachim Amm. 1992. *National capability data: Annual series, 1950–1988.* Ann Arbor, Mich.: Inter-University Consortium for Political and Social Research.

Terleckyj, Nestor E. 1974. Effects of R&D on the productivity growth of industries. Washington, D.C.: National Planning Association.

Thier, Herbert D., Robert Karplus, Chester Lawson, and Robert Knott. 1978. *Relative position and motion.* Chicago: Rand McNally.

Thomas, Keith. 1983. *Man and the natural world: Changing attitudes in England 1500–1899.* London: Allen Lane.

Thomas, George M., John W. Meyer, Francisco O. Ramirez, and John Boli. 1987. *Institutional structure: Constituting state, society, and the individual.* Newbury Park, Calif.: Sage.

Thomas, George, John Boli, and Young S. Kim. 1993. World culture and international non-governmental organization. Paper presented at American Sociological Association Annual Meeting, Miami, Fla., August 1993.

Thurow, Lester. 1974. *Generating inequality: Mechanisms of distribution in the U.S economy.* New York: Basic Books.

Tiffin, Scott, and Fola Osotimehin. 1992. Innovation of technology for industrial development. *Journal of Asian and African Studies* 27:94–113.

Tilly, Charles. 1985. War making and state making as organized crime. In *Bringing the state back in,* edited by Peter Evans, 169–91. Cambridge: Cambridge University Press.

Tolbert, Pamela, and Lynne Zucker. 1983. Institutional sources of change in the formal structure of organizations: The diffusion of civil service reform, 1880–1935. *Administrative Science Quarterly* 28:22–39.

Torney-Purta, Judith. 1990. International comparative research in education: Its role in educational improvement in the U.S. *Educational Researcher* 21:32–35.

Toulmin, Stephen. 1990. *Cosmopolis.* New York: Free Press.

Treiman, Donald. 1977. *Occupational prestige in comparative perspective.* New York: Academic Press.

Tuma, Nancy. 1992. *Invoking rate.* Palo Alto, Calif.: DMA Corporation.

———. 1994. Event History Analysis. In *Analyzing social and political change,* edited by A. Dale and R. B. Davies, 136–66. Thousand Oaks, Calif.: Sage.

Tuma, Nancy, and Michael Hannan. 1984. *Social dynamics: Models and methods.* Orlando: Academic Press.

Turner, Jack. 1998. In wilderness is the preservation of the world. In *The great new wilderness debate*, edited by J. Baird Callicott and Michael P. Nelson, 617–27. Athens, Ga.: University of Georgia Press.

Tyndall, J. 1883. *Albany elementary science readers, no. 2.* London: George Gill.

UNESCO. 1974. *UNESCO statistical yearbook.* Paris: UNESCO Publication.

———. 1979. *Science, technology, and government policy: A ministerial conference for Europe and North America.* Paris: UNESCO Publication.

———. 1987. *Migration of talent: Causes and consequences of brain drain.* Bangkok: UNESCO Principal Regional Office for Asia and the Pacific.

———. 1991. *Science for all and the quality of life.* Bangkok: UNESCO Principal Regional Office for Asia and the Pacific.

———. 1992. *UNESCO statistical yearbook.* Paris: UNESCO Publication.

———. 1993. *International forum on scientific and technological literacy for all (Project 2000+: Phase 2): Declaration.* Paris: UNESCO.

———. 1996. *World science report 1996.* Paris: UNESCO.

———. 1998. *UNESCO statistical yearbook.* Paris: UNESCO.

———. 1999a. *Declaration on science and the use of scientific knowledge.* Available at: <http://www.unesco.org/opi/science/>. Accessed May 2, 2002.

———. 1999b. *Science agenda—Framework for action.* Available at: <http://www.unesco.org/opi/science/>. Accessed May 2, 1999.

Union of International Associations (UIA). 1949–2000. *Yearbook of international organizations.* Munich: Saur.

———. 1949–2000. *Yearbook of international organizations.* Geneva: UIA.

United Kingdom. 1993. The rising tide: A report on women in science, engineering and technology. HMSO. Report of the UK Committee on Women in Science, Engineering and Technology to the Office of Science and Technology.

United Nations (UN). 1968. *Everyman's United Nations: A complete handbook of activities and evaluation of the UN during its first twenty years, 1945–1965.* 8th ed. New York: UN, Department of Public Information.

———. 1971. *World plan of action for the application of science and technology to development.* New York: UN Publication.

———. 1979. *Everyone's United Nations: A summary of activities of the UN during the five-year period.* 9th ed. New York: UN, Department of Public Information.

———. 1986. *Everyone's United Nations: A handbook of the work of the UN.* 10th ed. New York: UN, Department of Public Information.

———. 1995. *The world's women, 1995: Trends and statistics.* New York: United Nations.

United Nations Commission on Science and Technology for Development

(UNCSTD). 1995. *Missing links: Gender equity in science and technology for development.* Ottawa: International Development Research Center.

United Nations Development Fund for Women (UNIFEM). 1994. *Review of UN Agency Activities in the Field of Gender, Science and Technology.* New York: UNIFEM.

————. 1999. *Review of policies and activities of UN organizations in the field of gender science and technology.* New York: UNIFEM.

U.S. Bureau of Labor Statistics (BLS), Office of Employment Projections. 1997. *National industry-occupation employment projections 1996-2006.* Washington, D.C.: U.S. Department of Labor.

U.S. National Commission on Excellence in Education. 1984. *A nation at risk.* Cambridge, Mass.: USA Research.

Vas-Zoltan, Peter, ed. 1976. *The brain drain: An anomaly of international relations.* Budapest: Akademiai Kiado.

Velho, L. 1986. The "meaning" of citation in the context of the scientifically peripheral country. *Scientometrics* 2:74–89.

Ventresca, Marc. 1996. When states count: Institutional and political dynamics in modern census establishment, 1800–1993. Ph.D. diss., Stanford University.

Vogel, David. 1986. *National styles of regulation: Environmental policy in Great Britain and the United States.* Ithaca, N.Y.: Cornell University Press.

Wade, Robert. 1996. Japan, the World Bank, and the art of paradigm maintenance: The East Asian miracle in political perspective. *New Left Review* 217:3–36.

Wallerstein, Immanuel. 1974. *The modern world system.* Vol. 1. New York: Academic Press.

Walters, Pamela, and Richard Rubinson. 1983. Educational expansion and economic output in the U.S., 1890–1969: A production function analysis. *American Sociological Review* 48:480–93.

Walters, Pamela, and Philip J. O'Connell. 1990. Post–World War II higher educational expansion, the organization of work, and changes in labor productivity in the United States. *Research in Sociology of Education* 9:1–23.

Weber, Max. 1978. *Economy and society.* Berkeley: University of California Press.

————. 2002. *The Protestant ethic and the spirit of capitalism.* 3d Roxbury ed. Translated by Stephen Kalberg. Los Angeles: Roxbury Publishing Company.

Weeramantry, C. G. 1990. *Human rights and scientific and technological development.* Tokyo: UN University Press.

————. 1993. *The impact of technology of human rights: Global case-studies.* Tokyo: UN University Press.

Weick, Karl. 1976. Educational organizations as loosely coupled systems. *Administrative Science Quarterly* 21:1–19.

Welsh, Wayne N., and Henry N. Pontell. 1991. Counties in court: Interorganizational adaptations to jail litigation in California. *Law and Society Review* 25:73–101.

Williams, B. R. 1964. Research and economic growth: What should we expect? *Minerva* 3:57–71.

Williamson, John. 1996. What Washington means by policy reform. In *Latin American adjustment: How much has happened?*, edited by John Mendelsohn. Washington, D.C.: Institute for International Economics.

Windolf, Paul. 1997. *Expansion and Structural Change: Higher Education in Germany, the United States, and Japan, 1870–1990.* New York: Westview Press.

Woolgar, Steve. 1988. *Science: The very idea.* Chichester, U.K.: Ellis Harwood.

World Bank. 1950–1988. *World tables of economic and social indicators, 1950–1988.* Ann Arbor, Mich.: Inter-University Consortium for Political and Social Research.

————. 1992. *World tables.* Baltimore, Md.: Johns Hopkins University Press.

————. 1988. *Education in Sub-Saharan Africa.* Washington, D.C.: World Bank.

World Intellectual Property Organization (WIPO). 1983. *100 years protection of industrial property: Statistics, synoptic tables on patents, trademarks, designs, utility models, and plant varieties, 1883–1982.* Geneva: WIPO.

Wu, Lawrence. 1990. Simple graphical goodness-of-fit tests for hazard rates. In *Event history analysis in life course research*, edited by Karl Ulrich Mayer and Nancy B. Tuma, 184–202. Madison,Wis.: University of Wisconsin Press.

Wuthnow, Robert. 1980. The world-economy and the institutionalization of science in seventeenth century Europe. In *Studies of the modern world system*, edited by Albert Bergesen, 25–55. New York: Academic Press.

————, 1987. *Meaning and moral order.* Berkeley: University of California Press.

Yamaguchi, Kazuo. 1991. *Event history analysis.* Newbury Park, Calif.: Sage.

Yates, J. K., and Stalianos Aniftos. 1998. Developing standards and international standards organizations. *Journal of Management in Engineering* 14:57–64.

Zeger, S. L. 1988. A regression model for time series of counts. *Biometrika* 75:621–29.

Zeitler, William. 1981. Curriculum reviews. *Science and Children* 18:34–35.

Zils, Michael. 1978. *World guide to scientific associations and learned societies.* New York: Bowker.

———. 1998. *World guide to scientific associations and learned societies.* 7th ed. Munich: Saur.

Zuckerman, Harriet. 1989. The sociology of science. In *Handbook of sociology*, edited by N. Smelser, 511–74. Newbury Park, Calif.: Sage.

Zweekhorst, Marjolein B. M., and Joske F. G. Bunders. 1998. Strategies for globalization of a marginal national science system: The case of Bangladesh. Paper presented at the European Association for Studies of Science and Techology meeting, Lisbon, Portugal, October 1998.

# Index

Actor, 10, 13, 30, 46, 218–19, 267–68, 270–73, 329n1
Actorhood, 88, 267–69, 275, 290; construction of, 24, 30–34, 38, 137, 154, 253–55. *See also* Enactment
Africa, 89, 179–83, 204
Agency, 24
Air pollution, 258–60
Analysis, *see* Cross-sectional analysis; Event history analysis; Network analysis; Panel analysis; Pooled time-series analysis; Time-series analysis
Asia, 179–84, 247
Asian Tigers, 208, 212, 303
Australia, 210–11
Authority, *see* Science, authority of
Autocorrelation, 54–56, 59, 64–66
Autonomy: of nation-state, 228

Barinaga, Marcia, 189
Ben-David, Joseph, 27, 203
Biodiversity, 257
Bix, Amy Sue, 177
"Brain drain," 229
Brunsson, Nils, 32, 35
Bureaucratization, 292

Capitalism, 45
Carribean, *see* Latin America
Carson, Rachel, 222, 249–50
Catalonia, 163, 170–71
Chabbott, Colette, 9, 98
Christendom, 48. *See also* Religion
Civil society, 277
Classification, 148. *See also* Taxonomy
Cold War, 111, 199, 206

Colonialism, 48, 202, 204–5. *See also* Science, and colonial legacy
Communism, 1, 199–203, 205, 208, 275
Community: epistemic, 1, 38–39, 97, 99; imagined, 44
Convergence, 211, 281
Core countries, *see* Developed countries
Corporation: transnational, 236, 238. *See also* Research and development (R&D)
Cosmology: cosmological, 23, 113
Count models, 51–53, 55–57, 59
Coupling, 15–16, 18, 19, 118, 157, 278; conditions for 158–159; definition of, 159; measurement of, 166
Cross-sectional analysis, 52, 63–72
Cultural frames, 8–9, 12, 29–30, 61
Culture of science, 88

Decoupling, *see* Coupling
Democracy, 201–2, 235, 238, 276. *See also* Participatory politics
Democratization, 266, 276, 278
Density: organizational, 50, 125, 320n7
Dependency theory, *see* World system theory
Developed countries, 143, 150–51, 164–69, 199–203, 207–8, 281, 284
Developing countries, 104, 156, 158, 160–69, 199–200, 203–5, 208, 276, 281. *See also* Noncore countries; Periphery; Third world
Development: decade (*see* United Nations Development Decades); discourse (*see* Discourse, of development); economic, 125, 175; measures of, 128; and women, 175